Data Analysis with Excel®

Data analysis is of central importance in the education of scientists. This book offers a compact and readable introduction to techniques relevant to students of the physical sciences. The material presented has been thoroughly integrated with the popular and powerful spreadsheet package, Microsoft Excel®.

Excel® features of most relevance to the analysis of experimental data in the physical sciences are dealt with in some detail. Fully worked examples with a strong physical science bias reinforce basic principles. Underlying assumptions and the range of applicability of techniques are discussed, though detailed derivations of basic equations are mostly avoided or confined to the appendices. Exercises and problems are included, with answers to these at the back of the book. Further relevant topics may be accessed through the Internet.

Data Analysis with Excel® is suitable for intermediate and advanced undergraduate students in the physical sciences. It will also appeal to graduate students and researchers needing an introduction to statistical techniques and the use of spreadsheets in data analysis.

LES KIRKUP gained his PhD from Paisley College (now Paisley University) and has more than twenty years experience of working in tertiary education institutions. He held academic positions in England and Scotland before moving to Australia in 1990, and has been Associate Professor at the University of Technology, Sydney since 1997. A dedicated lecturer, he has focussed on improving the experimental skills of physical science and engineering undergraduates. One of his two books, *Experimental Methods* (Wiley, 1994), supports laboratory based experimentation. As an active researcher, his current research interests include the development of systems for the analysis of the electrical activity of the human brain. He is a member of the Institute of Physics and the Australian Institute of Physics and has written for a wide range of journals including: *Computer Applications in the Biosciences, Computers in Physics, Review of Scientific Instruments, Medical and Biological Engineering and Computing, Physics Education, European Journal of Physics, American Journal of Physics, Physiological Measurement and Measurement Science and Technology.*

Data Analysis with Excel®

An Introduction for Physical Scientists

Les Kirkup

University of Technology, Sydney

CAMBRIDGE UNIVERSITY PRESS

PUBLISHED BY THE PRESS SYNDICATE OF THE UNIVERSITY OF CAMBRIDGE
The Pitt Building, Trumpington Street, Cambridge, United Kingdom

CAMBRIDGE UNIVERSITY PRESS
The Edinburgh Building, Cambridge CB2 2RU, UK
40 West 20th Street, New York, NY 10011-4211, USA
477 Williamstown Road, Port Melbourne, VIC 3207, Australia
Ruiz de Alarcón 13, 28014 Madrid, Spain
Dock House, The Waterfront, Cape Town 8001, South Africa

http://www.cambridge.org

First published 2002

Printed in the United Kingdom at the University Press, Cambridge

Typeface Utopia 9.25/13.5pt.   *System* QuarkXPress® [SE]

A catalogue record for this book is available from the British Library

Library of Congress Cataloguing in Publication data

Kirkup, Les.
  Data analysis with Excel : an introduction for physical scientists / Les Kirkup.
    p. cm.
  Includes bibliographical references and index.
  ISBN 0-521-79337-8 – ISBN 0-521-79737-3 (pb.)
    1. Research–Statistical methods–Data processing. 2. Electronic spreadsheets.
  3. Microsoft Excel for Windows. I. Title.

  Q180.55.S7 K57 2002
  001.4′22′0285–dc21   2001037408

ISBN 0 521 79337 8 hardback
ISBN 0 521 79737 3 paperback

To Sarah and Amy.
Me canny lasses

Contents

Preface

Experiments and experimentation have central roles to play in the education of scientists. For many destined to participate in scientific enquiry through laboratory or field based studies, the ability to apply 'experimental methods' is a key skill that they rely upon throughout their professional careers. For others whose interests and circumstances take them into other fields upon completion of their studies, the experience of 'wrestling with nature' so often encountered in experimental work offers enduring rewards: skills developed in the process of planning, executing and deliberating upon experiments are of lasting value in a world in which some talents become rapidly redundant.

Laboratory and field based experimentation are core activities in the physical sciences. Good experimentation is a blend of insight, imagination, skill, perseverance and occasionally luck. Vital to experimentation is data analysis. This is rightly so, as careful analysis of data can tease out features and relationships not apparent at a first glance at the 'numbers' emerging from an experiment. This, in turn, may suggest a new direction for the experiment that might offer further insight into a phenomenon or effect being studied. Equally importantly, after details of an experiment are long forgotten, facility gained in applying data analysis methods remains as a highly valued and transferable skill.

My experience of teaching data analysis techniques at undergraduate level suggests that when the elements of content, relevance and access to contemporary analysis tools are sympathetically blended, students respond positively and enthusiastically. Believing that no existing text encourages or supports such a 'blend', I decided to write one. This text offers an introduction to data analysis techniques recognising the background and needs of students from the physical sciences. I have attempted to include those techniques most useful to students from the physical sciences and employ examples that have a physical sciences 'bias'.

It is natural to turn to the computer when the 'number crunching' phase of data analysis begins. Though many excellent computer based data analysis packages exist, I have chosen to exploit the facilities offered by spreadsheets throughout the text. In their own right, spreadsheets are powerful analysis tools which are likely to be familiar and readily accessible to students.

More specifically, my goals have been to:

- provide a readable text from which students can learn the basic principles of data analysis;
- ensure that problems and exercises are drawn from situations likely to be familiar and relevant to students from the physical sciences;
- remove much of the demand for manual data manipulation and presentation by incorporating the spreadsheet as a powerful and flexible utility;
- emphasise the analysis tools most often used in the physical sciences;
- focus on aspects often given less attention in other texts for scientists such as the treatment of systematic errors;
- encourage student confidence by incorporating 'worked' examples followed by exercises;
- provide access to extra material through generally accessible Web pages.

Computers are so much a part of professional and academic life that I am keen to include their use, especially where this aids the learning and application of data analysis techniques. The Excel® spreadsheet package by Microsoft has been chosen due to its flexibility, availability, longevity and the care that has been taken by its creators to provide a powerful yet 'user friendly' environment for the processing and presentation of data. At the time of writing, the most recent version of Excel® for PCs is Excel® 2002. This text does not, however, attempt a comprehensive coverage of the newest features of Excel®. Anyone requiring a text focussing on Excel®, and its many options, shortcuts and specialist applications must look elsewhere as only those features of most relevance to the analysis of experimental data are dealt with here.

While chapter 1 contains some material normally encountered at first year level, the text as a whole has been devised to be useful at intermediate and senior undergraduate levels. Derivations of formulae are mostly avoided in the body of the text. Instead, emphasis has been given to the assumptions underlying the formulae and range of applicability. Details of derivations may be found in the appendices. It is assumed that the reader is familiar with introductory calculus, graph plotting and the calculations of means and standard deviations. Experience of laboratory work at first year undergraduate level is also an advantage.

I am fortunate that many people have given generously of their time to help me during the preparation of this book. Their ideas, feedback and not least their encouragement are greatly appreciated. I also acknowledge many intense Friday night discussions with students and colleagues on matters relating to data analysis and their frequent pleadings with me to 'get a life'.

I would like to express my appreciation and gratitude to the following people:

From the University of Technology, Sydney (UTS): Geoff Anstis, Mark Berkhan, Graziella Caprarelli, Bob Cheary, Michael Dawson, Chris Deller, Sherri Hilario, Suzanne Hogg, Loraine Holley, Bob Jones, Ann-Marie Maher, Kendal McGuffie, Mary Mulholland, Matthew Phillips, Andrew Searle, Brian Stephenson, Mike Stevens, Paul Swift.
Formerly of UTS: Andreus Reuben, Tony Fisher-Cripps, Gary Norton.

The screen shots are reprinted with permission from Microsoft Corporation. I thank Simon Capelin and the staff at Cambridge University Press for their support, hardwork and professionalism. Finally, I thank Janet Sutherland for her encouragement and support during the preparation of this text.

March 2001   Les Kirkup
Sydney   Les.Kirkup@uts.edu.au

Chapter 1

Introduction to scientific data analysis

1.1 Introduction

'The principle of science, the definition almost, is the following: *The test of all knowledge is experiment.* Experiment is the *sole judge* of scientific "truth"'.

So wrote Richard Feynman, famous scientist and Nobel prize winner, noted for his contributions to physics.[1]

It is possible that when Feynman wrote these words he had in mind elaborate experiments devised to reveal the 'secrets of the universe', such as those involving the creation of new particles during high energy collisions in particle accelerators. However, experimentation encompasses an enormous range of more humble (but extremely important) activities such as testing the temperature of a baby's bath water by immersing an elbow into the water, or pressing on a bicycle tyre to establish whether it has gone 'flat'. The absence of numerical measures of quantities most distinguishes these experiments from those normally performed by scientists.

Many factors directly or indirectly influence the fidelity of data gathered during an experiment such as the quality of the experimental design, experimenter competence, instrument limitations and time available to perform the experiment. Appreciating and, where possible, accounting for such factors are key tasks that must be carried out by an experimenter. After every care has been taken to acquire the best data possible, it is time to apply techniques of data analysis to extract the most from the data. The

[1] See Feynman, Leighton and Sands (1963).

process of extraction requires qualitative as well as quantitative methods of analysis. The first steps require consideration be given to how data may be summarised numerically and graphically and this is the main focus of this chapter.[2] Some of the ideas touched upon in this chapter, such as those relating to error and uncertainty, will be revisited in more detail in later chapters.

1.2 Scientific experimentation

To find out something about the world, we experiment. A child does this naturally, with no training or scientific apparatus. Through a potent combination of curiosity and trial and error, a child quickly creates a viable model of the 'way things work'. This allows the consequences of a particular action to be anticipated. Curiosity plays an equally important role in the professional life of a scientist who may wish to know:

- the amount of contaminant in a pharmaceutical;
- the thickness of the ozone layer in the atmosphere;
- the surface temperature of a distant star;
- the stresses experienced by the wings of an aircraft;
- the blood pressure of a person;
- the frequency of electrical signals generated within the human brain.

In particular, scientists look for relationships between quantities. For example, a scientist may wish to establish how the amount of energy radiated from a body each second depends on the temperature of that body. In formulating the problem, designing and executing the experiment and analysing the results, the intention may be to extend the domain of applicability of an established theory, or to present strong evidence of the breakdown of that theory. Where results obtained conflict with accepted ideas or theories, a key goal is to provide an alternative and better explanation of the results. Before 'going public' with a new and perhaps controversial explanation, the scientist needs to be confident in the data gathered and the methods used to analyse those data. This requires that experiments be well designed. In addition, good experimental design helps anticipate difficulties that may occur during the execution of the experiment and encourages the efficient use of resources.

Successful experimentation is often a combination of good ideas, good planning, perseverance and hard work. Though it is possible to dis-

---

[2] This is sometimes referred to as 'exploratory data analysis'.

cover something interesting and new 'by accident', it is usual for science to progress by small steps. The insights gained by researchers (both experimentalists and theorists) combine to provide answers and explanations to some questions, and in the process create new questions that need to be addressed. In fact, even if something new *is* found by chance, it is likely that the discovery will remain a curiosity until a serious scientific investigation is carried out to determine if the discovery or effect is real or illusory. While scientists are excited by new ideas, a healthy amount of scepticism remains until the ideas have been subjected to serious and sustained scrutiny by others.

Though it is possible to enter a laboratory with only a vague notion of how to carry out a scientific investigation, there is much merit in planning ahead as this promotes the efficient use of resources, as well as revealing whether the investigation is feasible or overambitious.

1.2.1 Aim of an experiment

An experiment needs a focus, more usually termed an 'aim', which is something the experimenter returns to during the design and analysis phases of the experiment. Essentially the aim embodies a question which can be expressed as 'what are we trying to find out by performing the experiment?'

Expressing the aim clearly and concisely at the outset is important, as it is reasonable to query as the experiment progresses whether the steps taken are succeeding in addressing the aim, or whether the experiment has deviated 'off track'. Heading off on a tangent from the main aim is not necessarily a bad thing. After all, if you observe an interesting and unexpected effect during the course of an experiment, it would be quite natural to want to know more, as rigidly pursuing the original aim might cause you to bypass an important discovery. Nevertheless, it is likely that if a new effect has been observed, this effect deserves its own separate and carefully planned experiment.

Implicit in the aim of the experiment is an idea or hypothesis that the experimenter wishes to promote or test, or an important question that requires clarification. Examples of questions that might form the basis of an experiment include:

- Is a new spectroscopic technique better able to detect impurities in silicon than existing techniques?
- Does heating a glass substrate during vacuum deposition of a metal improve the quality of the thin films deposited onto the substrate?

- To what extent does a reflective coating on windows reduce the heat transfer into a motor vehicle?
- In what way does the cooling efficiency of a thermoelectric cooler depend on the amount of electrical current supplied to the cooler?

Such questions can be restated explicitly as aims of a scientific investigation. It is possible to express those aims in a number of different, but essentially equivalent, ways. For example:

(a) The aim of the experiment is to determine the change in heat transfer to a motor vehicle when a reflective coating is applied to the windows of that vehicle.

(b) The aim of the experiment is to test the hypothesis that a reflective coating applied to the windows of a motor vehicle reduces the amount of heat transferred into that vehicle.

Most physical scientists and engineers would recognise (a) as a familiar way in which an aim is expressed in their disciplines. By contrast, the explicit inclusion of a hypothesis to be tested, as stated in (b), is often found in studies in the biological, medical and behavioural sciences. The difference in the way the aim is expressed is largely due to the conventions adopted by each discipline, as all have a common goal of advancing understanding and knowledge through experimentation and observation.

1.2.2 Experimental design

Deciding the aim or purpose of an experiment 'up front' is important, as precious resources (including the time of the experimenter) are to be devoted to the experiment. Experimenting is such an absorbing activity that it is possible for the aims of an experiment to become too ambitious. For example, the aim of an experiment might be to determine the effect on the thermal properties of a ceramic when several types of atoms are substituted for (say) atoms of calcium in the ceramic. If a month is available for the study, careful consideration must be given to the number of samples of ceramic that can be prepared and tested and whether a more restricted aim, perhaps concentrating on the substitution of just one type of atom, might not be more appropriate.

Once the aim of an experiment is decided, a plan of how that aim might be achieved is begun. Matters that must be considered include:

- What quantities are to be measured during the experiment?
- Over what ranges should the controllable quantities be measured?

- What are likely to be the dominant sources of error?
- What equipment is needed and what is its availability?
- In what ways are the data to be analysed?
- Does the experimenter need to become skilled at new techniques (say, how to operate an electron microscope, or perform advanced data analysis) in order to complete the experiment?
- Does new apparatus need to be designed/constructed/acquired or does existing equipment require modification?
- Is there merit in developing a computer based acquisition system to gather the data?
- How much time is available to carry out the experiment?
- Are the instruments to be used performing within their specifications?

A particularly important aspect of experimentation is the identification of influences that can affect any result obtained through experiment or observation. Such influences are regarded as sources of 'experimental error' and we will have cause to consider these in this text. In the physical sciences, many of the experimental variables that would affect a result are easily identifiable and some are under the control of the experimenter. Identifying sources that would adversely influence the outcomes of an experiment may lead to ways in which the influence might be minimised. For example, the quality of a metal film deposited onto a glass slide may be dependent upon the temperature of the slide during the deposition process. By improving the temperature control of the system, so that the variability of the temperature of the slide is reduced to (say) less than 5 °C, the quality of the films may be enhanced.

Despite the existence of techniques that allow us to draw out much from experimental data, a good experimenter does not rely on data analysis to 'make up' for data of dubious worth. If large scatter is observed in data, a sensible option is to investigate whether improved experimental technique can reduce the scatter. For example, time spent constructing electromagnetic shielding for a sensitive electronic circuit in an experiment requiring the measurement of extremely small voltages can improve the quality of the data dramatically and is to be much preferred to the application of 'advanced' data analysis techniques which attempt to compensate for shortcomings in the data.

An essential feature of experiments in the physical sciences is that the measurement process yields numerical values for quantities such as temperature, pH, strain, pressure and voltage. These numerical values (often referred to as *experimental data*) may be algebraically manipulated, graphed, compared with theoretical predictions or related to values obtained by other experimenters who have performed similar experiments.

1.3 Units and standards

Whenever a value is recorded in a table or plotted on a graph, the unit of measurement must be stated, as numbers by themselves have little meaning. To encompass all quantities that we might measure during an experiment, we need units that are:

- comprehensive,
- clearly defined,
- internationally accepted,
- easy to use.

Reliable and accurate standards based on the definition of a unit must be available so that instruments designed to measure specific quantities may be compared with those standards. Without agreement between experimenters in, say, Australia and the United Kingdom as to what constitutes a metre or a second, a comparison of values obtained by each experimenter would be impossible.

A variety of instruments may be employed to measure quantities in the physical sciences, ranging from a 'low tech.' manometer to determine pressure in a chamber to a state of the art HPLC[3] to accurately determine the concentration of contaminant in a pharmaceutical. Whatever the particular details of a scientific investigation, we generally attach much importance to the 'numbers' that emerge from an experiment as they may provide support for a new theory of the origin of the universe, assist in monitoring damage to the earth's atmosphere or help save a life. Referring to the outcome of a measurement as a 'number' is rather vague and misleading. Through experiment we obtain *values*. A value is the product of a number and the unit in which the measurement is made. The distinction in scientific contexts between number and value is important. Table 1.1 includes definitions of number, value and other important terms as they are used in this text.

---

[3] HPLC stands for high pressure liquid chromatography

Table 1.1. *Definitions of commonly used terms in data analysis.*

Term	Definition
Quantity	An attribute or property of a body, phenomenon or material. Examples of quantities are: the temperature, mass or electrical capacitance of a body; the time elapsed between two events such as starting and stopping a stop watch; and the resistivity of a metal.
Unit	An amount of a quantity, suitably defined and agreed internationally, against which some other amount of the same quantity may be compared. As examples, the kelvin is a unit of temperature, the second is a unit of time and the ohm-metre is a unit of resistivity.
Value	The product of a number and a unit. As examples, 273 K is a value of temperature, 0.015 s is a value of time interval and 1.7×10^{-8} $\Omega \cdot$m is a value of resistivity.
Measurement	A process by which a value of a quantity is determined. For example, the measurement of water temperature using an alcohol-in-glass thermometer entails immersing a thermometer in the water followed by estimating the position of the top of a narrow column of alcohol against an adjacent scale.
Data	Values obtained through measurement or observation.

1.3.1 Units

The most widely used system of units in science is the SI system[4] which has been adopted officially by most countries around the world. Despite strongly favouring SI units in this text, we will also use some 'non-SI units' such as the minute and the degree, as these are likely to remain in widespread use in science for the foreseeable future.

The origins of the SI system can be traced to pioneering work done on units in France in the late eighteenth century. In 1960 the name 'SI system' was adopted and at that time it consisted of six fundamental or 'base' units. Since 1960 the system has been added to and refined and remains constantly under review. From time to time suggestions are made regarding how the definition of a unit may be improved. If this allows for easier or more accurate realisation of the unit as a standard (permitting, for

[4] SI stands for Système International.

Table 1.2. *SI base units, symbols and definitions.*

Quantity	Unit	Symbol	Definition
Mass	kilogram	kg	The kilogram is equal to the mass of the international prototype of the kilogram. (The prototype kilogram is made from an alloy of platinum and iridium and is kept under very carefully controlled environmental conditions near Paris.)
Length	metre	m	The metre is the length of the path travelled by light in a vacuum during a time interval of $\frac{1}{299\,792\,458}$ of a second.
Time	second	s	The second is the duration of 9 192 631 770 periods of the radiation corresponding to the transition between the two hyperfine levels of the ground state of the caesium 133 atom.
Thermodynamic temperature	kelvin	K	The kelvin is the fraction $\frac{1}{273.16}$ of the thermodynamic temperature of the triple point of water.
Electric current	ampere	A	The ampere is that current which, if maintained between two straight parallel conductors of infinite length, of negligible cross-section and placed 1 metre apart in a vacuum, would produce between these conductors a force of 2×10^{-7} newton per metre of length.
Luminous intensity	candela	cd	The candela is the luminous intensity, in a given direction, of a source that emits monochromatic radiation of frequency 540×10^{14} hertz and that has a radiant intensity in that direction of $\frac{1}{683}$ watt per steradian.
Amount of substance	mole	mol	The mole is the amount of substance of a system which contains as many elementary entities as there are atoms in 0.012 kilogram of carbon 12.

example, improvements in instrument calibration), then appropriate modifications are made to the definition of the unit. Currently the SI system consists of 7 base units as defined in table 1.2.

Other quantities may be expressed in terms of the base units. For example, energy can be expressed in units $kg \cdot m^2 \cdot s^{-2}$ and electric potential difference in units $kg \cdot m^2 \cdot s^{-3} \cdot A^{-1}$. The cumbersome nature of units expressed in this manner is such that other, so called *derived*, units are introduced which are formed from products of the base units. Some famil-

Table 1.3. *Symbols and units of some common quantities.*

Quantity	Derived unit	Symbol	Unit of quantity expressed in base units
Energy, work	joule	J	$kg \cdot m^2 \cdot s^{-2}$
Force	newton	N	$kg \cdot m \cdot s^{-2}$
Power	watt	W	$kg \cdot m^2 \cdot s^{-3}$
Potential difference, electromotive force (emf)	volt	V	$kg \cdot m^2 \cdot s^{-3} \cdot A^{-1}$
Electrical charge	coulomb	C	$s \cdot A$
Electrical resistance	ohm	Ω	$kg \cdot m^2 \cdot s^{-3} \cdot A^{-2}$

iar quantities with their units expressed in derived and base units are shown in table 1.3.

Example 1

The farad is the SI derived unit of electrical capacitance. With the aid of table 1.3, express the unit of capacitance in terms of the base units, given that the capacitance, C, may be written

$$C = \frac{Q}{V} \tag{1.1}$$

where Q represents electrical charge and V represents potential difference.

ANSWER

From table 1.3, the unit of charge expressed in base units is $s \cdot A$ and the unit of potential difference is $kg \cdot m^2 \cdot s^{-3} \cdot A^{-1}$. It follows that the unit of capacitance can be expressed with the aid of equation (1.1) as

$$\frac{s \cdot A}{kg \cdot m^2 \cdot s^{-3} \cdot A^{-1}} = kg^{-1} \cdot m^{-2} \cdot s^4 \cdot A^2$$

Exercise A

The henry is the derived unit of electrical inductance in the SI system of units. With the aid of table 1.3, express the unit of inductance in terms of the base units, given the relationship

$$E = -L\frac{dI}{dt} \tag{1.2}$$

where E represents emf, L represents inductance, I represents electric current and t represents time.

1.3.2 Standards

How do the definitions of the SI units in table 1.2 relate to measurements made in a laboratory? For an instrument to measure a quantity in SI units, the definitions need to be made 'tangible' so that an example or *standard* of the unit is made available. Only when the definition is realised as a practical and maintainable standard can values obtained by an instrument designed to measure the quantity be compared against that standard. If there is a difference between the standard and the value indicated by the instrument, then the instrument is adjusted or *calibrated* so that the difference is minimised.

Accurate standards based on the definitions of some of the units appearing in table 1.2 are realised in specialist laboratories. For example, a clock based on the properties of caesium atoms can reproduce the second to high accuracy.[5] By comparison, creating an accurate standard of the ampere based directly on the definition of the ampere appearing in table 1.2 is much more difficult. In this case it is common for laboratories to maintain standards of related derived SI units such as the volt and the ohm, which can be implemented to very high accuracy.

Most countries have a 'national standards laboratory' which maintains the most accurate standards achievable, referred to as *primary* standards. From time to time the national laboratory compares those standards with other primary standards held in laboratories around the world. In addition, a national laboratory creates and calibrates secondary standards by reference to the primary standard. Such secondary standards are found in some government, industrial and university laboratories. Secondary standards in turn are used to calibrate and maintain working standards and eventually a working standard may be used to calibrate (for example) a hand held voltmeter used in an experiment. If the calibration process is properly documented, it is possible to trace the calibration of an instrument back to the primary standard.[6] 'Traceability' is very important in some situations, particularly when the 'correctness' of a value indicated by an instrument is in dispute.

1.3.3 Prefixes and scientific notation

Values obtained through experiment are often much larger or much smaller than the base (or derived) SI unit in which the value is expressed.

---

[5] See appendix 2 of The International System of Units (English translation) 7th Edition, 1997, published by the Bureau International des Poids et Mesures (BIPM).
[6] See Morris (1997), chapter 3.

Table 1.4. *Prefixes used with the SI system of units.*

Factor	Prefix	Symbol	factor	Prefix	Symbol
10^{-24}	yocto	y	10^{1}	deka	da
10^{-21}	zepto	z	10^{2}	hecto	h
10^{-18}	atto	a	**10^{3}**	**kilo**	**k**
10^{-15}	femto	f	**10^{6}**	**mega**	**M**
10^{-12}	**pico**	**p**	**10^{9}**	**giga**	**G**
10^{-9}	**nano**	**n**	**10^{12}**	tera	T
10^{-6}	**micro**	**μ**	10^{15}	peta	P
10^{-3}	**milli**	**m**	10^{18}	exa	E
10^{-2}	centi	c	10^{21}	zetta	Z
10^{-1}	deci	d	10^{24}	yotta	Y

In such situations there are two widely used methods by which the value of the quantity may be specified. The first is to choose a multiple of the unit and indicate that multiple by assigning a *prefix* to the unit. So, for example, we might express the value of the capacitance of a capacitor as 47 μF. The symbol μ stands for the prefix 'micro' which represents a factor of 10^{-6}. A benefit of expressing a value in this way is the conciseness of the representation. A disadvantage is that many prefixes are required in order to span the orders of magnitude of values that may be encountered in experiments. As a result, several unfamiliar prefixes exist. For example, the size of the electrical charge carried by an electron is about 160 zC. Only dedicated students of the SI system would immediately recognise z as the symbol for the prefix 'zepto' which represents the factor 10^{-21}. Table 1.4 includes the prefixes currently used in the SI system. The prefixes shown in bold are the most commonly used.

Another way of expressing the value of a quantity is to give the number that precedes the unit in scientific notation. To express any number in scientific notation, we separate the first non-zero digit from the second digit by a decimal point, so for example, the number 1200 becomes 1.200. So that the number remains unchanged we must multiply 1.200 by 10^{3} so that 1200 is written as 1.200×10^{3}. Scientific notation is preferred for very large or very small numbers. For example, the size of the charge carried by the electron is written as 1.60×10^{-19} C. Though any value may be expressed using scientific notation, we should avoid taking this approach to extremes. For example, suppose the mass of a body is 1.2 kg. This *could* be written as 1.2×10^{0} kg, but this is possibly going too far.

Example 2

Rewrite the following values using: (a) commonly used prefixes and (b) scientific notation:

(i) 0.012 s; (ii) 601 A; (iii) 0.00064 J.

ANSWER

(i) 12 ms or 1.2×10^{-2} s; (ii) 0.601 kA or 6.01×10^2 A; (iii) 0.64 mJ or 6.4×10^{-4} J.

Exercise B

1. Rewrite the following values using prefixes:

(i) 1.38×10^{-20} J in zeptojoules; (ii) 3.6×10^{-7} s in microseconds; (iii) 43258 W in kilowatts; (iv) 7.8×10^8 m/s in megametres per second.

2. Rewrite the following values using scientific notation:

(i) 0.650 nm in metres; (ii) 37 pC in coulombs; (iii) 1915 kW in watts; (iv) 125 μs in seconds.

1.3.4 Significant figures

In a few situations, a value obtained in an experiment can be exact. For example, in an experiment to determine the wavelength of light using Newton's rings,[7] the number of rings can be counted exactly. By contrast, the temperature of an object cannot be known exactly and so we must be careful when we interpret values of temperature. Presented with the statement that '*the temperature of the water bath was 21°C*' it is unreasonable to infer that the temperature was 21.0000000 °C. It is more likely that the temperature of the water was closer to 21 °C than it was to either 20 °C or 22 °C. By writing the temperature as 21 °C, the implication is that the value of temperature obtained by a single measurement is known to two figures, often referred to as *two significant figures*.

Inferring how many figures are significant simply by the way a number is written can sometimes be difficult. If we are told that the mass of a body is 1200 kg, how many figures are significant? If the instrument measures mass to the nearest 100 kg, then the 'real' mass lies between 1150 kg and 1250 kg, so in fact only the first two figures are significant. On

---

[7] See Smith and Thomson (1988).

the other hand, if the measuring instrument is capable of measuring to the nearest kilogram, then all four figures are significant. The ambiguity can be eliminated if we express the value using scientific notation. If the mass of the body, m, is correct to two significant figures we would write

$$m = 1.2 \times 10^3 \text{ kg}$$

When a value is written using scientific notation, every figure preceding the multiplication sign is regarded as significant. If the mass is correct to four significant figures then we write

$$m = 1.200 \times 10^3 \text{ kg}$$

Though it is possible to infer something about a value by the way it is written, it is better to state explicitly the uncertainty in a value. For example, we might write

$$m = (1200 \pm 12) \text{ kg}$$

where 12 kg is the uncertainty in the value of the mass. Estimating uncertainty is considered in chapter 5.

It may be required to round a value to a specified number of significant figures. For example, we might want to round 1.752×10^{-7} m to three significant figures. To do this, we consider the fourth significant figure (which in this example is a '2'). If this figure is equal to or greater than 5, we increase the third significant figure by 1, otherwise we leave the figure unchanged. So, for example, 1.752×10^{-7} m becomes 1.75×10^{-7} m to three significant figures. Using the same convention, a mass of 3.257×10^3 kg becomes 3.3×10^3 kg to two significant figures.

Exercise C

1. How many significant figures are implied by the way each of the following values is written:

    (i) 1.72 m; (ii) 0.00130 mol/cm³; (iii) 6500 kg; (iv) 1.701×10^{-3} V; (v) 100°C;
       (vi) 100.0 °C?

2. Express the following values using scientific notation to two, three and four significant figures:

    (i) 775710 m/s²; (ii) 0.001266 s; (iii) −105.4°C; (iv) 14000 nH in henrys; (v) 12.400 kJ
       in joules; (vi) 101.56 nm in metres

1.4   Picturing experimental data

The ability possessed by humans to recognise patterns and trends is so good that it makes sense to exploit this talent when analysing experimental data. Though a table of experimental values may contain the same information as appears on a graph, it is very difficult to extract useful information from a table 'by eye'. To appreciate the 'big picture' it is helpful to devise ways of graphically representing the values.

When values are obtained through repeat measurements of a single quantity, then the histogram is used extensively to display data. When a single quantity or variable is being considered, the data obtained are often referred to as 'univariate' data. By contrast, if an experiment involves investigating the relationship between two quantities, then the x–y graph is a preferred way of displaying the data (such data are often referred to as 'bivariate' data).

1.4.1   Histograms

The histogram is a pictorial representation of data which is regularly used to reveal the scatter or distribution of values obtained from measurements of a single quantity. For example, we might measure the diameter of a wire many times in order to know something of the variation of the diameter along the length of the wire. A table is a convenient and compact way to present the numerical information. However, we are usually happy (at least in the early stages of analysis) to forego knowledge of individual values in the table for a broader overview of all the data. This should help indicate whether some values are much more common than others and whether there are any values that appear to differ greatly from the others. These 'extreme' values are usually termed *outliers*.

To illustrate the histogram, let us consider data gathered in a radioactive decay experiment. In an experiment to study the emission of beta particles from a strontium 90 source, measurements were made of the number of particles emitted from the source over 100 consecutive 1 minute periods. The data gathered are shown in table 1.5. Inspection of the table indicates that all the values lie between about 1100 and 1400, but little else can be discerned. Do some values occur more often than others and if so which values? A good starting point for establishing the distribution of the data is to count the number of values (referred to as the *frequency*) which occur in predetermined intervals of equal width. The next step is to plot a graph consisting of frequency on the vertical

Table 1.5. *Counts from a radioactivity experiment.*

1265	1196	1277	1320	1248	1245	1271	1233	1231	1207
1240	1184	1247	1343	1311	1237	1255	1236	1197	1247
1301	1199	1244	1176	1223	1199	1211	1249	1257	1254
1264	1204	1199	1268	1290	1179	1168	1263	1270	1257
1265	1186	1326	1223	1231	1275	1265	1236	1241	1224
1255	1266	1223	1233	1265	1244	1237	1230	1258	1257
1252	1253	1246	1238	1207	1234	1261	1223	1234	1289
1216	1211	1362	1245	1265	1296	1260	1222	1199	1255
1227	1283	1258	1199	1296	1224	1243	1229	1187	1325
1235	1301	1272	1233	1327	1220	1255	1275	1289	1248

Table 1.6. *Grouped frequency distribution for data shown in table 1.5.*

Interval (counts)	Frequency
$1160 < x \leq 1180$	3
$1180 < x \leq 1200$	10
$1200 < x \leq 1220$	7
$1220 < x \leq 1240$	24
$1240 < x \leq 1260$	25
$1260 < x \leq 1280$	16
$1280 < x \leq 1300$	6
$1300 < x \leq 1320$	4
$1320 < x \leq 1340$	3
$1340 < x \leq 1360$	1
$1360 < x \leq 1380$	1

axis versus interval on the horizontal axis. In doing this we create a histogram.

Table 1.6, created using the data in table 1.5, shows the number of values which occur in consecutive intervals of 20 counts beginning with the interval 1160 to 1180 counts and extending to the interval 1360 to 1380 counts. This table is referred to as a *grouped frequency distribution*. The distribution of counts is shown in figure 1.1. We note that most values are clustered between 1220 and 1280 and that the distribution is almost symmetric, with the suggestion of a longer 'tail' at larger counts. Other methods by which univariate data can be displayed include stem and leaf plots and

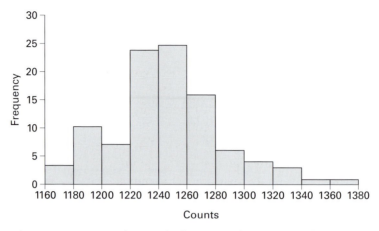

Figure 1.1. Histogram showing the frequency of counts in a radioactivity experiment.

pie charts,[8] though these tend to be used less often than the histogram in the physical sciences.

There are no 'hard and fast' rules about choosing the width of intervals for a histogram, but a good histogram:

- is easy to construct, so intervals are chosen to reduce the risk of mistakes when preparing a grouped frequency distribution. For example, an interval between 1160 and 1180 is preferable to one from (say) 1158 to 1178.
- reveals the distribution of the data clearly. If too many intervals are chosen then the number of values in each interval is small and the histogram appears 'flat' and featureless. At the other extreme, if the histogram consists of only two or three intervals, then all the values will lie in those intervals and the shape of the histogram reveals little.

In choosing the total number of intervals, N, a useful rule of thumb is to calculate N using

$$N = \sqrt{n} \tag{1.3}$$

where n is the number of values. Once N has been rounded to a whole number, the interval width, w, can be calculated using

$$w = \frac{\text{range}}{N} \tag{1.4}$$

---

[8] See Blaisdell (1998) for details of alternate methods of displaying univariate data.

where range is defined as

$$\text{range} = \text{maximum value} - \text{minimum value} \tag{1.5}$$

We should err on the side of selecting 'easy to work with' intervals, rather than holding rigidly to the value of w given by equation (1.4). If, for example, w were found using equation (1.4) to be 13.357, then a value of w of 10 or 15 should be considered, as this would make tallying up the number of values in each interval less prone to mistakes.

If there are many values then plotting a histogram 'by hand' becomes tedious. Happily, there are many computer based analysis packages, such as spreadsheets (discussed in chapter 2), which reduce the effort that would otherwise be required.

Exercise D

Table 1.7 shows the values of 52 'weights' of nominal mass 50 g used in an under-graduate laboratory. Using the values in table 1.7, construct

 (i) a grouped frequency distribution;
 (ii) a histogram.

Table 1.7. *Values of 52 weights.*

Mass (g)								
50.42	50.09	49.98	50.16	50.10	50.18	50.12	49.95	50.05
50.14	50.07	50.15	50.06	50.22	49.90	50.09	50.18	50.04
50.02	49.81	50.10	50.16	50.06	50.14	50.20	50.06	49.84
50.07	50.08	50.19	50.05	50.13	50.13	50.08	50.05	50.01
49.84	50.11	50.11	50.05	50.15	50.17	50.05	50.12	50.30
49.97	50.05	50.09	50.17	50.08	50.21	50.21		

1.4.2 Relationships and the *x–y* graph

A preoccupation of many scientists is to discover, and account for, the rela-tionship between quantities. This fairly innocent statement conceals the fact that a complex and sometimes unpredictable interplay between experiment and theory is required before any relationship can be said to be accounted for in a quantitative as well as qualitative manner. Examples of relationships that may be studied through experiment are:

 • the intensity of light emitted from a light emitting diode (LED) as the temperature of the LED is reduced;

- the power output of a solar cell as the angle of orientation of the cell with respect to the sun is altered;
- the change in electrical resistance of a humidity sensor as the humidity is varied;
- the variation of voltage across a conducting ceramic as the current through it changes;
- the decrease in the acceleration caused by gravity with depth below the earth's surface.

Let us consider the last example in a little more detail, in which the free-fall acceleration caused by gravity varies with depth below the earth's surface. Based upon considerations of the gravitational attraction between bodies, it is possible to predict a relationship between acceleration and depth when a body has uniform density. By gathering 'real data' this prediction can be examined. Conflict between theory and experiment might suggest modifications are required to the theory or perhaps indicate that some 'real' anomaly, such as the existence of large deposits of gold close to the site of the measurements, has influenced the values of acceleration.

As the acceleration in the example above depends on depth, we refer to the acceleration as the *dependent* variable, and the depth as the *independent* variable. (The independent and dependent variables are sometimes referred to as the predictor and response variables respectively.) A convenient way to record values of the dependent and independent variables is to construct a table. Though concise, a table of data is fairly dull and cannot assist efficiently with the identification of trends or patterns in data. A revealing and very popular way to display bivariate data is to plot an x–y graph (sometimes referred to as a scatter graph). The 'x' and the 'y' are the symbols used to identify the horizontal and vertical axes respectively of a Cartesian co-ordinate system.[9]

If properly prepared, a graph is a potent summary of many aspects of an experiment.[10] It can reveal:

- the quantities being investigated;
- the number and range of values obtained;
- gaps in the measurements;
- a trend between the x and y quantities;
- values that conflict with the trend shown by the majority of the data;
- the extent of uncertainty in the values (sometimes indicated by 'error bars').

[9] The horizontal and vertical axes are sometimes referred to as the abscissa and ordinate respectively.

[10] Cleveland (1994) discusses what makes 'good practice' in graph plotting.

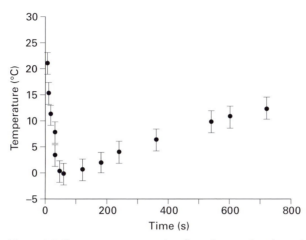

Figure 1.2. Temperature versus time for a thermoelectric cooler.

If a graph is well constructed, this qualitative information can be 'absorbed' in a few seconds. To construct a good graph we should ensure that:

- a caption describing the graph is included;
- axes are clearly labelled (and the label includes the unit of measurement);
- the scales for each axis are chosen so that plotting, if done by hand, is made easy so that values can be read easily from the graph;
- the graph is large enough to allow for the efficient extraction of information 'by eye';
- plotted values are clearly marked.

An x–y graph is shown in figure 1.2 constructed from data gathered in an experiment to establish the cooling capabilities of a thermoelectric cooler (TEC).[11] Attached to each point in figure 1.2 are lines which extend above and below the point. These lines are generally referred to as *error bars* and in this example are used to indicate the uncertainty in the values of temperature.[12] The 'y' error bars attached to the points in figure 1.2 indicate that the uncertainty in the temperature values is about 2 °C. As 'x' error bars are absent we infer that the uncertainty in values of time is too small to plot on this scale.

[11] A thermoelectric cooler is a device containing junctions of semiconductor material. When a current passes through the device, some of the junctions expel thermal energy (causing a temperature rise) while others absorb thermal energy (causing a temperature drop).

[12] Chapter 5 considers uncertainties in detail.

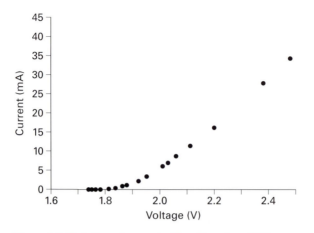

Figure 1.3. Variation of current with voltage for a LED.

If an x–y graph consists of many points, a convenient way to plot those points and attach error bars is to use a computer based spreadsheet (see section 2.7.1).

1.4.3 Logarithmic scales

The scales on the graph in figure 1.2 are linear. That is, each division on the x axis corresponds to a time interval of 200 s and each division on the y axis corresponds to a temperature interval of 5 °C. In some situations important information can be obscured if linear scales are employed. As an example, consider the current–voltage relationship for a LED as shown in figure 1.3. It is difficult to determine the relationship between current and voltage for the LED in figure 1.3 for values of voltage below about 2 V. As the current data span several orders of magnitude, the distribution of values can be more clearly discerned by replacing the linear y scale in figure 1.3 by a logarithmic scale. Though graph paper is available in which the scales are logarithmic, many computer based graph plotting routines, including those supplied with spreadsheet packages, allow easy conversion of the y or x or both axes from linear to logarithmic scales. Figure 1.4 shows the data from figure 1.3 replotted using a logarithmic y scale. As one of the axes remains linear, this type of graph is sometimes referred to as semi-logarithmic.

> **Exercise E**
> The variation of current through a Schottky diode is measured as the temperature of the diode increases. Table 1.8 shows the data gathered in the experiment. Choosing appropriate scales, plot a graph of current versus temperature for the Schottky diode.

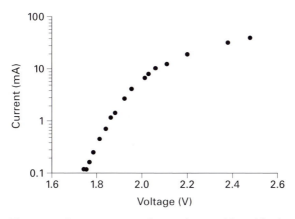

Figure 1.4. Current versus voltage using semi-logarithmic scales on the *x* and *y* axes.

Table 1.8. *Variation of current with temperature for a Schottky diode.*

Temperature (K)	Current (A)
297	2.86×10^{-9}
317	1.72×10^{-8}
336	6.55×10^{-8}
353	2.15×10^{-7}
377	1.19×10^{-6}
397	3.22×10^{-6}
422	1.29×10^{-5}
436	2.45×10^{-5}
467	9.97×10^{-5}
475	1.41×10^{-4}

1.5 Key numbers summarise experimental data

A significant challenge facing all experimenters is to find ways to express data in a concise fashion without obscuring important features. The histogram can give us the 'big picture' regarding the distribution of values and can alert us to important features such as lack of symmetry in the distribution, or the existence of outliers. This information, though very important, is essentially qualitative. What quantitative measures can we use to summarise all the data?

1.5.1 The mean and the median

It might seem surprising that many values may be faithfully summarised by a single statistic. But this is exactly what is done on a routine basis. Suppose that, as part of an experiment, we are required to measure the diameter of a wire. Upon making the measurements of diameter with a micrometer we find small but consistent variations in the diameter along the wire (these could be due to 'kinks', bends or scratches in the wire, or lack of experience in using the measuring instrument). Whatever the cause of the variations, there is unlikely to be any reason for favouring one particular value over another. What is required is to determine an 'average' of the values which is regarded as representative of all the values. Many types of average may be defined, but the most frequently used in the physical sciences is the *mean*, $\bar{x}$, which is defined as

$$\bar{x} = \frac{x_1 + x_2 + x_3 + \cdots + x_n}{n} = \frac{\sum_{i=1}^{i=n} x_i}{n} \tag{1.6}$$

where x_i denotes the ith value and n is the number of values.[13]

If values have been grouped so that the value x_i occurs f_i times, then the mean is given by

$$\bar{x} = \frac{\sum f_i x_i}{\sum f_i} \tag{1.7}$$

Another 'average' is the *median* of a group of values. Suppose data are ordered from the smallest to the largest value. The median separates the ordered values into two halves. As an example, consider the data in table 1.9 which shows values of the resonance frequency of an ac circuit. Table 1.10 shows the resonant frequencies arranged in ascending order. As the median divides the ordered values into two halves, it must lie between 2139 Hz and 2144 Hz. The median is taken to be the mean of these two values,

i.e. $\dfrac{2139 + 2144}{2} = 2141.5$ Hz

In general, if there is an even number of values, the median is the mean of the $\frac{1}{2}n$th value and the $(\frac{1}{2}n + 1)$th value. If there is an odd number of values, then the median is the $(\frac{1}{2}n + \frac{1}{2})$th value.

Although the mean and median differ little when there is a symmetric spread of data, they do differ considerably when the spread is asymmet-

---

[13] The limits of the summation are usually not shown explicitly, and we write $\bar{x} = \sum x_i / n$.

Table 1.9. *Resonance frequency of an ac circuit.*

Frequency (Hz)	2150	2120	2134	2270	2144	2156	2139	2122

Table 1.10. *Resonance frequency in ascending order from left to right.*

Frequency (Hz)	2120	2122	2134	2139	2144	2150	2156	2270

Median is the mean of the values 2139 Hz and 2144 Hz

ric, or if the group contains an outlier. As an example, for the data in table 1.9, the mean is 2154.4 Hz and the median is 2141.5 Hz. The difference between the mean and median is 12.9 Hz. The 'discrepancy' between mean and median is largely due to the outlier with value 2270 Hz. If the outlier is discarded, then the mean is 2137.9 Hz and the median is 2139 Hz, representing a difference of just over 1 Hz.

We are not suggesting that outliers should be discarded, as this is a matter requiring very careful consideration. However, this example does illustrate that the mean is more sensitive to outliers than the median. Despite this sensitivity, the mean is much more widely used than the median in the physical sciences for characterising the average of a group of values.

While the mean is arguably the most important number that can be derived from a group of repeat measurements, by itself it tells us nothing of the spread of the values. We need another number representative of the spread of the data.

Exercise F

Consider the data in table 1.11. Determine the mean and the median of the values of capacitance in this table.

Table 1.11. *Capacitance values.*

Capacitance (pF)	103.7	100.3	98.4	99.3	101.0	106.1	103.9	101.5	100.9	105.3

1.5.2 Variance and standard deviation

The starting point for finding a number which usefully describes the spread of values is to calculate the deviation from the mean of each value. If the ith value is written as x_i, then the deviation, d_i, is defined as[14]

$$d_i = x_i - \bar{x}$$ (1.8)

where $\bar{x}$ is the mean of the values.

At first inspection it appears plausible to use the mean of the sum of the deviations as representative of the spread of the values. In this case,

$$\text{mean deviation} = \frac{\sum d_i}{n} = \frac{1}{n}\sum (x_i - \bar{x})$$

and expanding the brackets gives

$$\text{mean deviation} = \frac{\sum x_i}{n} - \frac{\sum \bar{x}}{n}$$

$$= \bar{x} - \frac{n\bar{x}}{n} = 0$$

As the mean deviation is always zero, it is not a promising candidate as a number useful for describing the amount of spread in a group of values.

As a useful measure of spread we introduce the *variance*, σ^2, which is defined as the mean of the sum of the square of the deviations, so that

$$\sigma^2 = \frac{\sum (x_i - \bar{x})^2}{n}$$ (1.9)

One of the difficulties with using variance as a measure of spread of values is that its units are the square of the units in which the measurements were made. A new quantity based on the variance is therefore defined which is the *standard deviation* and is equal to the square root of the variance. Representing the standard deviation by σ, we have

$$\sigma = \left[\frac{\sum (x_i - \bar{x})^2}{n} \right]^{\frac{1}{2}}$$ (1.10)

Except in situations where we retain extra figures to avoid rounding errors in later calculations, we will express standard deviations to two significant figures.[15]

---

[14] d_i is sometimes referred to as the *residual*.

[15] See Barford (1985) for a discussion of rounding standard deviations.

Example 3

A rare earth oxide gains oxygen when it is heated at 600 °C in an oxygen-rich atmosphere. Table 1.12 shows the mass gain from twelve samples of the oxide which were heated to a temperature of 600 °C for 10 hours. Calculate (i) the mean, (ii) the standard deviation and (iii) the variance of the values in table 1.12.

ANSWER

(i) The mean of the values in table 1.12 = 5.9083 mg.

(ii) The standard deviation, as defined by equation (1.10), can be found on many scientific pocket calculators, such as those made by CASIO and Hewlett Packard. An alternative is to use a computer based spreadsheet, as most have built in functions for calculating σ. If neither of these options is available then it is possible to use equation (1.10) directly. To facilitate computation, equation (1.10) is rearranged into the form

$$\sigma = \left[\frac{\sum x_i^2}{n} - (\bar{x})^2 \right]^{\frac{1}{2}} \tag{1.11}$$

For the data in table 1.12, $\sum x_i^2 = 422.53$ (mg)^2 and $\bar{x} = 5.9083$ mg, so that

$$\sigma = \left[\frac{422.53}{12} - (5.9083)^2 \right]^{\frac{1}{2}} = 0.55 \text{ mg}$$

(iii) The variance $= \sigma^2 = 0.30$ (mg)^2.

Table 1.12. *Mass gain of twelve samples of ceramic.*

Mass gain (mg)	6.4	6.3	5.6	6.8	5.5	5.0	6.2	6.1	5.5	5.0	6.2	6.3

Exercise G

1 Show that equation (1.10) may be rewritten in the form given by equation (1.11).

2 When a hollow glass tube of narrow bore is placed in water, the water rises up the tube due to capillary action. The values of height reached by the water in a small bore glass tube are shown in table 1.13. For these values determine:

(i) the range;
(ii) the mean;
(iii) the median;
(iv) the variance;
(v) the standard deviation.

Table 1.13. *Heights to which water rises in a capillary tube.*

Height (cm)	4.15	4.10	4.12	4.12	4.32	4.20	4.18	4.13	4.15

1.6 Population and sample

In an experiment we must make decisions regarding the amount of time to devote to gathering data. This inevitably means that fewer measurements are made than would be 'ideal'. But what *is* ideal? This depends on the experiment being performed. As an example, suppose we want to know the mean and standard deviation of the resistance of a batch of 100 000 newly manufactured resistors. Ideally we would measure the resistance, R_i, of every resistor then calculate the mean resistance, $\bar{R}$, using

$$\bar{R} = \frac{\sum_{i=1}^{i=100\,000} R_i}{100\,000} \tag{1.12}$$

If the resistance of every resistor *is* measured, we regard the totality of values produced as the *population*. The standard deviation of the resistance values may be determined using equation (1.10).

Measuring the resistance of every resistor is costly and time consuming. Realistically, measurements are made of n resistors drawn at random from the population, where $1 < n \ll 100\,000$. The values of resistance obtained are regarded as a *sample* taken from a larger population. We hope (and anticipate) that the sample is representative of the whole population, so that the mean and the standard deviation of the sample are close to that of the population mean and standard deviation.

The population of resistors in the previous example, though quite large, is finite. There are other situations in which the size of the population is regarded as infinite. Suppose, for example, we choose a single resistor and measure its resistance many times. The values of resistance obtained are not constant but vary due to many factors including ambient temperature fluctuations, 50 Hz electrical interference, stability of the measuring instrument and (if we carried on for a very long time) ageing of the resistor. The number of repeat measurements of resistance that *could* be made on a single resistor is infinite and so we should regard the population as infinite. No matter the size of the population, we can estimate the mean and standard deviation by considering a sample drawn from the population. We will discover in chapter 3 that the larger the sample, the better are these estimates.

1.6.1 Population parameters

A population of values has a mean and a standard deviation. Any number that is characteristic of a population is referred to as a *population parameter*. One such parameter, the population mean, is usually denoted by the Greek symbol, μ. In situations in which the population is infinite, μ is defined as

$$\mu = \lim_{n \to \infty} \frac{\sum x_i}{n} \tag{1.13}$$

Similarly, for an infinite population, the standard deviation of the population, σ (most often referred to as the 'population standard deviation'), is given by

$$\sigma = \lim_{n \to \infty} \left[\frac{\sum (x_i - \mu)^2}{n} \right]^{\frac{1}{2}} \tag{1.14}$$

1.6.2 True value and population mean

The term 'true value' is often used to express the value of a quantity that would be obtained if no influences, such as shortcomings in an instrument used to measure the quantity, existed to 'interfere with' the measurement. In order to determine the true value of a quantity, we require that the following conditions hold:

- The quantity being measured does not change over the time interval in which the measurement is made.
- External influences that might affect the measurement, such as 50 Hz electrical interference or changes in room temperature and humidity, are absent.
- The instrument used to make the measurement is 'ideal'.[16]

As an example, suppose an experiment is devised to measure the charge carried by an electron. If only experimental techniques were good enough, and measuring instruments ideal, we would be able to know the value of the charge exactly. As the ideal instrument has yet to be devised, we must make do with values obtained using the best instruments at our disposal.

---

[16] An ideal instrument would have many (non-attainable) attributes such as freedom of influence on the quantity being measured (see section 5.9.3) and the capability of infinitely fine resolution.

By making many repeat measurements and taking the mean of the values obtained, we might expect that 'scatter' due to imperfections in the meas-urement process would cancel out, in which case the mean would be close to the true value. We might even go one step further and suggest that if we make an infinite number of measurements, the mean of the values (i.e. the population mean, μ) will coincide with the true value. Unfortunately, due to systematic errors,[17] not all imperfections in the measurement process 'average out' by taking the mean, and so we must be very careful when referring to the population mean as the 'true value'.

There are situations in which the term 'true value' may be mislead-ing. Returning to our example in which a population consists of the resis-tances of 100000 resistors, there is no doubt that this population has a mean, but in what sense, if any, is this population mean the 'true value'? Unlike the example of determination of the charge on an electron, in which variability in values is due to inadequacies in the measurement process, the variability in resistance values is mainly due to variations *between* resistors introduced during manufacture. Therefore no true value, in the sense used to describe an attribute of a single entity, such as the charge on an electron, exists for the group of resistors.

1.6.3 Sample statistics

As values of a quantity usually show variability, we take the mean, $\bar{x}$, of values obtained through repeat measurement as the best estimate of the true value (or the population mean). $\bar{x}$ is referred to as the *sample mean*, where

$$\bar{x} = \frac{\sum x_i}{n}$$

(1.15)

n is the number of values in the sample. As $\bar{x}$ is determined using a sample of values drawn from a population, it is an example of a *sample statistic*. As n tends to a large value, then $\bar{x} \to \mu$, as given by equation (1.13).

Another extremely important statistic determined using sample data is the estimate of the population standard deviation, s, given by

$$s = \left[\frac{\sum (x_i - \bar{x})^2}{n-1} \right]^{\frac{1}{2}}$$

(1.16)

As n tends to a very large number then $\bar{x} \to \mu$ and the difference between n and $n-1$ becomes negligible, so that $s \to \sigma$.

---

[17] Sections 1.7.2 and 5.9 deal with systematic errors.

Subtracting 1 from n in the denominator of equation (1.16) is some-times referred to as the 'Bessel correction'.[18] It arises from the fact that before we can use data to calculate s using equation (1.16), we must first use the same data to determine $\bar{x}$ using equation (1.15). In doing this we have reduced the number of *degrees of freedom* by 1. The degrees of freedom[19] is usually denoted by the symbol, ν.

Example 4

In a fluid flow experiment, the volume of water flowing through a pipe is determined by collecting water at 1 minute intervals in a measuring cylinder. Table 1.14 shows the volume of water collected in ten successive 1 minute intervals.

(i) Determine the mean of the values in table 1.14.
(ii) Estimate the standard deviation and the variance of the population from which these values were drawn.

ANSWER

With only ten values to deal with, it is quite easy to find $\bar{x}$ using equation (1.15). Determining s using equation (1.16) requires more effort. Most scientific calculators allow you to enter data and will calculate $\bar{x}$ and s. Perhaps an even better alternative is to use a spreadsheet (see chapter 2) as values entered can be inspected easily before proceeding to the calculations.

(i) Using equation (1.15), $\bar{x} = 251.3$ cm^3.
(ii) Using equation (1.16), the estimate of population standard deviation $s = 6.1$ cm^3. The variance $s^2 = 37$ (cm^3)$^2 = 37$ cm^6.

Table 1.14. *Volume of water collected in 1 minute intervals.*

Volume (cm^3)	256	250	259	243	245	260	253	254	244	249

Exercise H

In an experiment to study electrical signals generated by the human brain, the time was measured for a particular 'brain' signal to double in size when a person closes both eyes. The values obtained for 20 successive eye closures are shown in table 1.15. Using the values in the table, determine the sample mean and estimate the standard deviation of the population.

[18] See Kennedy and Neville (1986).
[19] In general, ν is equal to the number of observations, n, minus the number of constraints (in this situation there is one constraint, given by equation (1.15)).

Table 1.15. *Times for electrical signal to increase by a factor of 2.*

Time (s)									
4.51	2.33	1.51	1.91	2.54	1.91	1.51	1.52	2.71	3.03
2.12	2.61	0.82	2.51	2.07	1.73	2.34	1.82	2.32	1.92

1.6.4 Which standard deviation do we use?

A reasonable question to ask is 'which equation should be used to determine standard deviation?' (After all we have already introduced three, equations (1.10), (1.14) and (1.16).)

The first equation we can 'dismiss' is equation (1.14). The reason for this is that the parameter μ appears in the equation, and there is no way we can determine this parameter when the population is infinite. The best we can do is to estimate μ. As our goal is to identify the characteristics of the population from which the values are drawn, we choose equation (1.16) in preference to equation (1.10). This is because equation (1.16) *is* an estimate of the population standard deviation. By contrast, equation (1.10) gives the standard deviation of a population when *all* the values that make up that population are known. Since this rarely applies to data gathered in experiments in the physical sciences, equation (1.10) is not favoured.

Though equation (1.16) is preferred to equation (1.10), it is not difficult to demonstrate that, as the number of values, n, becomes large, the difference between the standard deviation calculated using each becomes negligible. To show this, we write the percentage difference, d, between the standard deviations given by equations (1.10) and (1.16) as

$$d = \left(\frac{s - \sigma}{s}\right) \times 100\% \tag{1.17}$$

By substituting equations (1.16) and (1.10) for s and σ respectively, we obtain

$$d = \left[1 - \left(\frac{n-1}{n}\right)^{\frac{1}{2}}\right] \times 100\% \tag{1.18}$$

Figure 1.5 shows the variation of d with n as given by equation (1.18), for n between 3 and 100. The graph indicates that as n exceeds 10, the percentage difference between standard deviations falls below 5%.

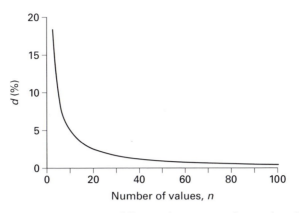

Figure 1.5. Percentage difference between s and σ as a function of n.

1.6.5 Approximating s

Calculating the standard deviation using equation (1.16) is time consuming, though using a spreadsheet or pocket calculator reduces the effort considerably. As the standard deviation, s, is usually not required to more than two significant figures (and often one figure is sufficient), we can use an approximation which works well so long as the number of values, n, does not exceed 12. The approximation is[20]

$$s \approx \frac{\text{range}}{\sqrt{n}} \qquad (1.19)$$

where the range is the difference between the maximum value and the minimum value in any given set of data (see equation (1.5)). We can regard equation (1.19) as a 'first order approximation' to equation (1.16) for small n. Equation (1.19) is extremely useful for determining s rapidly and with the minimum of effort.[21]

Example 5
Table 1.16 contains values of repeat measurements of the distance between a lens and an image produced using the lens. Use equations (1.16) and (1.19) to determine the standard deviation of the values in table 1.16.

---

[20] See Lyon (1980) for a discussion of equation (1.19).
[21] In situations in which n exceeds 12, the value of s given by equation (1.19) is consistently less than s as calculated using equation (1.16).

ANSWER
Using equation (1.16), $s = 2.5$ cm.
Using equation (1.19),

$$s \approx \frac{(42-35)\text{cm}}{\sqrt{7}} = 2.6 \text{ cm}$$

The difference between the two values of s is about 5% and would be regarded as unimportant for most applications.

Table 1.16. *Image distances.*

Image distance (cm)	35	42	40	38	39	36	36

Exercise I

The periods, T, of oscillation of a body on the end of a spring is measured and the values obtained are given in table 1.17.

  (i)   Use equation (1.16) to find the standard deviation, s.
 (ii)   Use equation (1.19) to find an approximate value for s.
(iii)   What is the percentage difference between the standard deviations calculated parts (i) and (ii)?

Table 1.17. *Periods of oscillation of a body on the end of a spring.*

T (s)	2.53	2.65	2.67	2.56	2.56	2.60	2.67	2.64	2.63

1.7 Experimental error

The elusive true value of a quantity sought through experiment could be established with a single measurement so long as 'interfering' factors (such as human reaction time when measuring an interval of time using a stopwatch) could be eliminated. Such factors are the cause of *errors* in values obtained through measurement. If a scale is misread on a meter, or a number transcribed incorrectly, this is not regarded as an error in the context of data analysis, but as a mistake which (in principle) can be avoided. Errors exist in all experiments and though their effects can be minimised in a number of ways, such as controlling the physical environment in which an experiment is performed, or by making repeat measurements, they cannot be eliminated completely. Chapter 5 deals in some detail with the identification and treatment of errors, but it is appropriate to introduce here two important categories of errors: random and systematic.

1.7.1 Random error

As a consequence of random errors, measured values are either above or below the true value. In fact, the most obvious 'signature' of random errors is a scatter or variability of values obtained through experiment.

  If we know the size of the error, we can correct the measured value by an amount equal to the error and hence obtain the true value. Regrettably, the size of the error is as difficult to determine as the true value. Nevertheless, by gathering many values we can quantify the scatter caused by random errors through determining the standard deviation of the values. The existence of random error introduces some *uncertainty* in the estimate of the true value of a quantity. Though the error cannot be determined, the uncertainty can. This is very important since, by calculating the uncertainty, we can express limits between which the true value of a quantity is expected to lie. When repeat measurements of a quantity are made, we can write

$$\text{true value} = \text{sample mean} \pm \text{uncertainty}$$

1.7.2 Systematic error

The existence of random errors is revealed through the variability in measured values. By contrast, systematic errors give no such clues. In an experiment, a poorly calibrated or faulty thermometer may consistently indicate water temperature as 25 °C, when the true value is close to 20 °C. The consistent *offset* or *bias* which causes the measured value to be consistently above or below the true value is most often referred to as a *systematic* error. From this example it might seem that all we need do is check the calibration of the instrument, correct the values appropriately and thereby eliminate the systematic error. Though we might reduce the systematic error in this way it cannot be eliminated as it requires the calibration procedure to be 'perfect' and that relies on some superior instrument determining the true value. No matter how 'superior' the instrument is, it still cannot determine the true value of the quantity, so some systematic error must remain.

Further systematic error may be introduced when dynamic measurements are made. For example, a thermometer may be used to measure the temperature of a water bath while heat is being transferred to the bath. Irrespective of how quickly a thermometer responds to a temperature rise, there will be some small difference between the water temperature and the temperature indicated by the thermometer.

1.7.3 Repeatability and reproducibility

A particular quantity, for example the time for a ball to fall through a viscous liquid, may be measured a number of times by the same observer using the same equipment where the conditions of the measurement (including the environmental conditions) remain unchanged. If the scatter of values produced is small, we say the measurement is *repeatable*. By contrast, if measurements are made of a particular quantity by various workers using a variety of instruments and techniques in different locations, we say that the measurements are *reproducible* if the values obtained by the various workers are in close agreement. The terms repeatability and reproducibility are qualitative only and in order to assess the degree of repeatability or reproducibility we must consider carefully the amount of scatter in the values as revealed by quantitative estimates, such as the standard deviation.

1.8 Modern tools of data analysis – the computer based spreadsheet

The availability of powerful computer based data analysis packages has reduced the amount of effort required to plot histograms and graphs, calculate standard deviations and perform sophisticated statistical analysis of data. A data set consisting of hundreds or thousands of values is no more difficult to analyse than one containing ten or twenty values. Though advanced statistical analysis tools are still the reserve of expensive and dedicated statistics packages, many spreadsheets incorporate analysis features useful to the scientist (and at less cost). One such spreadsheet package is Excel® by Microsoft. This spreadsheet is so useful and widely available that we have incorporated its use in this text where it complements the discussion of methods of analysis.

Perhaps it is worth sounding a note of caution even at this early stage. Spreadsheets (or any other computer based package) can return meaningless numbers with commendable speed. Using a spreadsheet places a great responsibility on the user to understand what is going on 'behind the scenes' when a calculation is performed. A preliminary evaluation of data (such as roughly estimating the mean of many values) can often help anticipate the number that the spreadsheet will return. If the number differs greatly from that expected, further investigation is warranted. Despite this concern, we recommend the use of spreadsheets for data analysis in science and recognise that spreadsheets have become so widely available that their use is taken as much for granted as the pocket calculator.

1.9 Review

Prior to the analysis of data, it is proper to consider issues relevant to good experimentation. Good experimentation begins with careful consideration of such matters as:

- What is the purpose of the experiment?
- Is the experimental design feasible?
- What factors might constrain the experiment (such as the availability of equipment)?
- Which 'exploratory' data analysis techniques should be used (such as the determination of means and standard deviations) and what display methods will allow the most insight to be drawn from the data?

In this chapter we have considered issues associated with experimentation and data analysis, such as experimental design and how to express values determined through experiment. Of special importance is the capacity to display data in ways that reveal the main features or trends in the data, as well as exposing anomalies or outliers. We suggested that graphs and histograms provide excellent visual summaries of data and, if carefully constructed, these can reveal features of interest that would be extremely difficult to identify through inspection of a table of 'raw' data.

Complementary to summarising data visually is to summarise data numerically. To this end we introduced the mean and standard deviation as the most frequently used measures of 'average' and 'scatter' respectively. In addition, we briefly considered units, standards, errors and uncertainty.

In the next chapter we will consider the spreadsheet as a powerful computer based tool that can assist in the efficient analysis of experimental data.

Problems

1. What are the derived SI units of the quantities:

    (i) specific heat capacity;

    (ii) pressure;

    (iii) thermal conductivity?

2. Express the units for the quantities in problem 1 in terms of the base units of the SI system.

3. The force, F, between two point charges Q_1 and Q_2, separated by a distance, r, in a vacuum is given by

$$F = \frac{Q_1 Q_2}{4\pi\varepsilon_0 r^2}$$

where ε_0 is the permittivity of free space. Use this equation and the information in table 1.3 to express the unit of ε_0 in terms of the base units in the SI system.

4. Write the following values in scientific notation to two significant figures:

    (i) 0.0000571 s; (ii) 13700 K; (iii) 1387.5 m/s; (iv) 101300 Pa;
    (v) 0.001525 Ω.

5. The lead content of river water was measured five times each day for 20 days. Table 1.18 shows the values obtained in parts per billion (ppb).

    (i) Construct a histogram beginning at 30 ppb and extending to 70 ppb with interval widths of 5 ppb.

    (ii) Determine the median value of the lead content.

6. Table 1.19 contains values of the retention time (in seconds) for pseudo-ephidrine using high pressure liquid chromatography (HPLC).

    (i) Construct a histogram using the values in table 1.19.

    (ii) Determine the mean retention time.

    (iii) Estimate the population standard deviation of the values in table 1.19.

7. The variation of electrical resistance of a humidity sensor with relative humidity is shown in table 1.20.

    (i) Plot a graph of resistance versus relative humidity. (Consider: should linear or logarithmic scales be employed?)

    (ii) Use your graph to estimate the relative humidity when the resistance of the sensor is 5000 Ω.

8. Consider table 1.21 which contains 50 values of nitrate ion concentration.

Table 1.18. *Lead content of river water (in ppb).*

43	58	53	49	60	48	48	49	49	45
57	35	51	67	49	51	59	55	62	59
40	52	53	70	48	51	44	51	46	47
42	41	53	56	40	42	47	54	54	56
46	40	57	57	47	54	48	61	51	56
56	57	45	42	54	56	66	48	48	52
64	56	49	54	66	45	63	49	65	40
54	54	50	56	51	49	51	46	52	41
43	46	57	51	46	68	58	69	52	49
52	53	46	53	44	36	54	60	61	67

Table 1.19. *Retention times for pseudoephidrine (in seconds).*

6.035	6.049	6.032	6.065	6.057	6.069	6.084	6.110
6.122	6.066	6.072	6.046	6.100	6.262	6.262	6.252
6.276	6.042	6.067	6.054	6.098	6.093	6.072	6.124
6.085	6.076	6.045	6.067	6.220	6.223	6.271	6.219

Table 1.20. *Variation of the resistance of a sensor with humidity.*

Resistance (Ω)	Relative humidity (%)
90.5	98
250	90
945	80
3250	70
8950	60
22500	50
63500	40
82500	30
124000	20

(i) Calculate the mean, $\bar{x}$, and standard deviation, s, of the nitrate concentration.

(ii) Draw up a histogram for the data in table 1.21.

Table 1.21. *Nitrate ion concentration (in μmol/mL).*

0.481	0.462	0.495	0.493	0.501	0.481	0.479	0.503	0.497
0.506	0.512	0.457	0.521	0.474	0.487	0.504	0.445	0.490
0.455	0.480	0.509	0.484	0.475	0.511	0.481	0.509	0.500
0.490	0.550	0.493	0.513	0.483	0.505	0.503	0.501	0.506
0.480	0.491	0.493	0.509	0.475	0.482	0.486	0.479	0.488

Table 1.22. *Vertical displacement of a GPS receiver.*

d (m)	120	108	132	125	118	106	115	103
	117	120	135	129	123	127	128	

9. Using accurate clocks in satellites orbiting the earth, the global position-ing system (GPS) may establish the position of a GPS receiver. Table 1.22 shows 15 values of the vertical displacement, d, (relative to mean sea level) of a receiver, as determined using the GPS. Using these values:

  (i) calculate the mean displacement and median displacement;
 (ii) calculate the range of values;
(iii) estimate the population standard deviation and the population vari-ance.

Chapter 2

Excel® and data analysis

2.1 Introduction

Thorough analysis of experimental data frequently requires extensive and time consuming numerical manipulation. Many tools exist to assist in the analysis of data, ranging from the pocket calculator to dedicated computer based statistics packages. Despite crude editing facilities and limited display options, the pocket calculator remains an effective tool for basic analysis due to its convenience, robustness and reliability. Intensive data analysis may require a statistics package such as Systat or Statistica.[1] As well as standard functions, such as those used to determine means and standard deviations, these packages possess advanced features required by researchers. Between the extremes of calculator and specialised statistics package is the spreadsheet. Originally designed for business users, spreadsheet packages have become popular with a wider audience due to their versatility and ease of use. The inclusion of advanced features into spreadsheets means that, in many situations, a spreadsheet is a viable alternative to a specialist statistics package. The most widely used spreadsheet package available for personal computers (PCs) is Excel® by Microsoft. Excel® appears throughout this book in the role of a convenient data analysis tool with short sections within most chapters devoted to describing specific features. Its clear layout, extensive help facilities, range of in built statistical functions and availability for both PCs and Macintosh computers make Excel® a popular choice for data analysis. This chapter introduces Excel® and describes some

---

[1] Systat is a product of Systat Inc, Illinois. Statistica is a product of Statsoft Inc, Oklahoma.

of its basic features using examples drawn from the physical sciences. Some familiarity with using a PC is assumed, to the extent that terms such as 'mouse', 'cursor', 'Enter key' and 'save' are assumed understood in the context of using a program such as Excel®.

2.2   What is a spreadsheet?

A computer based spreadsheet is a sophisticated and versatile analysis and display tool for numeric and text based data. As well as the usual arithmetic and mathematical functions found on a pocket calculator, spreadsheets offer other features such as data sorting and display of data in the form of an x–y graph. Some spreadsheet packages include more advanced analysis options such as linear regression and hypothesis testing. An attractive feature is the ability to 'import' data from other computer based applications, simplifying and speeding up data entry as well as avoiding mistakes caused by faulty transcription.

A spreadsheet consists of a two dimensional array of boxes, usually referred to as *cells*, into which text, symbols, numbers or formulae can be typed. An example of such an array is shown in sheet 2.1.[2] The data appearing in the array were gathered during a study of the performance of a vacuum system. The letters A and B in sheet 2.1 serve to identify the columns in the spreadsheet, and the numbers 1 to 5 identify the rows. So, for example, the number 95 appears in cell A3. Another way to identify a cell is to use a combination of letters and numbers for both rows and columns. As an example, the cell R4C2 would correspond to the cell in the fourth row, second column. In this text we confine ourselves to identifying a cell using a letter for the column and a number for the row, as illustrated by sheet 2.1.

Sheet 2.1. *Spreadsheet containing data on the variation of pressure in a vacuum system with time after 'switch on' of a vacuum pump.*

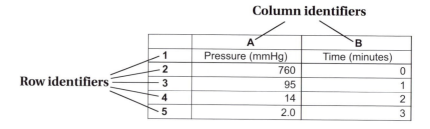

	A	B
1	Pressure (mmHg)	Time (minutes)
2	760	0
3	95	1
4	14	2
5	2.0	3

[2] We will use the term 'sheet', rather than 'table', in order to indicate that we are dealing with a spreadsheet.

As it stands, sheet 2.1 represents little more than a table constructed on paper into which values have been entered.[3] Where a computer based spreadsheet differs from a piece of paper is the dynamic way in which data in the spreadsheet can be manipulated, linked together and presented in the form of a graph. For example, the analysis of the data in sheet 2.1 may require that the pressure be expressed in pascals instead of mmHg, the time be expressed in seconds, and that a graph of pressure versus time be plotted. It is not a difficult task to perform the conversions or plot the graph 'by hand', but if the experiment consists of gathering pressure data at 1 minute intervals for an hour then the task of manipulating so much data and plotting a graph certainly becomes laborious and time consuming. This is where the spreadsheet comes 'into its own'. With a few clicks of a mouse button, formulae can be entered into the spreadsheet to perform the conversions. An extremely attractive feature is the ability to link cells dynamically so that, for example, where values have to be corrected, all calculations and graphs using those values are updated as soon as they are entered. This facility is extremely powerful and allows you to investigate the consequences, for example, of omitting, adding or changing one or more values in a set of data. This is sometimes referred to as a 'What if' calculation.

2.3 Introduction to Excel®

Designed originally with business users in mind, Excel® has evolved since the late 1980s into a powerful spreadsheet package capable of providing valuable support to users from science, mathematics, statistics, engineering and business disciplines. With regard to the analysis of experimental data, Excel® possesses 80 or so built in statistical functions which will evaluate such things as the standard deviation, maximum, minimum and mean of values. More advanced analysis facilities are available such as linear and multiple regression, hypothesis testing, histogram plotting and random number generation. Graphing options include pie and bar graphs and, perhaps the most widely used graph in the physical sciences, the 'x–y' graph.

At the time of writing, the latest version of Excel® for PCs is Excel® 2002. While recent versions of Excel® offer many short cuts to 'setting up'

---

[3] In chapter 1 we defined a 'value' as consisting of the product of number and a unit. When we refer in this and other chapters to 'entering a value' into a spreadsheet, it is taken for granted that we are considering only the numerical part of the value.

and using a spreadsheet,[4] they are not vital for solving data analysis problems and we will use them sparingly.

For convenience, sheets containing data referred to in this chapter can be found on the Internet at http://uk.cambridge.org/resources/0521 793378. Sheets in the file have the same name as the sheets referred to in this chapter. Files available can be read by versions of Excel® for PCs from Excel® 97 onwards. Some minor differences may be evident between how a spreadsheet appears in this book and in the corresponding Excel® file. For example, column widths or the number of figures displayed in each cell may differ between book and spreadsheet file.

2.3.1 Starting with Excel®

If the PC is running under the operating system Windows 95, 98 or 2000, then one way to start Excel® is to:

1. Using the left hand mouse button, click on [Start] at the bottom left hand corner of the screen.
2. Click on [Programs] then click on [Microsoft Excel].
3. After a few seconds the screen as shown in figure 2.1 appears.

Figure 2.1 shows toolbars[5] at the top of the screen. The icons on the toolbars allow the spreadsheet to be saved, printed out, a typing mistake to be undone or a list sorted into alphabetical order. A brief description of the function of each icon is obtained by moving the cursor onto the icon. Within a couple of seconds a short message appears describing the function the icon represents. Close to the top of the screen is a Menu bar offering the following options:

File Edit View Insert Format Tools Data Window Help

Moving the cursor to each word on the Menu bar and clicking the left hand mouse button causes an extensive 'pull down' menu to appear giving access to a large range of options. If an option is required that has not been used before, there may be a delay of a few seconds for that option to appear after clicking on the left hand mouse button. At the bottom and to the right of the screen there are 'scroll' bars which are useful for navigating around

---

[4] For example, see Blattner (2001).

[5] A toolbar consists of a number of icons that represent related functions. For example the Formatting toolbar includes icons that will allow numbers in cells to be centred, the font colour to be changed or the display precision to be modified.

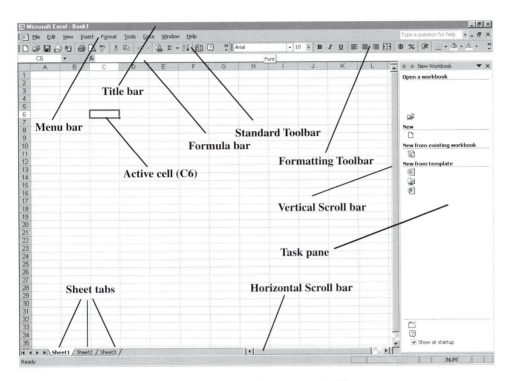

Figure 2.1. Appearance of screen after starting Excel®.

large spreadsheets. To the left of the horizontal scroll bar are sheet tabs which allow you to move from one sheet to another. Situated to the right of the screen (in Excel® 2002, but not in earlier versions) is a task pane which permits easy access to recent files used with Excel®, as well as other options such as Microsoft Excel® Help. The task pane may be closed (thereby allowing the cells to fill more of the screen) by clicking on the ✕ symbol next to the ▼ symbol in the top right corner of this pane.

2.3.2 Worksheets and Workbooks

When starting Excel®, a screen of empty cells appears as shown in figure 2.1. Sheet tabs, \Sheet1 ⧸ Sheet2 ⧸ Sheet3 ⧸, are visible near to the bottom of the screen. Switching between sheets is accomplished by clicking on each tab. Excel® refers to each of the sheets as a 'Worksheet'. These sheets (and others if they are added) constitute a 'Workbook'. A Workbook can contain many Worksheets, and incorporating several sheets into a Workbook is useful if interrelated data or calculations need to be kept together. For example, a Workbook might consist of three Worksheets, where:

- Worksheet 1 is used to store the 'raw' values as obtained directly from an experiment;
- Worksheet 2 contains values after conversion of units and other data manipulation steps have occurred;
- Worksheet 3 contains the values displayed in graphical form.

Worksheets can be renamed by moving the cursor to the Worksheet tab and double clicking on it. At this stage it is possible to overwrite (for example) 'Sheet1' with something more meaningful or memorable.

2.3.3 Entering and saving data

After starting Excel®, the majority of the screen is filled with cells. Sheet 2.2 shows cells which contain data from an experiment to establish the potassium concentration in a specimen of human blood. To begin entering text and data into the cells:

Sheet 2.2. *Potassium concentration in a blood sample.*

	A	B	C	D	E
1	Concentration (mmol/L)				
2	5.2				
3	4.7				
4	4.9				
5	4.6				
6	4.3				
7	3.9				
8	4.2				
9	5.2				
10	5.1				
11	4.6				
12					

1. Move the cursor to cell A1. Click the left hand mouse button to make the cell active. A conspicuous border appears around the active cell.
2. Type[6] **Concentration (mmol/L)**. Once the Enter key has been pressed, the active cell becomes A2.
3. Type **5.2** into cell A2 and press the Enter key. Continue this process until all the data are entered. If a transcription mistake occurs, use the mouse or the cursor key to activate the cell containing the mistake then retype the text or number.

Before moving on to manipulate the data, it is wise to save the data to disc. One way to do this is to:

1. Click on File on the Menu bar at the top of the screen.

---

[6] We adopt the convention that anything to be typed into a cell appears in **bold**.

2. Click on the Save As option.
3. Save the sheet by giving it an easy to remember name such as 'Conc1.xls', then clicking on the <u>S</u>ave button. The extension .xls iden-tifies the file as having been created by Excel®. If you do not add .xls to your filename, then Excel® does this for you.

2.3.4 Rounding, range and the display of numbers

The 'raw' values appearing in sheet 2.2 are likely to have been influenced by several factors: the underlying variation in the quantity being meas-ured, the resolution capability of the instrument used to make the meas-urement, as well as the experience, care and determination of the experimenter. As we begin to use a spreadsheet to 'manipulate' the values, we need to be aware that the spreadsheet introduces its own influences, such as those due to rounding errors or the accuracy of algorithms used in the calculation of the statistics. As with a pocket calculator, a spreadsheet is only able to store a value to a certain number of digits. However, in con-trast to a calculator, which might display up to ten digits and hold two 'in reserve' to ensure that rounding errors do not affect the least significant digit displayed, Excel® holds 15 digits internally. It is unlikely you will come across a situation in which 15 digits is insufficient.[7]

Most pocket calculators cannot handle values with magnitudes that fall outside the range 1×10^{-99} to $9.99999 \times 10^{+99}$. A value with magnitude less than 1×10^{-99} is rounded to zero, and a value with magnitude $10 \times 10^{+99}$ (or greater) causes an overflow error (displayed as -E-, or something similar, depending on the type of calculator). Excel® is able to cope with much larger and smaller values than the average scientific calculator, but cannot handle values with magnitudes that fall outside the range 2.23×10^{-308} to $9.99999 \times 10^{+307}$. A value with magnitude smaller than 2.23×10^{-308} is rounded to zero, and a value equal to or in excess of $1 \times 10^{+308}$ is regarded by Excel® as a string of characters and not a number. The limitation regard-ing the size of a value is unlikely to be of concern unless calculations are performed in which a divisor is close to zero. A division by zero causes Excel® to display the error message, #DIV/0! . If the result of a calculation is a value in excess of $9.99999 \times 10^{+307}$, then the error message #NUM! appears in the cell. Another way to provoke the #NUM! message to appear is to attempt to calculate the square root of a negative number.

[7] It is possible that if a process requires a very large number of iterations, rounding errors will become significant, especially if the calculations require the subtraction of two nearly equal numbers.

Values appear in cells as well as in the formula bar[8] as you type. Values such as 0.0023 or 1268 remain unchanged once they have been entered. However, if a value is very small, say, 0.000 000 000 0123, or very large, say, 165 000 000 000, then Excel® automatically switches to scientific notation. Thus 0.000 000 000 0123 is displayed as 1.23E − 11 and 165 000 000 000 as 1.65E + 11. The 'E' notation is interpreted as follows:

$$1.23\text{E} - 11 \equiv 1.23 \times 10^{-11}, \text{ and } 1.65\text{E} + 11 \equiv 1.65 \times 10^{11}$$

How a value appears in a cell can be modified. To illustrate some of the display options, consider sheet 2.3 containing data from an experiment in which a capacitor is discharged through a resistor. The voltage across the resistor is measured as a function of time. Cells A1 and B1 contain text indicating the quantities in the A and B columns. Neglecting to include such information risks the consequence of returning to the spreadsheet some time later with no idea of what the numbers relate to. Clear headings and formatting make an enormous difference to the readability and usefulness of any spreadsheet.

Scientific notation is often preferred for very small or very large values. We can present values in this manner using Excel®. As an example, consider the values in column B of sheet 2.3(a). These values are shown in scientific notation in column B of sheet 2.3(b). To convert values in column B to scientific notation:

Sheet 2.3. *Data from discharge of a 0.47 µF capacitor through a 12 MΩ resistor.*

(a) *voltage values in column B in 'general' format;*

(b) *voltage values in column B in scientific notation.*

	A	B
1	t(s)	V(volts)
2	0	3.98
3	5	1.58
4	10	0.61
5	15	0.24
6	20	0.094
7	25	0.035
8	30	0.016
9	35	0.0063
10	40	0.0031
11	45	0.0017
12	50	0.0011
13	55	0.0007
14	60	0.0006

	A	B
1	t(s)	V(volts)
2	0	3.98E+00
3	5	1.58E+00
4	10	6.10E-01
5	15	2.40E-01
6	20	9.40E-02
7	25	3.50E-02
8	30	1.60E-02
9	35	6.30E-03
10	40	3.10E-03
11	45	1.70E-03
12	50	1.10E-03
13	55	7.00E-04
14	60	6.00E-04

[8] The formula bar is indicated in figure 2.1.

1. Highlight cells B2 to B14 in sheet 2.3(a). This is done by moving the cursor to cell B2, pressing down on the left mouse button, then (with the button held down) 'dragging' the mouse from B2 to B14. When you reach cell B14, release the button. Every cell from B2 to B14 now lies within a clearly defined border.
2. From the Menu bar, click on Format. Click on the Cells submenu.
3. A box (usually referred to as a *dialog* box) appears headed Format Cells. Click on the Number tab. From the list that appears, choose the Scientific option.[9]
4. The dialog box shows how the first value in the column of highlighted numbers will appear in scientific notation. The number of decimal places to which the value is expressed can be modified at this stage. The default number is two decimal places. Click on the OK button. The values in column B now appear as in sheet 2.3(b).

Another option in the Format Cells dialog box, useful for science applications, is the Number category. With this option the number of decimal places to which a value is displayed may be modified (but the value is not forced to appear in scientific notation). Irrespective of how values are displayed by Excel® on the screen, they are stored internally to a precision of 15 digits.

2.3.5 Entering formulae

After entering data into a spreadsheet, the next step is usually to perform a mathematical, statistical or other operation on the data. This may be carried out by entering a formula into one or more cells. Though Excel® provides many advanced functions, very often only simple arithmetic operations such as multiplication or division are required. As an example, consider the capacitor discharge data in sheet 2.3. Suppose at each point in time we require both the current flowing through the discharge resistor and the charge remaining on the capacitor. The equations for the current, I, and charge, Q, are

$$I = \frac{V}{R} \tag{2.1}$$

$$Q = CV \tag{2.2}$$

where V is the voltage across the parallel combination of resistor and capacitor, R is the resistance and C is the capacitance.

[9] To save space, steps 2 and 3 are abbreviated as: Format ➤ Cells ➤ Number ➤ Scientific.

To enter a formula into a cell, we begin the formula with an equals sign. Sheet 2.4(a) shows a formula entered into cell C2. To calculate the current corresponding to the voltage value in cell B2, we divide the value in B2 (which is 3.98 V) by the resistance, 12 MΩ. To do this:

1. Make cell C2 active by moving the cursor to C2 and click on the left hand mouse button.
2. Type = **B2/12E6**.
3. Press the Enter key.
4. The value 3.31667E − 07 is returned[10] in cell C2, as shown in sheet 2.4(b).[11]

Sheet 2.4. *Calculation of current flowing through a 12 MΩ resistor:*

(a) *formula typed into cell C2;*

(b) *value returned in C2 when Enter key is pressed.*

	A	B	C
1	t(s)	V(volts)	I(amps)
2	0	3.98	=B2/12E6
3	5	1.58	
4	10	0.61	
5	15	0.24	
6	20	0.094	
7	25	0.035	
8	30	0.016	
9	35	0.0063	
10	40	0.0031	
11	45	0.0017	
12	50	0.0011	
13	55	0.0007	
14	60	0.0006	

	A	B	C
1	t(s)	V(volts)	I(amps)
2	0	3.98	3.31667E-07
3	5	1.58	
4	10	0.61	
5	15	0.24	
6	20	0.094	
7	25	0.035	
8	30	0.016	
9	35	0.0063	
10	40	0.0031	
11	45	0.0017	
12	50	0.0011	
13	55	0.0007	
14	60	0.0006	

If the current is required at other times between $t = 5$ s and $t = 60$ s, then it is possible to type = B3/12E6 into cell C3, = B4/12E6 into cell C4, and so on down to = B14/12E6 in cell C14. This is a lot of (unnecessary) work. Spreadsheets are designed to reduce the effort required to carry out such a task. To enter the required formulae:

1. Move the cursor to cell C2. With the left hand mouse button pressed down, move to cell C14 and release the button. The cells from C2 to C14 should be highlighted as shown in sheet 2.5(a).
2. Click on the Edit menu. Click on the Fill option, then click on the Down option.
3. Values now appear in cells C3 to C14, as shown in sheet 2.5(b).

[10] When Excel® performs a calculation, it 'returns' the result of the calculation into the cell in which the formula was typed.

[11] The number of figures can be increased or decreased using Format ➤ Cells ➤ Number ➤ Scientific, as described in section 2.3.4.

Sheet 2.5. *Entering formulae into a cell:*

(a) *highlighting cells;*   (b) *contents of cells after choosing*
Edit ➤ *Fill* ➤ *Down.*

	C
1	I(amps)
2	=B2/12E6
3	
4	
5	
6	
7	
8	
9	
10	
11	
12	
13	
14	

	C
1	I(amps)
2	3.31667E-07
3	1.31667E-07
4	5.08333E-08
5	0.00000002
6	7.83333E-09
7	2.91667E-09
8	1.33333E-09
9	5.25E-10
10	2.58333E-10
11	1.41667E-10
12	9.16667E-11
13	5.83333E-11
14	5E-11

The Edit ➤ Fill ➤ Down command copies the formula into the highlighted cells and automatically increments the cell reference in the formula so that the calculation is carried out using the value in the cell in the adjacent B column. Cell referencing is discussed in the next section.

Another common arithmetic operation is to raise a number to a power. If it is required that we find the square of the contents of, say, cell C2 in sheet 2.5, then we would type[12] in another cell = C2^2. To illustrate this, suppose we wish to calculate the power, P, dissipated in the 12 MΩ resistor. The equation required is

$$P = I^2 R \qquad (2.3)$$

where I is the current flowing through the resistance, R. The formula used to calculate the power dissipated in the resistor is shown in cell D2 of sheet 2.6.

Sheet 2.6. *Calculation of power dissipated in a resistor:*

(a) *formula to calculate power*   (b) *value returned in D2 when*
entered into cell D2   *Enter key is pressed.*

	C	D
1	I(amps)	P(watts)
2	3.31667E-07	=C2^2*12E6
3	1.31667E-07	
4	5.08333E-08	

	C	D
1	I(amps)	P(watts)
2	3.31667E-07	1.32003E-06
3	1.31667E-07	
4	5.08333E-08	

[12] The ^ symbol is found by holding down the shift key and pressing the '6' key.

Exercise A

1. Column B of sheet 2.4 shows the voltage across a 0.47 μF capacitor at times $t=0$ to $t=60$ s. Calculate the charge remaining on the capacitor (as given by equation (2.2)) at times $t=0$ to $t=60$ s. Tabulate values of charge in column D of the spreadsheet.
2. Enter a formula into cell E2 of sheet 2.4 to calculate the square root of the value of current in cell C2. Use Edit ➤ Fill ➤ Down to calculate the square root of the other values of current in sheet 2.4.

2.3.6   **Cell references and naming cells**

Relative referencing

Sheet 2.7. *Formula incorporating relative referencing of cells.*

	A	B	C
1	20	6.5	=A1*B1
2	30	7.2	
3	40	8.5	

How cells are referenced within other cells affects how calculations are performed. For example, consider the formula, =A1*B1, appearing in cell C1 in sheet 2.7. Excel® interprets the formula in cell C1 as 'starting from the current cell (C1) multiply the contents of the cell two to the left (the value in A1) by the contents of the cell one to the left (the value in B1)'. This is referred to as *relative* referencing of cells. When the Enter key is pressed, the value 130 appears in cell C1. If the Edit ➤Fill ➤Down command is used to fill cells C2 and C3 with formulae, relative referencing ensures that the correct cells in the A and B columns are used in the calculations. Specifically, =A2*B2 appears in cell C2, and =A3*B3 in cell C3. If cells are now moved around by 'cutting and pasting',[13] relative referencing ensures that calculations return the correct values irrespective of which cells contain the raw data. Excel® keeps track of where values or formulae are moved to and automatically updates the relative references.

Exercise B

Highlight cells C1 to C3 in sheet 2.7 and choose Edit ➤Fill ➤Down to calculate the product of values in adjacent cells in the A and B columns.

---

[13] To 'cut and paste', highlight the cells containing the values to be moved. Choose Edit ➤ Cut. Move the cursor to the cell where you want the first value to appear. Choose Edit ➤ Paste.

Absolute referencing

Sheet 2.8. *Formula using absolute referencing of cells.*

	A	B	C
1	20	6.5	=$A$1*$B$1
2	30	7.2	
3	40	8.5	

Another way in which cells may be referenced is shown in sheet 2.8. The formula in cell C1 is interpreted as 'multiply the value in cell A1 by the value in cell B1'. This is referred to as *absolute* referencing, and on the face of it, it doesn't seem very different from relative referencing. Certainly, when the Enter key is pressed, the value 130 appears in cell C1 just as in the previous example. The difference becomes more obvious by highlighting cells C1 to C3 and choosing Edit ➤ Fill ➤ Down. The consequences of these actions are shown in sheet 2.9. Cells C1 to C3 each contain 130. This is because the formulae in cells C1 to C3 use the contents of the cells which have been absolutely referenced, in this case cells A1 and B1, and no incrementing of references occurs when Edit ➤ Fill ➤ Down is used. This can be very useful, for example, if we wish to multiply values in a row or column by a constant.

Sheet 2.9. *Using* Edit ➤ Fill ➤ Down *with absolute referenced cells.*

	A	B	C
1	20	6.5	130
2	30	7.2	130
3	40	8.5	130

As an example, consider the values of distance, h, shown in sheet 2.10. The time, t, for an object to fall freely through a distance, h, is given by

$$t = \left(\frac{2h}{g}\right)^{\frac{1}{2}}$$

(2.4)

where g is the acceleration due to gravity ($=9.81$ m/s² on the earth's surface). Cell B2 contains a relative reference to cell A2. When cells B2 to B6 are highlighted and Edit ➤ Fill ➤ Down chosen, the cell references are automatically incremented. Cell B2 contains an absolute reference to cell D3 so that the value 9.81 is used in the formulae contained in cells in B2 to B6.

Sheet 2.10. *Calculation of time for object to fall a distance, h.*

	A	B	C	D
1	h (m)	t (s)		
2	2	=(2*A2/$D$3)^0.5		
3	4		g	9.81
4	6			
5	8			
6	10			

Exercise C

(i) Complete sheet 2.10 using the Edit ➤Fill ➤Down command to find the time of fall for all the heights given in the A column.

(ii) Use the spreadsheet to calculate the times of fall if the acceleration due to gravity is 1.6 m/s^2.

Naming cells

The use of absolute referenced cells for values that we might want to use again and again is fine, but it is possible to incorporate constants into a formula in a way that makes the formula easier to read. That way is to give the cell a name. Consider sheet 2.11.

Sheet 2.11. *Time of fall calculated using a named cell.*

	A	B	C	D
1	h (m)	t (s)		
2	2	=(2*A2/g)^0.5		
3	4		g	9.81
4	6			
5	8			
6	10			

This sheet is similar to sheet 2.10, the difference being that the absolute reference, $D$3, in cell B2 has been replaced by the symbol, g. Before proceeding, we must allocate the name g to the contents of cell D3. To allocate the name:

(i) Highlight cells C3 and D3.

(ii) Choose Insert ➤Name ➤Create. A dialog box appears with a cross in the Left Column box. This indicates that the value in cell D3 will be given the name appearing in C3. This is what we want, so click on OK.

Whenever we need to use the value of g in a formula, we use the symbol g instead of giving an absolute reference to the cell containing the value.[14]

A word of caution: some words or letters are reserved by Excel® and using them as a name will cause difficulties. For example, using the letter c (or C) as a name will appear acceptable until you try to use it in a formula. At this point Excel® indicates that there is an error in the formula, but does not assist in finding that error!

---

[14] We have omitted showing the units of g. If we type g(m/s^2) in cell C3, then g(m/s^2) becomes the name which we would need to type out in full in subsequent formulae.

Exercise D

The heat emitted each second, H, from a blackbody of surface area, A, at temperature, T, is given by

$$H = \sigma A T^4 \tag{2.5}$$

where σ is the Stefan–Boltzmann constant ($= 5.67 \times 10^{-8}\,\mathrm{W/(m^2 \cdot K^4)}$). In this problem take $A = 0.062\ \mathrm{m^2}$.

Create a spreadsheet with cells containing 5.67×10^{-8} and 0.062 and name the cells SB and A, respectively. Use Excel® to calculate H, for $T = 1000$ K to 6000 K in steps of 1000 K.

2.3.7 Operator precedence and spreadsheet readability

Care must be taken when entering formulae, as the order in which calculations are carried out affects the final values returned by Excel®. For example, suppose Excel® is used to calculate the equivalent resistance of two resistors of values 4.7 kΩ and 6.8 kΩ connected in parallel. The formula for the equivalent resistance, R_{eq}, of two resistors R_1 and R_2 in parallel is

$$R_{eq} = \frac{R_1 R_2}{R_1 + R_2} \tag{2.6}$$

Sheet 2.12 shows the resistor values entered into cells A1 and A2. The equation to calculate R_{eq} is entered in cell A3. When the Enter key is pressed, the value 13 600 appears in cell A3 as indicated in sheet 2.6(b). This value is incorrect, as R_{eq} should be 2.779 kΩ. Excel® interprets the formula in cell A3 as

$$= (A1*A2/A1) + A2, \quad \text{instead of} \quad = (A1*A2)/(A1 + A2)$$

Sheet 2.12. *Calculation of parallel equivalent resistance:*

(a) *formula entered into cell A3;*   (b) *value returned after pressing the Enter key.*

	A
1	4.70E3
2	6.80E3
3	=A1*A2/A1+A2
4	

	A
1	4.70E3
2	6.80E3
3	13600
4	

To avoid such mistakes, it is advisable to include parentheses in formulae, as Excel® carries out the operations within the parentheses first. Difficulties with calculations can often be avoided if a formula is divided into a number of smaller formulae, with each one entered into a different cell. As an example, consider an equation relating the velocity of water through a tube to the cross-sectional area of the tube:

$$v = \sqrt{\frac{2gh}{(A_1/A_2)^2 - 1}}$$
(2.7)

where v is the velocity of water through a tube, g is the acceleration due to gravity, h is the height through which the water falls as it moves through the tube whose cross-sectional area changes from A_1 to A_2. Values for these quantities are shown in sheet 2.13.

The formulae in cells B5, B6 and B7 return values which are subsequently used in the calculation of v in cell B8. This step by step approach is also useful if a mistake is suspected in the final value and troubleshooting is required. For completeness, symbols used for the quantities along with the appropriate units are shown in column A of the spreadsheet. The corresponding values and formulae are shown in column B of sheet 2.13(a). The values returned by the formulae are shown in sheet 2.13(b).

Sheet 2.13. *Calculation of the velocity of water:*

(a) *formulae entered into cells B5–B8;*   (b) *values returned in cells*

	A	B			A	B
1	g (m/s^2)	9.81		1	g (m/s^2)	9.81
2	h (m)	0.15		2	h (m)	0.15
3	A$_1$ (m^2)	0.062		3	A$_1$ (m^2)	0.062
4	A$_2$ (m^2)	0.018		4	A$_2$ (m^2)	0.018
5	2gh (m^2/s^2)	=2*B1*B2		5	2gh (m^2/s^2)	2.943
6	(A$_1$/A$_2$)^2	=(B3/B4)^2		6	(A$_1$/A$_2$)^2	11.8642
7	(A$_1$/A$_2$)^2 −1	=B6−1		7	(A$_1$/A$_2$)^2 −1	10.8642
8	v (m/s)	=(B5/B7)^0.5		8	v (m/s)	0.520471

Subscripts and superscripts have been added to quantities and units in column A. This further improves the readability of the spreadsheet. To add a super or subscript:

1. Click on the cell containing the symbol to be made into a super or subscript.
2. Highlight the symbol to be super or subscripted. This may be done in the formula bar. Hold down the left hand mouse button and drag across the symbol or number you wish to make a super or subscript.
3. From the menu bar choose Format ➤ Cells ➤ Font.
4. Click the mouse button in the Superscript or Subscript box. A tick should appear in the box selected.

5. Click on OK. The chosen symbol or number should appear as a superscript or a subscript.

Exercise E

1. The radius of curvature of a spherical glass surface may be found using the
Newton's rings method.[15] If the radius of the mth ring is r_m and the radius of the nth
ring is r_n, then the radius, R, of the spherical glass surface is given by

$$R = \frac{r_n^2 - r_m^2}{(n-m)\lambda} \tag{2.8}$$

where λ is the wavelength of the light incident on the surface. Table 2.1 contains data
from an experiment carried out to determine R. Using these data, create a spreadsheet to calculate R, as given by equation (2.8).

2. The velocity of sound, v_d, in a tube depends on the diameter of the tube, the frequency of the sound and the velocity of sound in free air. The equation relating the
quantities when the walls of the tube are made from smooth glass is

$$v_d = v\left(1 - \frac{3 \times 10^{-3}}{d\sqrt{f}}\right) \tag{2.9}$$

where v is the velocity of sound in free air in m/s, d is the diameter of the tube in
metres and f is the frequency of the sound in Hz. Taking $v = 344$ m/s and $f = 5$ Hz, use
Excel® to tabulate values of v_d when d varies from 0.1 m to 1 m in steps of 0.1 m.

Table 2.1. *Values used to
determine the radius of a
spherical surface.*

Quantity	value
r_m (m)	6.35×10^{-3}
r_n (m)	6.72×10^{-3}
m	52
n	86
λ (m)	6.02×10^{-7}

2.3.8 Verification and troubleshooting

In the last section we showed the ease with which a formula may be entered
into a cell that 'looks right', but without the appropriate parentheses the
formula returns values inconsistent with the equation upon which it is based.
Establishing or verifying that a spreadsheet is returning the correct values

[15] For details of the method refer to Daish and Fender (1970).

can sometimes be difficult, especially if there are many steps in the calculation. While no single approach can ensure that the spreadsheet is 'behaving'
as intended, there are a number of actions that can be taken which minimise
the chance of a mistake going unnoticed. If a mistake is detected, a natural
response is to suspect some logic error in the way the spreadsheet has been
assembled. However, it is easy to overlook the possibility that a transcription
error has occurred when entering data into the spreadsheet. In this situation
little is revealed by stepping through calculations performed by the spreadsheet in a 'step by step' manner. Table 2.2 offers some general advice intended
to help reduce the occurrence of mistakes.

Determining how the contents of various cells are 'brought together'
to calculate, for example, a mean and standard deviation is aided by using
some of Excel®'s in-built tools. These are the 'Auditing' tools and can assist
in identifying problems in a spreadsheet.

Table 2.2. *Troubleshooting advice.*

Suggestion	Explanation/Example
Make the spreadsheet do the work	Enter 'raw' values into a spreadsheet in the form in which they emerge from an experiment. For example, if the diameter of a ball bearing is measured using a micrometer, then it is unwise to convert the diameter to a radius 'in your head'. It is better to add an extra column (with a clear heading) and to calculate the radius in that column. This makes backtracking to find mistakes much easier.
Perform an order of magnitude calculation	If we have a feel for the size and sign of numbers emerging from the spreadsheet, then we are alerted when those numbers do not appear. For example, if we calculate the volume of a small ball bearing to be roughly 150×10^{-9} m^3 but the value determined by the spreadsheet is 143.8 m^3, this might point to an inconsistency in the units used for the volume calculation or that the formula entered is incorrect.
Use data that has already been analysed	On some occasions 'old' data are available that have already been analysed 'by hand' or using another computer package. The purpose of the spreadsheet might be to analyse similar data in a similar manner. By repeating the analysis of the 'old' data using the spreadsheet it is possible to establish whether the analysis is consistent with that performed previously.
Choose the appropriate built in function	Many built in functions in Excel® appear to be very similar, for example when calculating a standard deviation we could use STDEV(), STDEVA(), STDEVP() or STDEVPA(). Knowing the definition of each function (by consulting the help available

Table 2.2. (*cont.*)

Suggestion	Explanation/Example
	for each function if necessary) the appropriate function may be selected.
Graph the data and scan for outliers	If many values are entered by hand into a spreadsheet it is easy for a transcription mistake to occur. By using the x–y graphing feature of Excel® we are able to plot out the values in the order in which they appear in a column or row of the spreadsheet. Any gross mistakes can usually be identified 'by eye' within a few moments. As an example, most y values in figure 2.2 are about 3.6. As one value is close to 6.3 we should consider the possibility that a transcription error has occurred.

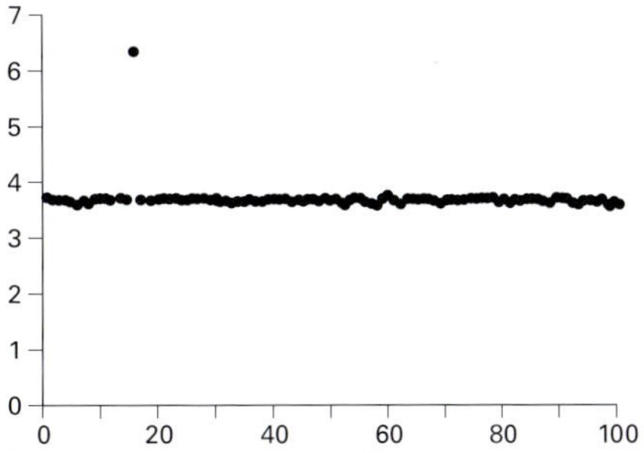

Figure 2.2. x–y graph of data indicating a possible transcription error.

Suggestion	Explanation/Example
Be alert to error messages	#DIV/0!, #NAME?, #REF!, #NUM! and #N/A! are some of the error messages that Excel® may return into a cell due to a variety of causes. As examples:

- A cell containing the #DIV/0! error indicates that the calculation being attempted in that cell requires Excel® to divide a value by zero. A common cause of this error is that a cell such as B1 contains no value, yet this cell is referenced in a formula such as $=32.45/B1$.
- The #NAME? error occurs when a cell contains reference to an invalid name. For example if we type $=$averag(A1:A10) into a cell, Excel® does not recognise averag() as a function (we probably meant to type $=$average() which is a valid Excel® function, but spelled it incorrectly). Excel® assumes that averag(A1:A10) is a name that has not been defined and so returns the error message.
- A cell containing the #NUM! error indicates that some invalid mathematical operation is being attempted. For example, if a cell contains $=$LN(-6) then the #NUM! error is returned into that cell as it is not possible to take the logarithm of a negative number.

2.3.9 Auditing tools

Sheet 2.14 contains nine values of temperature as well as the mean and the standard deviation of the values. Calculating the mean and standard deviation 'independently' (say, by using a pocket calculator) we find that the mean = 24.74 °C (to four significant figures), consistent with the mean calculated in cell B11 of sheet 2.14. By contrast, the standard deviation found using a pocket calculator is 2.008 °C. To trace the calculation of the standard deviation appearing in sheet 2.14, we can use Excel®'s auditing tools which give a graphical representation of how values in cells are calculated. Specifically, by using the auditing tools, we can establish which cells contribute to the calculation of a value in a cell. To access the auditing tools go to the Menu bar and choose Tools ➤ Auditing ➤ Show Formula Auditing Toolbar. At this point the formula auditing toolbar appears as shown in figure 2.3. To use the auditing tools we proceed as follows:

Figure 2.3. The Formula Auditing toolbar.

Sheet 2.14. *Calculation of mean and standard deviation.*

	A	B
		Temperature (°C)
1		Temperature (°C)
2		23.5
3		24.2
4		26.4
5		23.1
6		22.8
7		22.5
8		25.4
9		28.3
10		26.5
11	mean	24.74444444
12	standard deviation	1.893328117

1. Click on the cell which contains the calculation we wish to trace (in the above example that would be cell B12).

2. Click on the ![icon] icon on the formula auditing toolbar. The blue line and arrow that appear indicate which cells are used in the calculation of the value in cell B12. If a range of cells contributes to the calculation, then this range is outlined in blue.

Figure 2.4 shows the formula auditing tool used to trace the calculation of the standard deviation in sheet 2.14. The formula auditing tool indicates that cells in the range B2 to B11 have been used in the calculation of the standard deviation in cell B12. The mistake that has been made has been

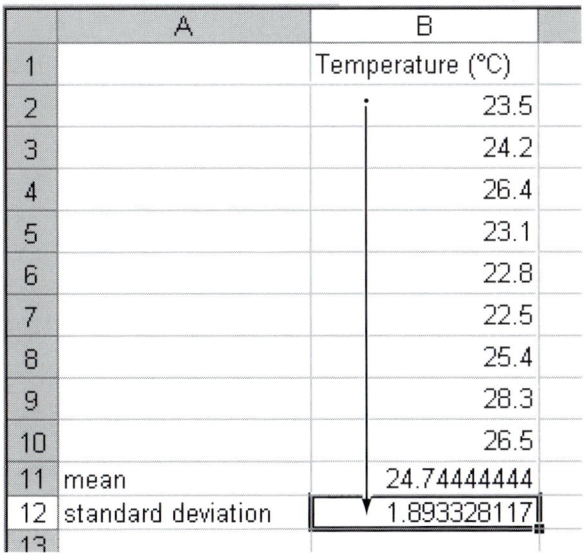

	A	B
1		Temperature (°C)
2		23.5
3		24.2
4		26.4
5		23.1
6		22.8
7		22.5
8		25.4
9		28.3
10		26.5
11	mean	24.74444444
12	standard deviation	1.893328117
13		

Figure 2.4. Tracing a calculation using the formula auditing tool.

the inclusion of the cell B11 in the range, as B11 does not contain 'raw data', but the mean of the contents of cells B2 to B10. This mistake could have been detected by examining the range appearing in the formula in cell B12. However, as the relationships between cells becomes more complex, the pictorial representation of those relationships as provided by the formula auditing tool can help identify mistakes that would be otherwise difficult to find. The formula auditing toolbar contains other facilities to assist in tracking mistakes and allows for the inclusion of comments to help document calculations being carried out by Excel®. Details of other formula auditing facilities can be found either by using Excel®'s Help or by referring to a standard text.[16]

2.4 Built in mathematical functions

Excel® possesses the mathematical and trigonometrical functions normally present on a scientific pocket calculator. In addition, Excel®

[16] See, for example, Blattner (2001).

possesses other functions rarely found on a calculator such as a function to calculate the sum of a geometric series. A full list of mathematical and trigonometrical functions provided by Excel® is found by going to the Menu bar and choosing Insert ►Function. From the category box choose Math and Trig.

In many situations, data may need to be transformed before further analysis or graphing can take place. For example, in the capacitor discharge experiment discussed in section 2.3.4, there is every likelihood that the decay of the current with time follows a logarithmic relationship. The first step that would normally be taken to verify this would be to calculate the natural logarithm of all of the values of current. In Excel®, the natural logarithm of a number is calculated using the LN() function. To calculate the natural logarithm of the value appearing in cell C2:

1. Type the function =**LN(C2)** into cell D2[17], as shown in sheet 2.15(a). (LN can be in either upper or lower case letters) and press the Enter key. The number –14.9191 is returned in cell D2 as indicated in sheet 2.15(b).

Sheet 2.15. *Calculation of natural logarithm of current:*

	C	D
1	I(amps)	ln(I)
2	3.31667E-07	=LN(C2)
3	1.31667E-07	
4	5.08333E-08	
5	0.00000002	
6	7.83333E-09	
7	2.91667E-09	
8	1.33333E-09	
9	5.25E-10	
10	2.58333E-10	
11	1.41667E-10	
12	9.16667E-11	
13	5.83333E-11	
14	5E-11	

	C	D
1	I(amps)	ln(I)
2	3.31667E-07	-14.9191
3	1.31667E-07	
4	5.08333E-08	
5	0.00000002	
6	7.83333E-09	
7	2.91667E-09	
8	1.33333E-09	
9	5.25E-10	
10	2.58333E-10	
11	1.41667E-10	
12	9.16667E-11	
13	5.83333E-11	
14	5E-11	

(a) *formula entered into cell D2;*   (b) *value returned in cell D2.*

2. Highlight cells D2 to D14 and choose Edit ►Fill ►Down to calculate the natural logarithms of all the values in the C column.

[17] As you begin typing =**LN(C2)** in Excel® 2002, a 'tooltip' appears which advises on the argument(s) that appear in the function. Excel® 2002 provides tooltips for all of its built-in functions.

Exercise F

1. Calculate the logarithms to the base 10 of the values shown in column C of sheet 2.15.

2. An equation used to calculate atmospheric pressure at height h above sea level, when the air temperature is T, is

$$P = P_0 \exp\left(\frac{-3.39 \times 10^{-2} h}{T}\right) \tag{2.10}$$

where P is the atmospheric pressure in pascals, P_0 is equal to 1.01×10^5 Pa, h is the height in metres above sea level and T is the temperature in kelvins.

Assuming $T = 273$ K, use the EXP() function to calculate P for $h = 1 \times 10^3$ m, 2×10^3 m, and so on, up to 9×10^3 m. Express the value of P in scientific notation to three significant figures.

2.4.1 Trigonometrical functions

The usual sine, cosine and tangent functions are available in Excel®, as are their inverses. Note Excel® expects angles to be entered in radians and not degrees. An angle in degrees can be converted to radians by multiplying the angle by $\pi/180$. In Excel® π is entered as PI(). Sheet 2.16 shows a range of angles and the functions for calculating the sine, cosine and tangent of each angle.

Sheet 2.16. *Calculation of sine, cosine and tangent.*

	A	B	C	D	E
1	x(degrees)	x (radians)	sin(x)	cos(x)	tan(x)
2	10	=A2*PI()/180	=SIN(B2)	=COS(B2)	=TAN(B2)
3	30				
4	56				
5	125				

In section 2.3.5 we used Edit ➤ Fill ➤ Down to fill a single column of Excel® with formulae. We can complete sheet 2.16 by filling all the cells from B2 to E5 with the required formulae. To do this:

1. Enter the numbers, headings and formulae as shown in sheet 2.16.
2. Move the cursor to cell B2 and hold down the left mouse button.
3. With the button held down, drag across and down to cell E5, then release the button.
4. From the Menu bar choose Edit ➤ Fill ➤ Down.
5. The cells from B1 to E5 should now contain the values shown in sheet 2.17.

Sheet 2.17. *Values returned by Excel®'s trigonometrical functions.*

	A	B	C	D	E
	x(degrees)	x (radians)	sin(x)	cos(x)	tan(x)
1					
2	10	0.174533	0.173648	0.984808	0.176327
3	30	0.523599	0.5	0.866025	0.57735
4	56	0.977384	0.829038	0.559193	1.482561
5	125	2.181662	0.819152	-0.57358	-1.42815

Exercise G

Use the Excel® functions ASIN(), ACOS() and ATAN() to calculate the inverse sine, inverse cosine and inverse tangent of the values of x in table 2.3. Express the inverse values in degrees.

Table 2.3. *Data for exercise G.*

x	0.0	0.1	0.2	0.3	0.4	0.5	0.6	0.7	0.8	0.9	1.0

2.5   Built in statistical functions

In addition to mathematical and trigonometrical functions, Excel® possesses about 80 functions directly related to statistical analysis of data. Some of these functions are found on scientific pocket calculators. For example, the AVERAGE() function calculates the arithmetic mean of a list of values and so performs the same function as the $\bar{x}$ button on most pocket calculators. Other statistical functions are more specialised, such as NORMDIST(), which is related to the normal distribution. We consider such functions in later chapters.

2.5.1   SUM(), MAX() and MIN()

The summation of values occurs frequently in data analysis, for example it is a necessary step when finding the mean of a column of numbers. Summation can be accomplished by Excel® in several ways. Consider the values in sheet 2.18. To sum the values in the B column:

Sheet 2.18. *Example of summing numbers.*

	A	B
1		y
2		23
3		34
4		42
5		42
6		65
7		87
8		

1. Click on the cell B8.
2. Move the cursor to the Standard toolbar at the top of the screen and click on the summation sign, Σ.
3. A flashing dotted line appears around cells B2 to B7 and =SUM(B2:B7) appears in cell B8. Excel® has made an 'intelligent guess' that we want to sum the numbers appearing in cells B2 to B7 inclusive.
   =SUM(B2:B7) can be read as 'sum the values in cells B2 through to B7'.
4. Pressing the Enter key causes the number 293 to be returned in cell B8. The function can be cancelled by clicking on the symbol, ✗, on the formula bar.

Another way to calculate the sum of the values in cells B2 to B7 is to make cell B8 active then type =SUM(B2:B7). On pressing the Enter key, the number 293 appears in cell B8.

Exercise H

Consider the *x* and *y* values in sheet 2.19.

Sheet 2.19. *x and y values for exercise H.*

	A	B	C	D
1	x	y	xy	x^2
2	1.75	23		
3	3.56	34		
4	5.56	42		
5	5.85	42		
6	8.76	65		
7	9.77	87		
8				

(i) Use Excel® to calculate the product of *x* and *y* in column C.
(ii) Calculate x^2 in column D.
(iii) Use the SUM() function in cells A8 to D8 to calculate Σx, Σy, Σxy, Σx^2.

The MAX() and MIN() functions return the maximum and minimum values respectively in a list of values. Such functions are useful, for example, if we wish to scale a list of values by dividing each value by the maximum or minimum value in that list, or if we require the range (i.e. the maximum value – the minimum value) as the first step to plotting a histogram.[18] To illustrate these functions, consider the values in sheet 2.20. The formula for finding the maximum value in cells A1 to E6 is shown in cell A8. When the Enter key is pressed, 98 is returned in cell A8.

Sheet 2.20. *Identifying the maximum value in a group of values.*

	A	B	C	D	E
1	23	23	13	57	29
2	65	22	45	87	76
3	34	86	76	79	35
4	45	55	89	34	43
5	45	61	56	43	12
6	98	21	87	56	34
7					
8	=MAX(A1:E6)				

Exercise I

Incorporate the MIN() function into sheet 2.20 so that the smallest value appears in cell B8. Show the range of the values in cell C8.

2.5.2   AVERAGE(), MEDIAN() and MODE()

The AVERAGE(), MEDIAN() and MODE() functions calculate three numerical descriptions of the centre of a set a values. Briefly,

> AVERAGE() returns the arithmetic mean of a group or list of values.
> MEDIAN() orders values in a list from lowest to highest and returns the middle value in the list. The middle value is the $(\frac{1}{2}n + 1)$th value if the list has an odd number of values. If there is an even number of values in the list then the median value is the mean of the $\frac{1}{2}n$th value and the $(\frac{1}{2}n+ 1)$th value.
> MODE() determines the frequency of occurrence of values in a list and returns the value which occurs most often. If two or more values occur equally often then Excel® gives only the first value that it completed tallying. Of the three measures of centre of data discussed in this section, the mode is used least frequently in the physical sciences.

---

[18] See section 2.8.1 for a description of how to use Excel® to plot a histogram.

Consider sheet 2.21 containing 48 integer values. The mean of the values is found by entering the AVERAGE() function into cell A9. When the Enter key is pressed, the value 23.60417 is returned in cell A9.

Sheet 2.21. *Use of the AVERAGE() function.*

	A	B	C	D	E	F
1	27	1	49	2	39	11
2	27	29	40	8	28	5
3	18	5	25	0	4	33
4	26	30	20	14	22	10
5	23	28	33	30	28	16
6	23	5	27	9	48	4
7	39	41	46	22	25	25
8	35	49	13	8	40	43
9	=AVERAGE(A1:F8)					

Exercise J

Calculate the median and mode of the values in sheet 2.21 by entering the MEDIAN() and MODE() functions into cells B9 and C9 respectively.

2.5.3 Other useful functions

A full list of functions in Excel® may be obtained by making any cell active then going to the Menu bar and choosing Insert ➤ Function. In the category box choose 'All'. The functions appear in the dialog box ordered alphabetically. Using the scroll bar you can scroll down to, say, STDEV. To obtain information on the functions, you can click on [Help on this function] .

Other useful statistical functions are given in table 2.4 along with brief descriptions of each.

Exercise K

Repeat measurements made of the pH of river water are shown in sheet 2.22. Use the built in functions in Excel® to find the mean, harmonic mean, average deviation and estimate of the population standard deviation for these data.

Sheet 2.22. *Data for exercise K.*

	A	B	C	D	E	F	G	H	I	J
1	6.6	6.4	6.8	7.1	6.9	7.4	6.9	6.4	6.3	7.0
2	6.8	7.2	6.4	6.7	6.8	6.1	6.9	6.7	6.4	7.1
3	6.8	6.7	6.3	6.6	7.0	6.7	6.4	6.7	6.7	6.4
4	6.6	6.7	7.4	7.1	7.0	6.8	7.0	6.8	6.7	6.2
5	7.1	6.4	6.7	6.9	6.9	6.6	7.2	6.8	6.4	6.5

Table 2.4. *Useful statistical functions in Excel®.*

Function	What it does	Defining equation	Example of use	Value returned
AVEDEV() (abbreviated to AD)	Calculates the mean of the absolute deviations of values from the mean	$AD = \frac{1}{n}\sum|x - \bar{x}|$	=AVEDEV(14, 23, 12, 34, 36)	8.96
HARMEAN() (abbreviated to HM)	Calculates the reciprocal of the mean of reciprocals of values	$HM = \dfrac{n}{\sum\left(\dfrac{1}{x}\right)}$	=HARMEAN(2.4, 4.5, 6.7, 6.4, 4.3)	4.248266
STDEV() (symbol, s)	Calculates s which is the estimate of population standard deviation	$s = \left[\dfrac{\sum(x - \bar{x})^2}{n-1}\right]^{\frac{1}{2}}$	=STDEV(2.1, 3.5, 4.5, 5.6)	1.488568
STDEVP() (symbol, σ)	Calculates the population standard deviation, σ (all values in the population are required)	$\sigma = \left[\dfrac{\sum(x - \mu)^2}{n}\right]^{\frac{1}{2}}$	=STDEVP(2.1, 3.5, 4.5, 5.6)	1.289137

2.6 Presentation options

A spreadsheet consisting largely of numbers and formulae can appear incomprehensible to the user, even if that user is the person who created it. For clarity it is important to consider the layout of the spreadsheet carefully.

It may be necessary to move the contents of cells to allow space for titles or labels. This does not pose a problem, as whole sections of a spreadsheet can be moved without 'messing up' the calculations. For example, when columns containing values are moved, the formulae in cells that use those values are automatically updated to include the references to the new cells containing the values.

Sheet 2.23 contains data from the measurement of the coefficient of static friction between two flat wooden surfaces. To improve the readability of the spreadsheet we might:

Sheet 2.23. *Coefficient of static friction data.*

	A	B	C	D	E	F
1	0.34	0.34	0.40	0.49	0.42	0.33
2	0.41	0.39	0.38	0.39	0.44	0.49
3	0.42	0.50	0.31	0.36	0.48	0.26
4	0.46	0.42	0.32	0.37	0.40	0.36
5	0.44	0.40	0.36	0.38	0.45	0.37
6	0.38	0.30	0.29	0.39	0.49	0.55

- give the sheet a clear heading;
- indicate what quantities (with units, if any) the values in the cells represent.

To allow some space at the top of sheet 2.23 for labels and titles to be inserted:

1. Highlight the cells containing values by moving the cursor to cell A1 and, with the left hand mouse button held down, drag down and across to cell F6. Release the mouse button.

2. Move the cursor to the outline of the highlighted cells. The cursor should change from an open cross, ✛, to an arrow, ↖. Holding the left hand mouse button pressed down, move the mouse around the screen. The outline of the cells should move with the mouse.

3. Release the mouse button to transfer the contents of the cells to a new location.

By moving the contents of sheet 2.23 down by three rows, and adding labels, the spreadsheet can be made to look like sheet 2.24. A title (in bold font) has been added to the sheet and the quantity associated with the values in the cells is clearly indicated.

Sheet 2.24. *Improved spreadsheet layout for static friction data.*

	A	B	C	D	E	F
1	**Coefficient of static friction for two wooden surfaces in contact**					
2						
3	Coefficient of static friction (no units)					
4	0.34	0.34	0.40	0.49	0.42	0.33
5	0.41	0.39	0.38	0.39	0.44	0.49
6	0.42	0.50	0.31	0.36	0.48	0.26
7	0.46	0.42	0.32	0.37	0.40	0.36
8	0.44	0.40	0.36	0.38	0.45	0.37
9	0.38	0.30	0.29	0.39	0.49	0.55

The impact of a spreadsheet can be improved by formatting cells. This is done by choosing Format ➤ Cells. With this option, the colour of digits or text contained within cells may be changed, borders can be drawn around one or more cells, fonts can be altered and background shading added. A word of caution: too much formatting and too many bright colours can be distracting and, instead of making the spreadsheet easy to read, actually has the opposite effect.

2.7 Charts in Excel®

A spreadsheet consisting solely of 'numbers' is sometimes difficult to decipher, even when it is well laid out. Often the most efficient way to 'take in' data and to identify trends, anomalies and features of interest is to create a picture or graph. The graphical features of Excel® are extensive and include options such as bar, pie and x–y graphs. Excel® refers to graphs as *charts*. Using the facility referred to as the Chart Wizard, graphs may be constructed quite quickly. Among the many graphing alternatives is the x–y graph. This is extensively used in the physical sciences and we will concentrate on this graph in the next section.

2.7.1 The x–y graph

The x–y graphing facility in Excel® (referred to in Excel® as an x–y scatter chart) offers many options, including:

- conversion of scales from linear to logarithmic;
- addition of error bars to data points;
- addition of 'line of best fit' using a variety of models;
- automatic updating of points and trendline(s) on the graph when values within cells are changed.

The creation of a graph is simplified using Excel®'s built in Chart Wizard. This is accessed by clicking on the ▦ icon on the standard toolbar. We illustrate x–y graphing features using data gathered from an experiment in which the electrical resistance of a tungsten wire is measured as a function of temperature. We consider:

- plotting an x–y graph;
- adding a line of best fit (referred to in Excel® as a 'trendline') to the graph;
- adding error bars to each point.

We begin by entering the resistance–temperature data as shown in sheet 2.25. To simplify the construction of an x–y graph, values to be plotted are placed in adjacent columns. The values in the left of the two columns are interpreted by Excel® as the x values and those in the other column as the y values. To plot the data shown in sheet 2.25:

Sheet 2.25. *Resistance versus temperature data for a tungsten wire.*

	A	B
1	t (°C)	R (ohm)
2	20	15.9
3	30	15.1
4	40	16.3
5	50	17.5
6	60	18.2
7	70	18.5
8	80	18.9
9	90	19.7
10	100	20.1
11	110	20.2

1. To highlight cells A2 to B11, move the cursor to cell A2. With the left hand mouse button held down, drag across and down from cell A2 to cell B11. Release the mouse button.
2. Click on the Chart Wizard icon 📊 .
3. A dialog box appears identified as Chart Wizard - Step 1 of 4 - Chart Type. Click on the XY (Scatter) option. Click on Next.
4. Step 2 of 4. An x–y graph appears. At this point the data series can be named and other data series added. We do not want to do this, so click on Next.
5. Step 3 of 4. A chart title and axes labels can be added. Click on the Chart Title box and type **Resistance versus temperature for a tungsten wire**. In the Value (X) box type **Temperature (°C).** In the Value (Y) type **Resistance (ohms)**. Click on Next.
6. Step 4 of 4. You are asked where you want to place the chart. The default is to embed the chart into the sheet containing the data used in the plotting of the graph. Take the default. Click on Finish.
7. To raise the 'o', in front of the symbol C, which appears in the x axis label, click on the x axis label and with the left hand mouse button held down, drag across the 'o', then release the button. Next, choose Format ➤ Selected Axis Title. In the dialog box click on the Superscript Effects box, then click OK.

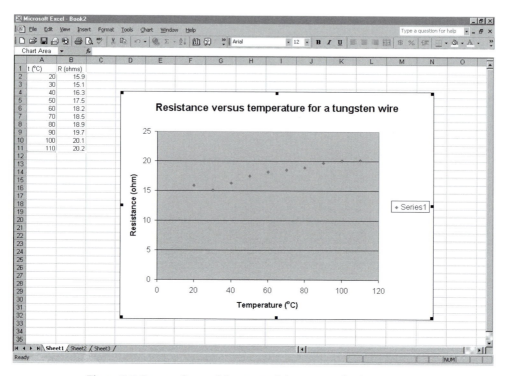

Figure 2.5. Screen of spreadsheet containing *x–y* graph of resistance versus temperature.

The graph that appears embedded in the sheet is quite small. To enlarge it, move the cursor to one of the corners of the outline around the graph, then with the left hand mouse button pressed down, drag the outline until the graph size has increased till it fills about 2/3 of the sheet as shown in figure 2.5.

Adding a trendline

If there is evidence to suggest that there is a linear relationship between quantities plotted on the *x* and *y* axes, as there is in this case, then a 'line of best fit', or trendline may be added to the data.[19] To add a trendline to the resistance versus temperature data shown in sheet 2.25:

1. Single click on the chart to make it active.
2. From the Menu bar choose Chart ➤Add Trendline. A dialog box should appear with the default set to the Linear option.

---

[19] Fitting a line to data is dealt with in detail in chapter 6.

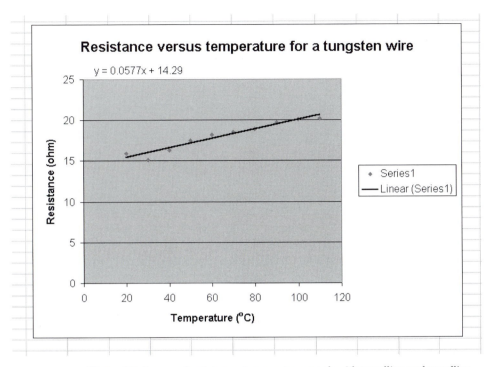

Figure 2.6. Screen of resistance–temperature graph with trendline and trendline equation added.

3. Click on the Options tab and click in the box beside Display equation on chart. Click on OK.
4. A line of best fit should now appear, along with the equation of the line (y = 0.0577x + 14.29).

Figure 2.6. shows the line of best fit attached to the resistance versus temperature data.[20] If data appearing in the cells A2 to B11 are changed, then the points on the line, the trendline and the trendline equation are immediately updated.

Adding error bars

When drawing x–y graphs by hand, error bars are often omitted due to the effort required to plot them, especially if the graph contains many points. Error bars can be added to all points in matter of seconds using Excel®.

To add error bars to data shown in figure 2.6:

---

[20] The legend to the right of the screen can be removed by making the chart active, clicking on the legend, then pressing the delete key.

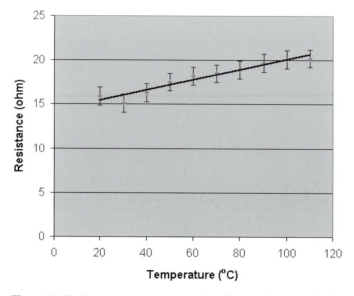

Figure 2.7. Resistance–temperature graph with error bars attached.

1. Make the chart active.
2. Move the cursor arrow to one of the data points and double click on the point. A dialog box appears which contains a number of tabs.
3. Click on the Y Error Bars tab.
4. If the uncertainty in the resistance is estimated to be, say ± 1 Ω, then type 1 into the Custom + box, and type 1 in the Custom − box.
5. Click on OK. Error bars appear attached to each data point as shown in figure 2.7.

Exercise L

1. The graph in figure 2.6 would be improved by better filling the graph with the data. This can be accomplished if the *y* axis were to begin at 14 ohms, rather than at zero. Open the file Chapter2 which can be found at http://uk.cambridge.org/ resources/0521793378 and select the tab 2.25. This sheet contains the chart shown in figure 2.6. Make the chart active by clicking on it, move the cursor close to the *y* axis and double click. A dialog box appears, with a tab labelled Scale. Click on Scale and change the Minimum value from 0 to 14 then click OK.

2. The intensity of light emitted from a red light emitting diode (LED) is measured as a function of current through the LED. Table 2.5 shows data from the experiment. Use Excel®'s *x*–*y* graphing facility to plot a graph of intensity versus current. Attach a straight line to the data using the Trendline option, and show the equation of the line on the graph.

Table 2.5. *Variation of light intensity with current for an LED.*

Current (mA)	1	2	3	4	5	6	7	8	9	10
Light intensity (arbitrary units)	71	97	127	134	159	175	197	203	239	251

2.7.2 Plotting multiple sets of data on an *x–y* graph

There are situations in which it is helpful to show more than one set of data on an *x–y* graph, as this allows comparisons to be made between data. Two basic approaches may be adopted.

- Several sets of data can be plotted 'simultaneously' by arranging the left most column of the spreadsheet to contain the *x* values and adjacent columns to contain the corresponding *y* values for each set of data.
- If an *x–y* graph has already been created with one set of data, then another set of data may be added to the same graph.

For example, consider the five sets of data shown in sheet 2.26. Columns B to F contain *y* values for each set corresponding to the *x* values in column A. To plot the data, highlight the contents of cells A1 through to F11, then follow the instructions in section 2.7.1. The graph produced is shown embedded in the Worksheet in figure 2.8. Note that by including the headings of the columns when selecting the range of data to be plotted, Excel® has used those headings in the legend at the right hand side of the graph.

Sheet 2.26. *Multiple data sets to be plotted on an x–y graph.*

	A	B	C	D	E	F
1	x	y1	y2	y3	y4	y5
2	0.2	71	77	83	89	96
3	0.4	53	61	70	81	93
4	0.6	42	50	61	74	91
5	0.8	36	44	54	69	89
6	1.0	34	40	50	65	87
7	1.2	33	39	48	62	86
8	1.4	34	39	47	61	85
9	1.6	36	40	47	60	85
10	1.8	39	42	48	60	85
11	2.0	42	44	49	60	85

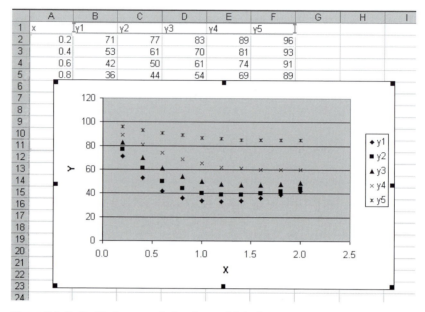

	A	B	C	D	E	F	G	H	I
1	x	y1	y2	y3	y4	y5			
2	0.2	71	77	83	89	96			
3	0.4	53	61	70	81	93			
4	0.6	42	50	61	74	91			
5	0.8	36	44	54	69	89			

Figure 2.8. Embedded x–y graph showing multiple data sets.

If another set of data is to be added to an existing Excel® x–y graph, then:

1. Place the new x–y data in adjacent columns in a Worksheet.
2. Highlight the x–y data (but not the headings) and choose Edit ➤ Copy.
3. Click on the existing x–y graph to make it active.
4. Choose Edit ➤ Paste Special.
5. In the dialog box choose 'Add cells as New series'.
6. Make sure that the box described as 'Categories (X Values) in First Column' has been ticked and that the option 'Values (Y) in Columns' has been selected.
7. Click on OK.

The new data should be added to the graph and the legend updated.

Exercise M

The intensity of radiation (in units W/m³) emitted from a body can be written

$$I = \frac{A}{\lambda^5 (e^{B/\lambda T} - 1)} \tag{2.11}$$

where T is the temperature of the body in kelvins, λ is the wavelength of the radiation in metres, $A = 3.75 \times 10^{-16}$ W·m² and $B = 1.4435 \times 10^{-2}$ m·K.

(i) Use Excel® to determine I at $T = 1250$ K for wavelengths between 0.2×10^{-6} m and 6×10^{-6} m in steps of 0.2×10^{-6} m.

(ii) Repeat part (i) for temperatures 1500 K, 1750 K and 2000 K.

(iii) Plot I versus λ at temperatures 1250 K, 1500 K, 1750 K and 2000 K on the same x–y graph.

2.8 Data analysis tools

Besides many built in statistical functions, there exists an 'add in' within Excel® that offers more powerful data analysis facilities. This is referred to as the Analysis ToolPak. To establish whether the ToolPak is available, choose Tools from the Menu bar. If the option 'Data Analysis' does not appear near to the bottom of the Tools menu, then the Analysis ToolPak must be added.

To add the Analysis ToolPak, choose Tools ➤ Add-Ins. A dialog box appears listing the available add-ins. Near to the top of the list is Analysis ToolPak. Click in the box next to Analysis ToolPak, then click OK. This adds the utility to the bottom of the Tools menu. If we go to the Menu bar again and choose Tools ➤ Data Analysis, the dialog box shown in figure 2.9 appears.[21]

We discuss other analysis tools in chapter 9, but for the moment we consider just two: Histogram and Descriptive Statistics.

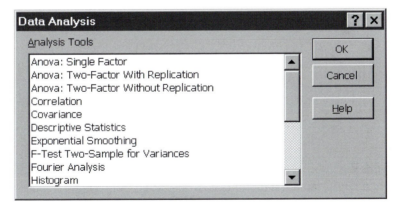

Figure 2.9. Dialog box for Data Analysis tools.

[21] Most 'dialog boxes' in the Data Analysis tool allow for the easy entry of cell ranges into the tool and often allow you to modify parameters.

2.8.1 Histograms

As discussed in section 1.4.1, a histogram is a very revealing way to display repeated values of a single quantity or variable. The Histogram tool within the Analysis ToolPak takes much of the tedium out of creating a histogram. To illustrate this tool, consider the data shown in sheet 2.27 consisting of 60 repeat measurements of the lead content (in ppb) in river water.[22]

Sheet 2.27. *Lead content of river water.*

	A	B	C	D	E	F	G	H
1	Lead content of river water (ppb)							
2	45	71	47	55	44	59	46	49
3	46	45	57	44	53	46	45	66
4	52	54	48	53	49	40	41	46
5	43	63	35	42	33	33	44	57
6	51	41	40	43	37	40	32	37
7	39	60	34	47	58	37	40	57
8	48	50	32	40	47	37	41	61
9	49	53	37	58	40	45	21	43
10								

Before using the Histogram tool, consideration should be given to what intervals, or 'bins' we want to appear in the histogram. If this is not specified, Excel® divides up the data range into bins of equal width, but may not choose a sensible number of bins. For example, applying the histogram tool to the data in sheet 2.27 without specifying the bins produces a histogram with only three bins. Such a histogram reveals little about the distribution of the data.

Before selecting the bins, it is helpful to find the range of the data by using the MAX() and MIN() functions on the data in sheet 2.27. The maximum value is 71 ppb and the minimum value is 21 ppb. It seems reasonable to choose the limits of the bins to be 20, 30, 40, 50, 60, 70, 80. These numbers are entered into the spreadsheet as shown in sheet 2.28. We are now ready to create the histogram.

Sheet 2.28. *Bin limits for the lead content histogram.*

	A	B
11	Bin limits	
12	20	
13	30	
14	40	
15	50	
16	60	
17	70	
18	80	

[22] ppb stands for 'parts per billion'.

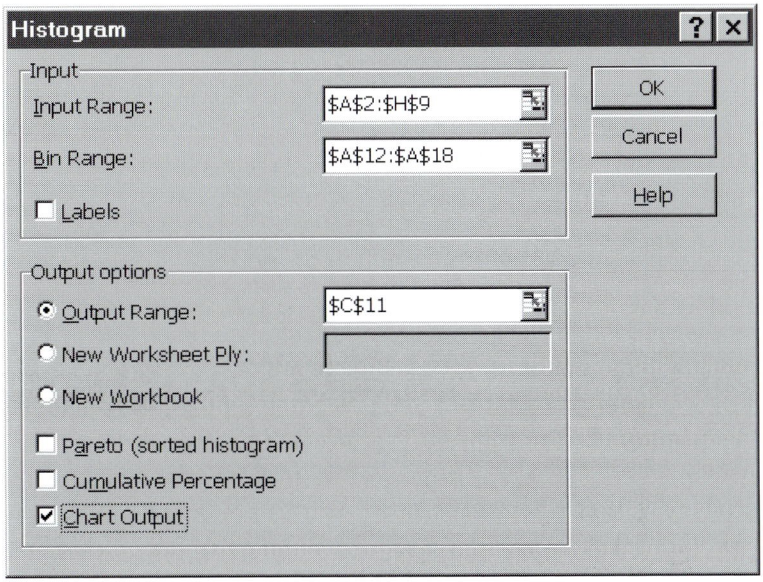

Figure 2.10. Dialog box for the Histogram tool.

1. From the Menu bar, choose Tools ➤ Data Analysis ➤ Histogram.
2. A dialog box appears into which we must enter references to the cells containing the values, references to the cells containing the bin information and indicate where we want the histogram frequencies to appear. Enter the cell references into the dialog box, as shown in figure 2.10.
3. To obtain a graphical output of the histogram, tick the Chart output box in the dialog box. Click on the OK button.

Sheet 2.29. *Numerical output from Histogram tool.*

	A	**B**	**C**	**D**
11	Bin limits		*Bin*	*Frequency*
12	20		20	0
13	30		30	1
14	40		40	18
15	50		50	27
16	60		60	14
17	70		70	3
18	80		80	1
19			More	0

The values returned in columns C and D are shown in sheet 2.29. It is not immediately clear how the bin limits in the C column of sheet 2.29 relate to the frequencies appearing in the D column. The way to interpret the values in cells D12 to D18 is shown in table 2.6, where x represents the lead

Table 2.6. *Relating frequencies to intervals in the histogram output shown in sheet 2.29.*

Excel® bin label	Actual interval (ppb)	Frequency
20	$x \leq 20$	0
30	$20 < x \leq 30$	1
40	$30 < x \leq 40$	18
50	$40 < x \leq 50$	27
60	$50 < x \leq 60$	14
70	$60 < x \leq 70$	3
80	$70 < x \leq 80$	1
More	$x > 80$	0

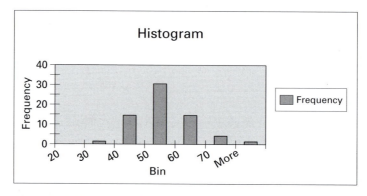

Figure 2.11. Histogram of lead content.

concentration in ppb. Figure 2.11 shows the chart created using the bins and frequencies appearing in sheet 2.29. A close inspection of figure 2.11 reveals that the horizontal axis is not labelled correctly. It appears that in the interval between 20 and 30 there are no values. This conflicts with table 2.6 which indicates that one value lies in this interval. To remedy this problem:

1. Make the chart active, then double click on the *x* axis.
2. From the Format Axis dialog box choose the Alignment tab.
3. Move to the graphic which shows the alignment of the text and with the left hand mouse button pressed down, rotate the text to –45 Degrees. Click OK.

The *x* axis labels should now appear in positions consistent with the frequencies in the histogram.

2.8.2 Descriptive statistics

The mean, median, mode, maximum and minimum of a list of numbers can be determined using the built in functions of Excel®, as described in sections 2.5.1 to 2.5.3. However, if all of these quantities are required, then it is more efficient to use the Descriptive Statistics tool within the Analysis ToolPak. Consider the data in sheet 2.30 which were gathered in an experiment to measure the density of rock specimens. To use the Descriptive Statistics tool:

Sheet 2.30. *Density data of rock samples.*

	A	B	C	D	E	F	G	H	I	J	K
1	Density (g/cm³)	7.3	6.4	7.7	8.6	8.5	9.0	5.7	7.3	8.4	6.6

1. From the Menu bar choose Tools ➤ Data Analysis ➤ Descriptive Statistics.
2. Type the values shown in figure 2.12 into the dialog box.
3. All the values are in row 1, so click on the option 'Grouped By Rows'.
4. Tick the Summary statistics box.
5. Click on OK.

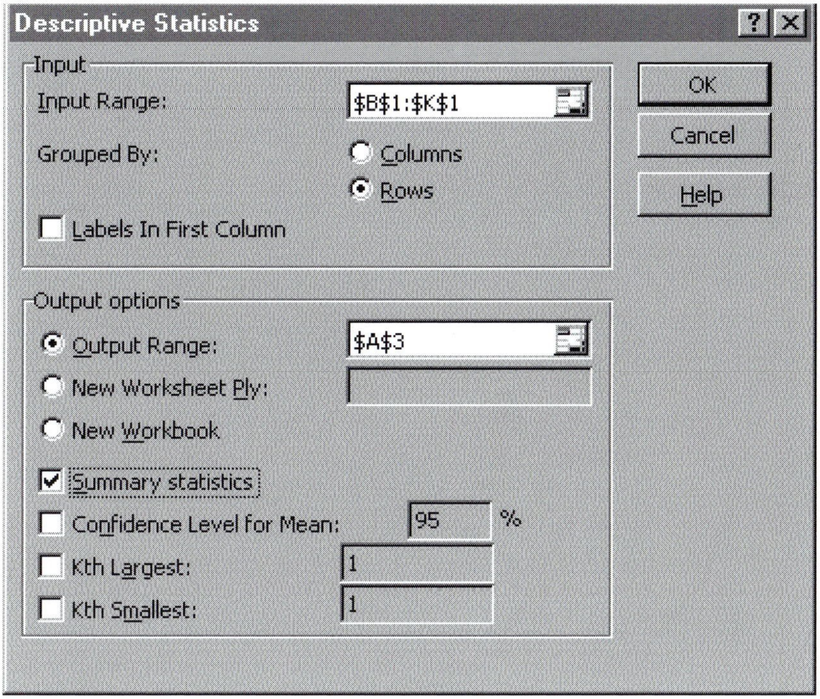

Figure 2.12. Dialog box for Descriptive Statistics.

Excel® returns values for the mean, median and the other statistics as shown in sheet 2.31.

Sheet 2.31. *Values returned when applying the Descriptive Statistics tool to density data.*[23]

	A	B
3	*Row1*	
4		
5	Mean	7.55
6	Standard Error	0.343592
7	Median	7.5
8	Mode	7.3
9	Standard Deviation	1.086534
10	Sample Variance	1.180556
11	Kurtosis	-1.01032
12	Skewness	-0.33133
13	Range	3.3
14	Minimum	5.7
15	Maximum	9
16	Sum	75.5
17	Count	10

As with most of the tools in the Analysis ToolPak, if raw data are changed, the values returned by the Descriptive Statistics tool are *not* immediately updated. To update the values, the sequence beginning at step 1 above must be repeated. If the consequences need to be considered of changing one or more values when, for example, a standard deviation is to be calculated, it may be better to forego use of the Descriptive Statistics tool in favour of using Excel®'s built in functions whose outputs are updated as soon as the contents of a cell have been changed. The trade off between ease of use offered by a tool in the Analysis ToolPak as compared to the increased flexibility and responsiveness of a 'user designed' spreadsheet needs to be considered before analysis of data using the spreadsheet begins.

2.9 Review

The analysis of experimental data may require something as straight-forward as calculating a mean, to much more extensive numerical manip-

[23] Standard error is discussed in chapter 3. Kurtosis is a statistic relating to the 'flatness' of a distribution of data. Skewness is a statistic which indicates the extent to which a distribution of data is asymmetric. For details see Spiegel and Stephens (1998).

ulation, involving the use of advanced statistical functions. The power and flexibility of computer based spreadsheets make them attractive alternatives for this purpose to, say, pocket calculators. Pocket calculators have limited analysis and presentation capabilities and become increasingly cumbersome to use when a large amount of data is to be analysed.

The use of spreadsheets for analysing and presenting experimental data continues to become more common as access to these tools at college, university or home increases. In this chapter we have considered some of the features of the Excel® spreadsheet package which are particularly useful for the analysis of scientific data, including use of functions such as AVERAGE() and STDEV() as well as the more extensive features offered by the Analysis ToolPak.

Ease of use, flexibility and general availability make Excel® a popular tool for data analysis. Other features of Excel® will be treated in later chapters, in situations where they support or clarify the basic principles of data analysis under discussion. In the next chapter we consider those basic principles and, in particular, the way in which experimental data are distributed and how to summarise the main features of data distributions.

Problems

1. An equation relating the speed, v, of a wave to its wavelength, λ, for a wave travelling along the surface of a liquid is

$$v = \sqrt{\frac{\lambda g}{2\pi} + \frac{2\pi\gamma}{\rho\lambda}} \qquad\qquad (2.12)$$

where g is the acceleration due to gravity, γ is the surface tension of the liquid, and ρ is the density of the liquid. Given that

$g = 9.81$ m/s^2,
$\gamma = 73 \times 10^{-3}$ N/m for water at room temperature,
$\rho = 10^3$ kg/m^3 for water,

use Excel® to calculate the speed of a water wave, v, for $\lambda = 1$ cm to 9 cm in steps of 1 cm.

2. The relationship between the mass of a body, m, and its velocity, v, can be written

$$m = \frac{m_0}{\left(1 - \dfrac{v^2}{c^2}\right)} \qquad\qquad (2.13)$$

where m_0 is the rest mass of the body and c is the velocity of light. Excel® is used to calculate m, for values of v close to the speed of light. Note that $c=3.0\times10^8$ m/s and we take $m_0=9.1\times10^{-31}$ kg. A formula has been entered into cell B2 of sheet 2.32 to calculate m. Cell D2 has been given the name mo and cell D3 has been given the name c_ (see section 2.3.6 for restrictions on naming cells).

When the Enter key is pressed, the value returned is −9.34E−1. This is inconsistent with equation (2.13). Find the mistake in the formula in cell B2, correct it, and complete column B of sheet 2.32.

Sheet 2.32. *Mass of a body moving with velocity,* v.

	A	B	C	D
1	v	m		
2	2.90E+08	=mo/1-A2^2/c_^2	mo	9.10E-31
3	2.91E+08		c_	3.00E+08
4	2.92E+08			
5	2.93E+08			
6	2.94E+08			
7	2.95E+08			
8	2.96E+08			
9	2.97E+08			
10	2.98E+08			
11	2.99E+08			

3. Sheet 2.33 contains 100 values of relative humidity (%).

Sheet 2.33. *100 relative humidity values (%).*

	A	B	C	D	E	F	G	H	I	J
1	70	62	58	76	60	55	56	60	68	59
2	69	54	61	58	62	71	68	63	73	72
3	67	68	68	63	63	64	75	65	64	67
4	68	57	72	69	61	70	73	58	63	66
5	72	63	64	65	59	65	63	63	66	58
6	73	61	72	71	69	65	58	66	77	61
7	72	66	67	65	62	70	65	67	66	72
8	64	72	61	69	70	66	64	60	61	68
9	63	70	65	62	70	75	64	79	68	62
10	63	63	72	75	70	66	72	79	78	69

(i) Use the built in functions on Excel® to find the mean, standard deviation, maximum, minimum and range of these values.

(ii) Use the Histogram tool to plot a histogram of the values.

4. A diaphragm vacuum pump is used to evacuate a small chamber. Table 2.7 shows the pressure, P, in the chamber at a time, t, after the pump starts. Use Excel® to:

Table 2.7. *Variation of
pressure with time for a
diaphragm pump.*

P (Torr)	t (s)
750	1.9
660	4.2
570	6.8
480	10.7
390	16.2
300	25.2
210	49.2
120	69.0

Table 2.8. *Variation of vapour pressure with temperature for three volatile
liquids.*

Temperature (°C)	Pressure (Torr)		
	Halothane	Chloroform	Trichloroethylene
0	102	61	20
5	120	68	24
10	148	82	30
15	188	98	38
20	235	133	51
25	303	182	65
30	382	243	92
35	551	352	119

(i) Convert the pressures in torrs to pascals, given that the relationship
between torrs and pascals is 1 Torr = 133.3 Pa.

(ii) Plot a graph of pressure in pascals versus time in seconds. The graph
should include fully labelled axes.

5. The vapour pressure of halothane, chloroform and trichloroethylene
were measured over the temperature range 0 °C to 35 °C. The data obtained
are shown in table 2.8. Show the variation of vapour pressure with temperature for the three liquids on an x–y graph.

6. Equation (2.14) relates the electrical resistivity, ρ, of blood cells to the
packed cell volume H (where H is expressed as a percentage):

Table 2.9. *Variation of force with displacement.*

F(N)	y (cm)
2	2.4
4	4.2
8	6.9
10	7.6
12	9.0
14	10.4
16	11.6

$$\frac{\rho}{\rho_P} = \frac{1 + (f - 1)\,\dfrac{H}{100\%}}{1 - \dfrac{H}{100\%}} \tag{2.14}$$

where ρ_p is the resistivity of the plasma surrounding the cells and f is a form factor which depends on the shape of the cells.

   (i) Use Excel® to determine ρ/ρ_p for H in the range 0% to 80% in steps of 5% when $f = 4$.
   (ii) Repeat part (i) for $f = 3, 2.5, 2.0$ and 1.5.
   (iii) Show the ρ/ρ_p versus H data determined in parts (i) and (ii) on the same x–y graph.

7. A force, F, is applied to the string of an archer's bow, causing the string to displace a distance y. Table 2.9 contains the values of F and y.

   (i) Plot a graph of displacement versus force.
   (ii) Assuming a linear relationship between the displacement and the force, add a trendline and a trendline equation to the graph.
   (iii) If the uncertainty in the y values is ± 0.2 cm, attach error bars to all the points on the graph.

8. An experiment is performed to study the effectiveness of reflective coatings on a glass window for reducing the transmission of infra-red radiation into a room. Sheet 2.34 in the Excel® file named Chapter2 at http://uk.cambridge.org/resources/0521793378 contains temperature versus time data for coated and uncoated glass. Use the x–y graphing option to plot temperature versus time for each coating

Chapter 3

Data distributions I

3.1 Introduction

It is tempting to believe that the laws of 'chance' that come into play when we toss a coin or roll dice have little to do with experiments carried out in a laboratory. Rolling dice and tossing coins are the stuff of games. Surely, well planned and executed experiments provide precise and reliable data, immune from the laws of chance. Not so. Chance, or what we might call more formally *probability*, has rather a large role to play in every experiment. This is true whether an experiment involves counting the number of beta particles detected by nuclear counting apparatus in one minute, measuring the time a ball takes to fall a distance through a liquid or determining the values of resistance of 100 resistors supplied by a component manufacturer. Because it is not possible to predict with certainty what value will emerge when a measurement is made of a quantity, say of the time for a ball to fall through liquid, we are in a similar position to a person throwing dice, who cannot know in advance which numbers will appear 'face up'. If we are not to give up in frustration at our inability to discover the 'exact' value of a quantity experimentally, we need to find out more about probability and how it can assist rather than impede our experimental studies.

In many situations a characteristic pattern or *distribution* emerges in data gathered when repeat measurements are made of a quantity. A distribution of values indicates that there is a probability associated with the occurrence of any particular value. Related to any distribution of 'real' data there is a *probability distribution* which allows us to calculate the probability of the occurrence of any particular value. Real probability distributions can often be approximated by a 'theoretical' probability distribution. Though it is possible to devise many theoretical probability distributions, it is the so called '*normal*' probability distribution[1] (also referred to as the 'Gaussian' distribution) that is most widely used. This is because histograms of data in many experiments have shapes that are very similar to that of the normal distribution. An attraction of the normal and other distributions is that they provide a way of describing data in a quantitative manner which complements and extends diagrammatic representations of data such as the histogram. Using the properties of the normal distribution we are usually able to summarise a whole data set, which may consist of many values, by one or two carefully chosen numbers. Consideration of the normal distribution allows us to define a *confidence interval* for data, so that, though we cannot say with certainty what the next measured value will be in a 'time of fall' experiment, we *will* be able to determine the probability that the value lies between (say) 3.2 s and 3.6 s.

Due to the reliance placed on the normal distribution when analysing experimental data, we need to be aware in what circumstances it is inappropriate to use this distribution and what the alternatives are. In particular, we need to consider a way of assessing the 'normality' of data.

Probability tables are often used in the calculation of probabilities and confidence intervals. Excel® has several built in functions which relieve much of the tedium connected with using such tables. These functions and their applications will be discussed after the basic principles of probability have been considered.

3.2 Probability

In science it is common to speak of 'making measurements' and accumulating 'values' which might also be referred to as 'data'. The topic of probability has its own distinctive vocabulary with which we should be familiar. If an experiment in probability requires that a die[2] is thrown 150 times, then it is usual to say that the experiment consists of 150 *trials*. The outcome of a trial is referred to as an *event*.[3] So, if on the first throw of a die a '6' occurs, we speak of the event being a '6'. We can draw equivalences between trial and measurement, and between event and value.

The record of many events constitutes the data gathered in an experiment.

---

[1] Usually referred to simply as the normal distribution.
[2] The singular of dice is die, so you can throw two dice, but only one die.
[3] Also sometimes referred to as an 'outcome'.

Example 1
A die is thrown once; what are the possible events?

ANSWER
The possible events on a single throw of a die (i.e. a single trial) are 1 or 2 or 3 or 4 or 5 or 6.

The possible outcomes of a trial, or a number of trials, are often referred to as the *sample space*. This is written as S{}. The sample space for a single throw of a die is S{1,2,3,4,5,6}.

3.2.1 Rules of probability

We will use some of the rules of probability as ideas are developed regarding the variability in data. It is useful to illustrate the rules of probability by reference to a familiar 'experiment' which consists of rolling a die one or more times. It is emphasised that, though the outcomes of rolling a die are considered, the rules are applicable to the outcomes of many experiments.

Rule 1
All probabilities lie in the range 0 **to** 1
Notes:

(i) Example of probability equal to zero: the probability of an event being a '7' on the roll of a die is zero as only whole numbers from 1 to 6 are possible.

(ii) Example of probability equal to 1: the probability of an event being from 1 to 6 (inclusive) on the roll of a die is 1, as all the possible events lie between these inclusive limits.

(iii) Due to the symmetry of the die, it is reasonable to suppose that each number will have the same probability of occurring. Writing the probability of obtaining a 1 as $P(1)$ and the probability of obtaining a 2 as $P(2)$ and so on, it follows that

$$P(1) + P(2) + P(3) + P(4) + P(5) + P(6) = 1.$$

If all the probabilities are equal, then $P(1) = P(2) = P(3)$ etc. $= \frac{1}{6}$.

(iv) Another way to find $P(1)$ is to do an experiment consisting of n trials, where n is a large number – say 1000 and to count the number of times a 1 appears – call this number, N. The ratio N/n (sometimes called the *relative frequency*) can be taken as the probability of obtaining a 1. Assuming we use a die that has not been tampered with, it is reasonable to expect N/n to be about $\frac{1}{6}$. This can be looked at another way: If the probability of a particular event A is $P(A)$, then the expected number of times that event A will occur in n trials is equal to $nP(A)$.

Rule 2

When events are *mutually exclusive*, the probability of event A occurring, $P(A)$, or event B occurring, $P(B)$, is

$$P(A \text{ or } B) = P(A) + P(B)$$

Notes

(i) *Mutually exclusive events:* this means that the occurrence of one event excludes other events occurring. For example, if you roll a die and obtain a 6 then no other outcome is possible (you cannot obtain, for example, a 6 *and* a 5 on a single roll of a die).

(ii) Example: What is the probability of obtaining a 2 or a 6 on a single roll of a die?

$$P(2 \text{ or } 6) = P(2) + P(6) = \tfrac{1}{6} + \tfrac{1}{6} = \tfrac{1}{3}.$$

Rule 3

When events are *independent*, the probability of event A, $P(A)$, *and* event B, $P(B)$, occurring is $P(A) \times P(B)$. This is written[4] as,

$$P(A \text{ and } B) = P(A) \times P(B)$$

Notes

(i) *Independent events:* this means that the occurrence of an event has no influence on the probability of the occurrence of succeeding events. For example, on the first roll of a die the probability of obtaining a 6 is $\frac{1}{6}$. The next time the die is rolled, the probability that a 6 will occur remains $\frac{1}{6}$.

(ii) Example: Using the rule, the probability of throwing two 6s in succession is

$$P(6 \text{ and } 6) = P(6) \times P(6) = \tfrac{1}{6} \times \tfrac{1}{6} = \tfrac{1}{36}$$

We now have sufficient probability rules for our purposes.[5] Next we consider how functions which describe the probability of particular outcome

---

[4] $P(A \text{ and } B)$ is often written as $P(AB)$.

[5] Adler and Roessler (1972) discuss the other rules of probability.

Table 3.1. *One hundred random numbers in the interval 0 to 1.*

0.632	0.328	0.696	0.166	0.665	0.157	0.010	0.391	0.454	0.396
0.322	0.454	0.087	0.540	0.603	0.138	0.021	0.203	0.272	0.763
0.055	0.095	0.410	0.422	0.109	0.713	0.834	0.029	0.577	0.984
0.575	0.932	0.772	0.043	0.464	0.112	0.234	0.062	0.657	0.839
0.600	0.894	0.421	0.186	0.213	0.676	0.504	0.028	0.916	0.809
0.798	0.841	0.927	0.335	0.505	0.549	0.352	0.430	0.984	0.853
0.803	0.302	0.389	0.814	0.175	0.309	0.607	0.198	0.569	0.177
0.711	0.445	0.279	0.091	0.469	0.572	0.719	0.901	0.993	0.034
0.571	0.277	0.345	0.119	0.688	0.512	0.437	0.141	0.903	0.453
0.048	0.597	0.532	0.864	0.936	0.040	0.553	0.129	0.077	0.706

occurring can be useful when describing variability observed in 'real' data. We begin by discussing probability distributions.

3.3 Probability distributions

In the last section we considered the probability that a particular outcome or event will occur when a die is rolled. A single roll of a die has only six possible outcomes. What do we do if our data consists of many events or, as we usually prefer to call them, values – is it possible to assign a probability to each value observed and is it sensible to do this? For illustrative purposes we consider a computer based 'experiment' which consists of generating 1000 random numbers[6] with values between 0 and 1. Table 3.1 shows the first 100 random numbers generated.

In contrast to the situation in which a die is rolled, we do not know the probability that a particular value appearing in table 3.1 will occur.[7] Is it possible to establish the probability of obtaining a random number, x, between, say, 0.045 and 0.621, using the data that appear in table 3.1? A good starting point is to view the values in picture form to establish whether any values occur more frequently than others. The histogram shown in figure 3.1 indicates the frequency of occurrence of the 1000 random numbers.

---

[6] Random numbers between 0 and 1 can be generated using the RAN function found on many calculators such as those manufactured by CASIO. The RAND() function on Excel® can also be used to generate random numbers.

[7] Unless we know the details of the algorithm that produced the random numbers.

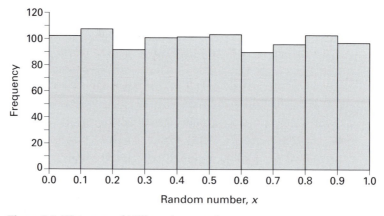

Figure 3.1. Histogram of 1000 random numbers.

The histogram provides evidence to support the statements:

- All values of x lie in the range 0 to 1.
- The number of values (i.e. the frequency) appearing in any given interval, say from 0.0 to 0.1, is about the same as that appearing in any other interval of the same width.

The histogram in figure 3.1 can be transformed into a probability distribution. This allows the probability of obtaining a particular random number in any interval to be calculated. First, the bars in figure 3.1 are merged together and the frequencies scaled so that the total area under the curve created by merging the bars is 1. That is, for this distribution, the probability is 1 that x lies between 0 and 1. Taking the length of each bar to be the same, the graph shown in figure 3.2 is created.

In going from figure 3.1 to 3.2 we have made quite an important step. Figure 3.1 shows the distribution of the 1000 random numbers. These

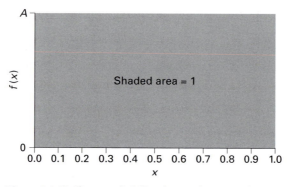

Figure 3.2. Uniform probability density function, $f(x)$ versus x.

numbers may be regarded as a *sample* of all the random numbers that *could* have been generated. The number of values that could have been generated is infinite in size, so the population is, in principle, infinite. By contrast, the graph in figure 3.2 allows us to calculate the probability of obtaining a value of x within a particular interval. What we have done by innocently smoothing out the 'imperfections' in the histogram shown in figure 3.1 prior to scaling is to *infer* that the population of random numbers behaves in this manner, i.e. it consists of values that are evenly spread between 0 and 1. Inferring something about a population by considering a sample drawn from that population is a central goal in data analysis. Note that the label on the vertical axis in figure 3.2 differs from that in figure 3.1. The vertical axis is labelled $f(x)$, where $f(x)$ is referred to as the *probability density function*, or pdf for short.

As the area under any $f(x)$ versus x curve is 1 and, in general, x may have any value between $-\infty$ and $+\infty$, we write

$$\int_{-\infty}^{\infty} f(x)\,dx = 1 \tag{3.1}$$

With reference to the graph in figure 3.2,

$f(x) = 0$ for $x < 0$
$f(x) = 0$ for $x > 1$
$f(x) = A$ for $0 \le x \le 1$

Using equation (3.1),

$$\int_{0}^{1} A\,dx = 1$$

therefore,

$$A|x|_{0}^{1} = A|1 - 0| = 1$$

It follows that $A = 1$, so $f(x) = 1$ from $x = 0$ to $x = 1$.

A word of caution here: $f(x)$ is *not* the probability of the occurrence of the value x. If that were the case then, as $f(x) = 1$ for all values of x between 0 and 1, we would conclude that the probability of observing, say, the number 0.476 is 1. This is not so. It is the *area* under the curve that is interpreted as a probability. If we want to know the probability of observing a value between x_1 and x_2, where $x_2 > x_1$, we must integrate $f(x)$ between these limits,[8] i.e.

---

[8] Read $P(x_1 \le x \le x_2)$ as 'the probability that x lies between the limits x_1 and x_2'.

$$P(x_1 \le x \le x_2) = \int_{x_1}^{x_2} f(x)\,dx \tag{3.2}$$

For this probability distribution, $f(x)$ is 1 between $x=0$ and $x=1$, and the following relationship emerges

$$P(x_1 \le x \le x_2) = \int_{x_1}^{x_2} 1 \cdot dx = x_2 - x_1 \tag{3.3}$$

Example 2

What is the probability of observing a random number, generated in the manner described in this section, between 0.045 and 0.621?

ANSWER

Substituting $x_2 = 0.621$ and $x_1 = 0.045$ into equation (3.3) gives

$$P(0.045 \le x \le 0.621) = 0.621 - 0.045 = 0.576$$

In performing the integration given by equation (3.3), the implication is that the variable, x, represents values of a continuous quantity.[9] That is, x can take on any value between the limits x_1 and x_2. Mass, length and time are quantities regarded as continuous[10] but some values measured through experiment are discrete. An example is the counting of particles emerging from a radioactive substance. In this case the recorded values must be a whole number of counts. The probability distributions used to describe the distribution of discrete quantities differ from those that describe continuous quantities. For the moment we focus on continuous distributions as we tend to encounter these more often than discrete distributions in the physical sciences.

The probability distribution given by figure 3.2 and represented mathematically by equation (3.3) is convenient for illustrating some of the basic features common to all probability distributions. However, few, if any, 'real' scientific experiments generate data distributed in such a manner. In section 3.4 we consider data gathered in more conventional experiments and look at a probability distribution with more features in common with real data distributions.

---

[9] x is referred to as a *continuous random variable*, or *crv*.

[10] We can argue (with justification) that on an atomic scale, mass and length are no longer continuous; however, few measurements we make in the laboratory are likely to be sensitive to the 'graininess' of matter on this scale.

Exercise A

1. A particular probability density function is written $f(x) = Ax$ for the range $0 \le x \le 4$ and $f(x) = 0$ outside this range.

    (i)   Sketch a graph of $f(x)$ versus x.

    (ii) Use equation (3.1) to find A.

    (iii) Calculate the probability that x lies between $x = 3$ and $x = 4$.

    (iv) If 1000 measurements are made, how many values would you expect to fall between $x = 3$ and $x = 4$, assuming this probability function to be valid?

2. The exponential probability density function[11] may be written

$$f(x) = \lambda e^{(-\lambda x)}$$

where $\lambda > 0$ and $x \ge 0$.

    (i)   Show that, for $\lambda > 0$, $\int_{0}^{\infty} f(x)\,dx = 1$.

    (ii) If $\lambda = 0.3$, calculate the probability that x lies between $x = 0$ and $x = 2$.

    (iii) Calculate the probability that $x > 2$.

3.3.1 Limits in probability calculations

When calculating a probability using a probability density function, does it matter whether the limits of the interval, x_1 and x_2, are included in the calculation? In other words, does $P(x_1 \le x \le x_2)$ differ from $P(x_1 < x < x_2)$? The answer to this is no, and to justify this we can write

$$P(x_1 \le x \le x_2) = P(x_1) + P(x_1 < x < x_2) + P(x_2) \qquad (3.4)$$

$P(x_1)$ is the area under the probability curve at $x = x_1$. Using equation (3.2), this probability is written

$$P(x_1) = P(x_1 \le x \le x_1) = \int_{x_1}^{x_1} f(x)\,dx \qquad (3.5)$$

Both limits of the integral are x_1 and so the right hand side of equation (3.5) must be equal to zero. Therefore, $P(x_1) = P(x_2) = P(x) = 0$, and equation (3.4) becomes

$$P(x_1 \le x \le x_2) = P(x_1 < x < x_2) \qquad (3.6)$$

Equation (3.6) is valid for continuous random variables.

[11] This distribution is sometimes used to predict the lifetime of electrical and mechanical components. For details see Kennedy and Neville (1986).

The fact that $P(x) = 0$ can be disturbing. Let us return to table 3.1 and consider the first value, $x = 0.632$. If we are dealing with the distribution of a random variable, then $P(0.632) = 0$. However, the value 0.632 *has* been observed, so how are we able to reconcile this with the fact that $P(0.632) = 0$? The explanation is that $x = 0.632$ is a *rounded* value[12] and the 'actual' value could lie anywhere between $x = 0.63150$ and $x = 0.63250$. That is, though it is not obvious, we are dealing with an implied range by the way the number is written. If we now ask what is the probability that a value of x (for this distribution) lies between $x = 0.63150$ and $x = 0.63250$, we have, using equation (3.3),

$$P(x_1 \leq x \leq x_2) = 0.63250 - 0.63150 = 0.001$$

If 1000 measurements are made, then the number of times 0.632 is expected to occur is $1000 \times 0.001 = 1$.

3.4 Distributions of real data

We now consider some real data gathered in laboratory based experiments with a view to identifying some general features of data distributions.

Figure 3.3 shows data from an experiment to determine the viscosity of an oil. In the experiment, a small metal sphere was allowed to fall many times through the oil and the time of fall was recorded. Figure 3.4 shows the resistances of a sample of 100 resistors of nominal resistance 4.7 kΩ which were measured with a high precision multimeter. Figure 3.5 shows the number of counts detected by a scintillation counter in 100, 10-second time intervals, when the counter was placed in close proximity to the β particle emitter, strontium 90. As the experiments described here are so diverse and the underlying causes of variability in each experiment likely to be quite different,[13] we might expect the distribution of data in each experiment as shown in figures 3.3 to 3.5 to be markedly different. This is not the case, and we are struck by the similarity in the shapes of the histograms, rather than the differences between them. Note that:

---

[12] Rounding could have been carried out by the experimenter, or the value recorded could have been limited by the precision of the instrument used to make the measurement.

[13] For example, a major source in the variability of time measured in the viscosity experiment is due inconsistency in hand timing the duration of the fall of the ball, whereas in the resistance experiment the dominant factor is likely to be variability in the process used to manufacture the resistors.

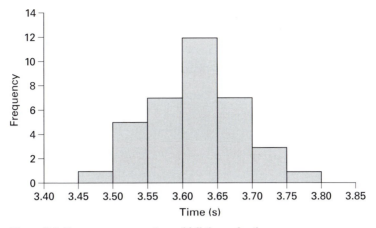

Figure 3.3. Frequency versus time of fall through oil.

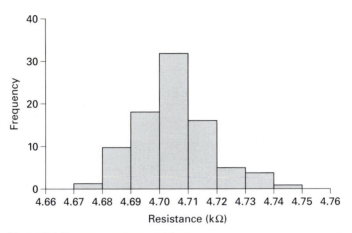

Figure 3.4. Frequency versus resistance.

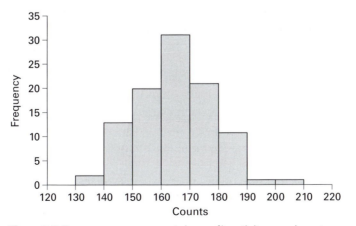

Figure 3.5. Frequency versus counts in a radioactivity experiment.

- the distribution of each data set is approximately symmetric;
- there is a single central 'peak';
- most data are clustered around that peak (and consequently few data are found in the 'tails' of the distribution).

A commonly used shorthand way of expressing the main features of histograms, such as those in figures 3.3 to 3.5, is to say that the data they represent follow a 'bell shaped' distribution. To illustrate this, figure 3.6 shows a bell shaped curve superposed on the time of fall data shown in figure 3.3.

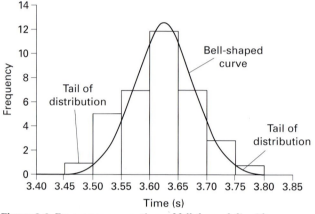

Figure 3.6. Frequency versus time of fall through liquid.

We shift our focus to the mathematical function that generates the bell shaped curve in figure 3.6. This is because it appears reasonable to argue that the distribution of data in this sample and, more fundamentally, the distribution of the population from which the sample was drawn, can be described well by this function. A consideration of this function and its properties provides further insights into how the variability in data can be expressed concisely in numeric form.

It is perhaps worth sounding a note of caution at this point. By no means do all data follow the shape given by figures 3.3 to 3.5. However, approximations to this shape are found so frequently in experimental data that it makes sense to focus attention on this distribution ahead of any other. Additionally, even if the spread of data *is* distinctly asymmetric, a bell shaped distribution can be constructed from the raw data in a way that assists in expressing the variability or uncertainty in the *mean* of the data.[14] This is discussed in section 3.7.

---

[14] The term 'raw' data refers to data that has been obtained directly through experiment or observation and has not been manipulated in any way, such as combining values to calculate a mean.

3.5 The normal distribution

The bell shaped curve appearing in figure 3.6 is generated using the probability density function

$$f(x) = \frac{1}{\sigma\sqrt{2\pi}} e^{-(x-\mu)^2/2\sigma^2} \tag{3.7}$$

where μ and σ are the population mean and the population standard deviation respectively introduced in chapter 1 and which we used to describe the centre and spread respectively of a data set. Equation (3.7) is referred to as the *normal probability density function*. It is a 'normalised' equation which is another way of saying that when it is integrated between $-\infty \leq x \leq +\infty$, the value obtained is 1.

Using equation (3.7) we can generate the bell shaped curve associated with any combination of μ and σ and hence, by integration, find the probability of obtaining a value of x within a specified interval. Figure 3.7 shows $f(x)$ versus x for the cases in which $\mu = 50$ and $\sigma = 2$, 5 and 10. The population mean, μ, is coincident with the centre of the symmetric distribution and the standard deviation, σ, is a measure of the spread of the data. A larger value of σ results in a broader, flatter distribution, though the total area under the curve remains equal to 1. On closer inspection of figure 3.7 we see that $f(x)$ is quite small for x outside the interval $\mu - 2.5\sigma$ to $\mu + 2.5\sigma$.

The normal distribution may be used to calculate the probability of obtaining a value of x between the limits x_1 and x_2. This is given by

$$P(x_1 \leq x \leq x_2) = \frac{1}{\sigma\sqrt{2\pi}} \int_{-\infty}^{x_1} e^{-(x-\mu)^2/2\sigma^2} dx \tag{3.8}$$

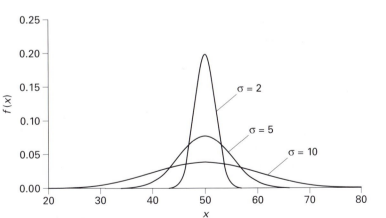

Figure 3.7. Three normal distributions with the same mean, but differing standard deviations.

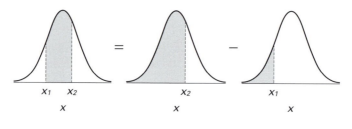

$$P\,(x_1 \le x \le x_2) \quad = \quad P\,(-\infty \le x \le x_2) \quad - \quad P\,(-\infty \le x \le x_1)$$

Figure 3.8. Finding the area under the normal curve.

The steps to determine the probability are shown symbolically and picto-
rially in figure 3.8. The function given by

$$F(x \le x_1) = P(-\infty < x < x_1) = \frac{1}{\sigma\sqrt{2\pi}} \int_{-\infty}^{x_1} e^{-(x-\mu)^2/2\sigma^2} dx \tag{3.9}$$

is usually referred to as the *cumulative distribution function*, cdf, for the
normal distribution. The cdf is often used to assist in calculating probabilities, such as those shown in figure 3.8.

As an example of the use of the normal distribution, we calculate the
probability that x lies between $\pm\,\sigma$ of the mean, μ. Finding this probability
requires that $f(x)$ in equation (3.2) be replaced by the expression in equa-
tion (3.7). The limits of the integration are $x_1 = \mu - \sigma$, and $x_2 = \mu + \sigma$. We have

$$P(\mu - \sigma \le x \le \mu + \sigma) = \frac{1}{\sigma\sqrt{2\pi}} \int_{\mu-\sigma}^{\mu+\sigma} e^{-(x-\mu)^2/2\sigma^2} dx \tag{3.10}$$

To assist in evaluating the integral in equation (3.10), it is usual to change
the variable from x to z, where z is given by

$$z = \frac{x - \mu}{\sigma} \tag{3.11}$$

z is a random variable with mean of zero and standard deviation of 1.
Equation (3.7) reduces to

$$f(z) = \frac{1}{\sqrt{2\pi}} e^{-z^2/2} \tag{3.12}$$

Equation (3.10) becomes

$$P(-1 \le z \le 1) = \frac{1}{\sqrt{2\pi}} \int_{-1}^{1} e^{-z^2/2} dz \tag{3.13}$$

Figure 3.9 shows the variation of $f(z)$ with z. The integral appearing in
equation (3.13) cannot be evaluated analytically and so a numerical

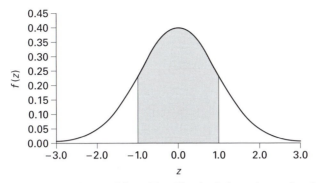

Figure 3.9. Variation of $f(z)$ with z. The shaded area is equal to the probability that z lies between -1 and $+1$.

method for solving for the area under the curve is required. It turns out that the shaded area indicated in figure 3.9 is about 0.68. We conclude that the probability that a value obtained through measurement lies within $\pm \sigma$ of the population mean is 0.68. Or, looked at another way, that 68% of all values obtained through experiment are expected to lie within $\pm \sigma$ of the population mean.[15]

3.5.1 Excel®'s NORMDIST() function

Traditionally, 'probability tables' have been used to assist in determining probabilities associated with areas under the normal probability distribution.[16] However, spreadsheet packages such as Excel® have built in functions which offer an attractive alternative for determining probabilities.

The NORMDIST() function in Excel® calculates the value of the probability density function (pdf) given by equation (3.7) and the value of the cumulative distribution function[17] (cdf) as given by equation (3.9).

The syntax of the function is

NORMDIST(x,mean,standard deviation,cumulative)

In order to select the cdf option, the cumulative parameter in the function is set to TRUE. To choose the pdf option, the cumulative parameter is set to FALSE.

---

[15] As long as the values are normally distributed.

[16] It remains important to know how to use such tables, as a computer will not always be 'to hand'. We consider one such table in section 3.5.2.

[17] The cdf in Excel® is referred to as the *normal cumulative distribution*.

Example 3

Calculate the value of the pdf and the cdf when $x = 46$ for normally distributed data with mean $= 50$ and standard deviation $= 4$.

ANSWER

The NORMDIST() function for the pdf and cdf is shown entered in cells A1 and A2 respectively of sheet 3.1(a). Sheet 3.1(b) shows the values returned in cells A1 and A2 after the ENTER key has been pressed. We conclude that

$$f(x = 46) = 0.06049 \text{ and } P(-\infty \le x \le 46) = 0.15866$$

Sheet 3.1. *Using the* NORMDIST() *function.*

(a) *values entered into* NORMDIST();

(b) *values returned by the* NORMDIST() function.

	A	B	C
1	=NORMDIST(46,50,4,FALSE)		
2	=NORMDIST(46,50,4,TRUE)		

	A	B	C
1	0.06049		
2	0.15866		

Example 4

Calculate the area under the curve between $x = 46$ and $x = 51$ for normally distributed data with mean $= 50$ and standard deviation $= 4$.

ANSWER

$$P(46 \le x \le 51) = P(-\infty \le x \le 51) - P(-\infty \le x \le 46)$$

In cell A1 of sheet 3.2(a) we calculate $P(-\infty \le x \le 51)$ and in cell A2 we calculate $P(-\infty \le x \le 46)$. Subtraction takes place in cell A3 to give the required result. Sheet 3.2(b) shows the values returned by Excel®. The area under the curve between $x = 46$ and $x = 51$ is 0.44005.

Sheet 3.2. *Example of the application of the* NORMDIST() *function.*

(a) *numbers entered into* NORMDIST() *function;*

(b) *result of calculations.*

	A	B	C
1	=NORMDIST(51,50,4,TRUE)		
2	=NORMDIST(46,50,4,TRUE)		
3	=A1-A2		

	A	B	C
1	0.59871		
2	0.15866		
3	0.44005		

> **Exercise B**
>
> The masses of small metal spheres are known to be normally distributed with $\mu = 8.1$ g and $\sigma = 0.2$ g. Use the NORMDIST() function to find the values of the cdf for values of mass given in table 3.2 to four significant figures.

Table 3.2. *Values of mass for exercise B.*

x (g)	7.2	7.4	7.6	7.8	8.0	8.2	8.4	8.6	8.8	9.0

3.5.2 The standard normal distribution

In order to assist in integrating the normal pdf in section 3.5, a mathematical 'trick' was played consisting of changing variables using equation (3.11). This has the effect of creating a new distribution with a mean of 0 and a standard deviation of 1. This distribution is usually referred to as the *standard normal distribution*. It is particularly useful as it can represent any normal distribution, whatever its mean and standard deviation. So, if we wish to evaluate the area (and hence the probability) under a normal distribution curve between the two limits, x_1 and x_2, we need to integrate the pdf not as given by equation (3.7), which differs from one normal distribution to another, but as given by equation (3.12) which is the same for all normal distributions.

To find the probability of obtaining a value of x between x_1 and x_2, first transform x values to z values, so that

$$z_1 = \frac{x_1 - \mu}{\sigma} \qquad z_2 = \frac{x_2 - \mu}{\sigma}$$

and find the probability that a value of z lies between the limits z_1 and z_2 by evaluating the integral

$$P(z_1 \le z \le z_2) = \frac{1}{\sqrt{2\pi}} \int_{z_1}^{z_2} e^{-z^2/2} dz \tag{3.14}$$

Table 3.3, as well as table 1 of appendix 1, may be used to assist in the evaluation of equation (3.14).

If we do not have access to Excel® or some other spreadsheet package that is able to evaluate the area under a normal curve directly, we can use 'probability' tables such as those found in appendix 1. To explain the use of table 3.3, first consider figure 3.10. For any value of z_1, say $z_1 = 0.75$, go down

Table 3.3. *Normal probability table giving the area under the standard normal curve between $z=-\infty$ and $z=z_1$.*

z_1	0.00	0.01	0.02	0.03	0.04	0.05	0.06	0.07	0.08	0.09
0.00	0.5000	0.5040	0.5080	0.5120	0.5160	0.5199	0.5239	0.5279	0.5319	0.5359
0.10	0.5398	0.5438	0.5478	0.5517	0.5557	0.5596	0.5636	0.5675	0.5714	0.5753
0.20	0.5793	0.5832	0.5871	0.5910	0.5948	0.5987	0.6026	0.6064	0.6103	0.6141
0.30	0.6179	0.6217	0.6255	0.6293	0.6331	0.6368	0.6406	0.6443	0.6480	0.6517
0.40	0.6554	0.6591	0.6628	0.6664	0.6700	0.6736	0.6772	0.6808	0.6844	0.6879
0.50	0.6915	0.6950	0.6985	0.7019	0.7054	0.7088	0.7123	0.7157	0.7190	0.7224
0.60	0.7257	0.7291	0.7324	0.7357	0.7389	0.7422	0.7454	0.7486	0.7517	0.7549
0.70	0.7580	0.7611	0.7642	0.7673	0.7704	**0.7734**	0.7764	0.7794	0.7823	0.7852
0.80	0.7881	0.7910	0.7939	0.7967	0.7995	0.8023	0.8051	0.8078	0.8106	0.8133
0.90	0.8159	0.8186	0.8212	0.8238	0.8264	0.8289	0.8315	0.8340	0.8365	0.8389
1.00	0.8413	0.8438	0.8461	0.8485	0.8508	0.8531	0.8554	0.8577	0.8599	0.8621

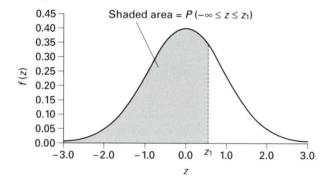

Figure 3.10. Relationship between area under curve and z_1.

the first column as far as $z_1=0.70$, then across to the column headed 0.05; this brings you to the entry in the table with value 0.7734. We conclude that the probability of obtaining a value of $z\leq0.75$ is

$$P(-\infty\leq z\leq0.75)=\frac{1}{\sqrt{2\pi}}\int_{-\infty}^{0.75}e^{-z^2/2}dz=0.7734$$

Using the knowledge that:

- the area under the curve $=1$,
- the curve is symmetric about $z=0$,

it is possible to calculate the probabilities in many other situations.

Example 5

What is the probability that a value of z is *greater* than 0.75?

ANSWER

We have

$$P(z \leq 0.75) + P(z \geq 0.75) = 1$$

so

$$P(z \geq 0.75) = 1 - P(z \leq 0.75) = 1 - 0.7734 = 0.2266$$

Example 6

What is the probability that z lies between the limits $z_1 = -0.60$ and $z_2 = 0.75$?

ANSWER

Solving this type of problem is assisted by drawing a picture representing the area (and hence the probability) to be determined. Figure 3.11 indicates the area required. The shaded area in figure 3.11 is equal to (area to the left of z_2) – (area to the left of z_1). Table 3.3 gives $P(-\infty \leq z \leq 0.75) = 0.7734$.

Probability tables do not usually include negative values of z, so we must make use of the symmetry of the distribution and recognise that[18]

$$P(z \leq -|z_1|) = 1 - P(z \leq |z_1|) \tag{3.15}$$

Now,

$$P(z \leq 0.60) = 0.7257$$

so

$$P(z \leq -0.60) = 1 - 0.7257 = 0.2743$$

The shaded area in figure 3.11 is therefore

$$0.7734 - 0.2743 = 0.4991 \text{ (to four significant figures)}$$

Exercise C

With reference to figure 3.11, if $z_1 = -0.75$ and $z_2 = -0.60$, calculate the probability $P(-0.75 \leq z \leq -0.60)$.

---

[18] Table 1 in appendix 1 is slightly unusual as it gives the area under the normal curve for negative values of z.

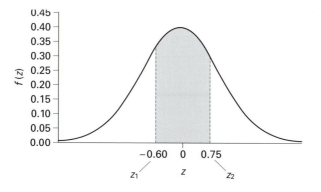

Figure 3.11. Shaded area under curve is equal to probability that z lies between -0.6 and 0.75.

3.5.3 Excel®'s NORMSDIST() function[19]

This function calculates the area under the standard normal curve between $z=-\infty$ and, say, $z=z_1$. This is given as

$$P(-\infty \leq z \leq z_1) = \frac{1}{\sqrt{2\pi}} \int_{-\infty}^{z_1} e^{-z^2/2} dz$$

and is referred to in Excel® as the *standard normal cumulative distribution function*

The syntax of the function is

NORMSDIST(z)

Example 7

Calculate the area under the standard normal distribution between $z=-\infty$ and $z=-1.5$.

ANSWER

Sheet 3.3(a) shows the NORMSDIST() function entered into cell A1. Sheet 3.3(b) shows the value returned in cell A1 after the Enter key is pressed.

---

[19] The NORMSDIST() function is very similar to the NORMDIST() function discussed in section 3.5.1 and so care must be taken to select the correct function.

Sheet 3.3. *Use of* NORMSDIST() *function.*

(a) *number entered into* NORMSDIST();   (b) *result of calculation.*

	A	B	C
1	=NORMSDIST(-1.5)		
2			

	A	B	C
1	0.06681		
2			

Example 8

Calculate the area under the standard normal distribution between $z=-2.5$ and $z=-1.5$.

ANSWER

The area between $z=-2.5$ and $z=-1.5$ may be written as

$$P(-2.5 \leq z \leq -1.5) = P(-\infty \leq z \leq -1.5) - P(-\infty \leq z \leq -2.5)$$

In cell A1 of sheet 3.4(a) we calculate $P(-\infty \leq z \leq -1.5)$ and in cell A2 $P(-\infty \leq z \leq -2.5)$. The two probabilities are subtracted in cell A3 to give the required result. Sheet 3.4(b) shows the values returned in cells A1 to A3 after the Enter key is pressed. We conclude that $P(-2.5 \leq z \leq -1.5) = 0.06060$.

Sheet 3.4. *Example of use of* NORMSDIST() *function.*

(a) *numbers entered into* NORMSDIST() *function.*   (b) *result of calculations.*

	A	B	C
1	=NORMSDIST(-1.5)		
2	=NORMSDIST(-2 5)		
3	=A1-A2		

	A	B	C
1	0.06681		
2	0.00621		
3	0.06060		

Exercise D

Use the NORMSDIST() function to calculate the values of the standard cdf for the z values given in table 3.4. Give the values of the standard cdf to four significant figures.

Table 3.4. *z values for exercise D.*

z	0.0	0.1	0.2	0.3	0.4	0.5	0.6	0.7	0.8	0.9	1.0

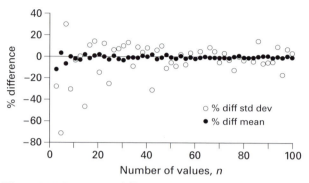

Figure 3.12. Percentage difference between population parameters and sample statistics.

3.5.4 $\bar{x}$ and s as approximations to μ and σ

Before proceeding to consider the analysis of real data using the normal distribution, we should admit once again that, in the majority of experiments, we can never know the population mean, μ, and the population standard deviation, σ, as they can only be determined if all the values that make up the population are known. If the population is infinite, as it often is in science, this is impossible. We must 'make do' with the best estimates of μ and σ, namely $\bar{x}$ and s respectively. μ and $\bar{x}$ are given by equations (1.13) and (1.6) respectively. σ and s are given by equations (1.14) and (1.16) respectively. $\bar{x}$ tends to μ and s tends to σ as the number of points, n, tends to infinity. It is fair to ask for what sample size, n, does the approximation of $\bar{x}$ for μ and s for σ become 'reasonable'?

Figure 3.12 shows the percentage difference between $\bar{x}$ and μ, and between σ and s for samples of size n, where n is between 2 and 100. Samples were drawn from a population consisting of simulated values known to be normally distributed.[20] The percentage difference between the means is defined as $[(\bar{x} - \mu)/\mu] \times 100\%$ and the percentage difference between the standard deviations is defined as $[(s - \sigma)/\sigma] \times 100\%$. Figure 3.12 indicates that the magnitude of the percentage difference between μ and $\bar{x}$ for these data is small (under 5%) for $n > 10$. Figure 3.12 also shows that, as n increases, s converges more slowly to σ than $\bar{x}$ converges to μ. The percentage difference between σ and s is only consistently[21] below 20% for $n > 30$. As

---

[20] Normally distributed values used in the simulation were generated with known μ and σ using the 'Random Number Generation' facility in Excel®.

[21] However, note that one sample of $n \approx 40$ produced a percentage difference between σ and s of about -30%.

a rule of thumb we assume that if a sample consists of n values, where n is in excess of 30, we can take s as a reasonable approximation[22] to σ.

Exercise E

1. Figure 3.6 shows a histogram of the time of fall of a small sphere through a liquid. For these data, $\bar{x}=3.624$ s and $s=0.068$ s.

   (i) Calculate the probability that a value lies in the interval 3.65 s to 3.70 s.

   (ii) If the data consist of 36 values, how many of these would you expect to lie in the interval 3.65 s to 3.70 s?

2. With reference to the standard normal distribution, what is the probability that z is

   (i) Less than -2?

   (ii) Between -2 and $+2$?

   (iii) Between -3 and -2?

   (iv) Above $+2$?

Sketch the standard normal distribution for each of parts (i) to (iv) above, indicating clearly the probability required.

3. For normally distributed data, what is the probability that a value lies further than 2.5σ from the mean?

4. A large sample of resistors supplied by a manufacturer is tested and found to have a mean resistance of 4.70 kΩ and a standard deviation of 0.01 kΩ. Assuming the distribution of resistances is normal:

   (i) What is the probability that a resistor chosen from the population will have a resistance

      (a) Below 4.69 kΩ?

      (b) Between 4.71 kΩ and 4.72 kΩ?

      (c) Above 4.73 kΩ?

   (ii) If 10 000 resistors are measured, how many would be expected to have resistance in excess of 4.73 kΩ?

3.6 Confidence intervals and confidence limits

Instead of asking the question 'what is the area under the normal curve between limits x_1 and x_2?', we ask the closely related question 'between

---

[22] We will discuss this approximation further in section 3.8.1.

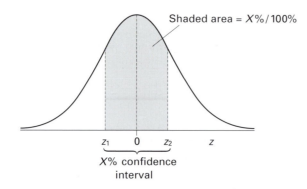

Figure 3.13. Standard normal distribution showing the $X\%$ confidence interval.

what limits (symmetrical about the mean) is the area under the normal curve equal to, say, 0.5?' Obtaining limits which define an interval is useful if we want to answer the question:

> Between what limits (symmetric about the mean) does $X\%$ of the population lie?

$X\%$ is sometimes referred to as the *confidence level* and the interval within which $X\%$ of the data lie is the *confidence interval.*

In order to determine the confidence interval:

1. Choose the confidence level $X\%$ (levels commonly chosen are 50%, 68%, 95% or 99%).
2. Calculate $X\%/100\%$.
3. Sketch a graph of the standard normal distribution. Shade the area $X\%/100\%$ under the curve, symmetric about the mean as shown in figure 3.13.
4. Use the normal probability tables in appendix 1 (or the NORMSINV() function in Excel®) to determine z_1 and z_2 (note that $z_1 = -z_2$).
5. Transform back to 'x values' using the relationship $x = \mu + z\sigma$.

In section 3.8.1 we extend the idea of confidence intervals to include the confidence interval for the true value (or population mean) of a quantity. This is extremely useful as we are able to make 'concrete' the sometimes vague notion of 'uncertainty' in a quantity, i.e. we can specify limits for the true value and say what the probability is that the true value of the quantity lies between those limits.

Example 9

A distribution of data consisting of many repeat measurements of the flow of water through a tube has a mean of 3.7 mL/s and a standard deviation of 0.41 mL/s. What is the 50% confidence interval for these data?

ANSWER

We assume that the question provides good estimates of the population parameters, μ and σ, so that $\mu = 3.7$ mL/s and $\sigma = 0.41$ mL/s. To find the 50% confidence interval it is necessary to consider the standard normal distribution and indicate the required area under the curve. The area is distributed symmetrically about $z = 0$ and is shown shaded in figure 3.14.

Table 1 in appendix 1 can be used to find z_2 so long as we can determine the area to the left of z_2, i.e. the area between $-\infty \le z \le z_2$. As the normal distribution is symmetrical about $z = 0$, half of the total area under the curve lies between $z = -\infty$ and $z = 0$. This area is equal to 0.5, as the total area under the curve is equal to 1. As the required area is symmetrical about the mean, half the shaded area in figure 3.14 must lie between $z = 0$ and $z = z_2$. The total area under the curve between $-\infty$ and z_2 is therefore $0.5 + 0.25 = 0.75$.

The next step is to refer to the normal probability integral tables and look in the table for the probability 0.75 (or the nearest value to it). Referring to table 1 in appendix 1, a probability of 0.75 corresponds to a z value of 0.67, so $z_2 = 0.67$. As the confidence limits are symmetrical about $z = 0$, it follows that $z_1 = -0.67$.

The final step is to apply equation (3.11) which is used to transform z values to x values. Rearranging equation (3.11) we obtain

$$x = \mu + z\sigma \tag{3.16}$$

Substituting $z_1 = -0.67$ and $z_2 = 0.67$ gives $x_1 = 3.43$ mL/s and $x_2 = 3.97$ mL/s.

In summary, the 50% confidence interval (i.e. the interval expected to contain 50% of the values) lies between 3.43 mL/s and 3.97 mL/s. Equivalently, if a measurement is made of water flow, the probability is 0.5 that the value obtained lies between 3.43 mL/s and 3.97 mL/s.

3.6.1 The 68% and 95% confidence intervals

As the normal distribution extends between $+\infty$ and $-\infty$, it follows that the 100% confidence interval lies between these limits. It is hardly very discriminating to say that all data lie between $\pm\infty$. Instead, it is useful to quote a confidence interval between which the majority of the data are expected to lie. We discovered in section 3.5 that about 68% of normally distributed

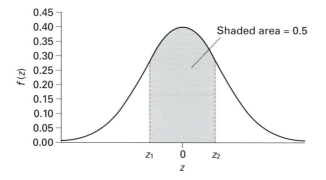

Figure 3.14. Standard normal curve indicating area of 0.5 distributed symmetrically about $z=0$.

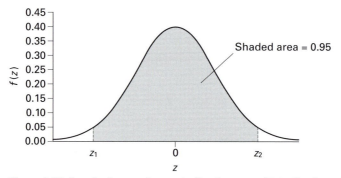

Figure 3.15. Standard normal curve indicating area of 0.95 distributed symmetrically about $z=0$.

data lie within $\pm\sigma$ of the population mean. It follows that the 68% confidence interval for such data can be written

$$\mu - \sigma \leq x \leq \mu + \sigma \tag{3.17}$$

In terms of the standard normal variable, z, the 68% confidence interval is

$$-1 \leq z \leq 1 \tag{3.18}$$

Another often quoted confidence interval is that which includes 95% of the data (or equivalently, where the area under the normal curve is equal to 0.95). This is shown in figure 3.15. As stated in the previous section, the area between $z=-\infty$ and $z=0$ is 0.5. The area between $z=0$ and $z=z_2$ is half the shaded area in figure 3.15. Now the $\frac{1}{2} \times$ shaded area $= \frac{1}{2} \times 0.95 = 0.475$, so the area between $z=-\infty$ and $z=z_2$ is $0.5 + 0.475 = 0.975$.

The z value corresponding to a probability of 0.975, found using table 1 in appendix 1 is equal to 1.96, so that $z_2 = 1.96$. As the distribution is symmetric about $z=0$, it follows that $z_1 = -z_2 = -1.96$. So the 95% confidence interval lies between $z=-1.96$ and $z=+1.96$.

Example 10

The current gain of many BC107 transistors is measured.[23] The mean gain is found to be 209 with the standard deviation of 67. What are the 68% and 95% confidence intervals for the distribution of the gain of the transistors?

ANSWER

Taking $\mu = 209$ and $\sigma = 67$, rearrange equation (3.11) to give

$$x = \mu + z\sigma$$

Considering the 68% confidence interval first, the lower limit occurs for $z = -1$, giving the lower limit of x, x_1, as

$$x_1 = 209 - 67 = 142$$

Similarly, the upper limit of x, x_2, is

$$x_2 = 209 + 67 = 276$$

We conclude that the 68% confidence interval lies between 142 and 276.

For the 95% confidence interval, the lower limit occurs for $z = -1.96$, giving the lower limit of x, x_1, as

$$x_1 = 209 + (-1.96) \times 67 = 78$$

The upper limit of z is 1.96, giving x_2 as

$$x_2 = 209 + 1.96 \times 67 = 340$$

We conclude that the 95% confidence interval for the current gain of the transistors lies between 78 and 340. The z values for other confidence intervals may be calculated in a similar manner. Table 3.5 shows a summary of confidence levels and the corresponding z values.

Table 3.5. *z values corresponding to X% confidence levels.*

X%	50%	68%	90%	95%	99%
$z_{X\%}$	0.67	0.99	1.64	1.96	2.58

[23] Current gain has no units.

Exercise F

A water cooled heat sink is used in a thermal experiment. The variation of the temperature of the heat sink is recorded at 1 minute intervals over the duration of the experiment. Forty values of heat sink temperature (in °C) are shown in table 3.6. Assuming the data in table 3.6 are normally distributed, use the data to estimate the 50%, 68%, 90%, 95% and 99% confidence intervals for the population of heat sink temperatures.

Table 3.6. *Heat sink temperature data.*

18.2	17.4	19.0	18.7	17.8	19.4	18.5	18.4
18.4	18.6	18.6	17.1	17.3	19.4	18.9	17.7
19.8	17.6	18.2	19.3	18.7	18.1	18.4	18.6
18.9	18.9	18.4	18.5	19.0	18.6	18.3	19.4
17.9	18.8	18.0	18.1	18.0	18.0	18.2	17.3

3.6.2   Excel®'s NORMINV() function

If the area is known under the normal distribution curve between $x = -\infty$ and $x = x_1$, what is the value of x_1? For a given probability, the NORMINV() function in Excel® will calculate x_1 so long as the mean and standard deviation of the normal distribution are given. The syntax of the function is

NORMINV(probability, mean, standard deviation)

Example 11

Normally distributed data have a mean of 126 and a standard deviation of 18. If the area under the normal curve between $-\infty$ and x_1 for these data is 0.75, calculate x_1.

ANSWER

Sheet 3.5(a) shows the function required to calculate x_1 in cell A1. Sheet 3.5(b) shows the value returned in cell A1 after the Enter key is pressed.

Sheet 3.5. *Use of the NORMINV() function:*

(a) *number entered into* NORMINV() *function;*   (b) *result of calculation.*

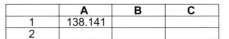

	A	B	C
1	=NORMINV(0.75,126,18)		
2			

	A	B	C
1	138.141		
2			

Example 12

Values of density of a saline solution are found to be normally distributed with a mean of 1.150 g/cm^3 and a standard deviation of 0.050 g/cm^3. Use the NORMINV() function to find the 80% confidence limits for the density data.

ANSWER

It is helpful to sketch a diagram of the normal distribution and to indicate the area under the curve and the corresponding limits, x_1 and x_2. The total area to the left of x_2 in figure 3.16 is $0.8 + 0.1 = 0.9$. Now we can use the Excel® function NORMINV() to find the value of x_2.

Sheet 3.6(a) shows the NORMINV() function entered into cell A1. Sheet 3.6(b) shows the value returned in cell A1 after the Enter key is pressed. The upper limit of the confidence interval is 1.214 g/cm^3, which is more than the mean by an amount 1.214 g/cm$^3 - 1.150$ g/cm$^3 = 0.064$ g/cm^3. The lower limit is less than the mean by the same amount. It follows that the lower limit is 1.150 g/cm$^3 - 0.064$ g/cm$^3 = 1.086$ g/cm^3.

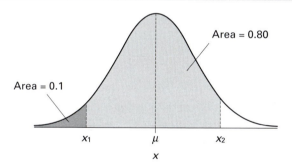

Figure 3.16. Normal curve for example 12.

Sheet 3.6. *Use of* NORMINV() *function.*

(a) *numbers entered into* NORMINV() *function;*

(b) *result of calculation.*

	A	B	C
1	=NORMINV(0.9,1.15,0.05)		
2			

	A	B	C
1	1.21408		
2			

3.6.3 Excel®'s NORMSINV() function

If the area under the standard normal distribution between $-\infty$ and z_1 is known, then the NORMSINV() function returns the value of z_1 corresponding to that area. The syntax of the function is

NORMSINV(probability)

Example 13

If the area under the standard normal curve between $-\infty$ and z_1 is equal to 0.85, what is the value of z_1?

ANSWER

Sheet 3.7(a) shows the NORMSINV() function entered into cell A1. Sheet 3.7(b) shows the value returned in cell A1 after the Enter key is pressed, i.e. $z_1 = 1.03643$.

Sheet 3.7(a). *Use of* NORMSINV() *function.*   Sheet 3.7(b). *Result of calculation.*

	A	B	C
1	=NORMSINV(0.85)		
2			

	A	B	C
1	1.03643		
2			

Exercise G

If the area under the standard normal curve between z_1 and $+\infty$ is 0.2, use the NORMSINV() function to find z_1.

3.7 Distribution of sample means

The normal distribution would be useful enough if it only described patterns observed in 'raw' data. However, the use of the normal distribution can be extended to data that are 'non-normally' distributed. Suppose, instead of drawing up a histogram of raw data, we group data which do not follow a normal distribution and take the mean of the group. We repeat this many times until we are able to plot a histogram representative of the distributions of the means. Perhaps surprisingly, the distribution of the *means* takes on the characteristics of the normal distribution. To illustrate this we revisit the situation discussed in section 3.3 in which 1000 random numbers are generated and where the underlying probability distribution is uniform, with all values falling between 0 and 1. Figure 3.17 shows a histogram of the raw data which exhibits none of the characteristics associated with the normal distribution. Figure 3.18 is constructed by taking the data (in the order in which they were generated) as pairs of numbers, then calculating the mean of each pair. A histogram of 500 means is now constructed in the usual way. The main characteristics of the normal distribution are beginning to emerge in figure 3.18: a clearly defined centre to the distribution, symmetry about that centre and an approximately bell shaped distribution. What, if instead of plotting a histogram consisting of

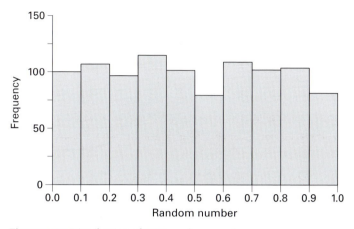

Figure 3.17. Distribution of 1000 random numbers.

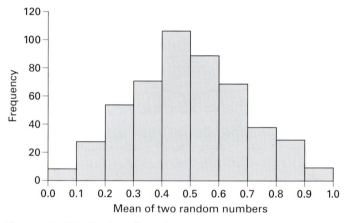

Figure 3.18. Distribution of sample means.

the means of pairs of values, we had chosen larger groupings of raw data, say containing ten values, and plotted the means of these groupings? Figure 3.19 shows a histogram consisting of 100 means. These means were determined using the same data as appears in figure 3.17. The distribution of means does not look as convincingly normal as, say, figure 3.18. This is due to the fact that it consists of only 100 means. The distribution in figure 3.19 is narrower than those shown in figures 3.17 and 3.18 to the extent that there are no means below 0.25, or above 0.75. In the next section we quantify the 'narrowing effect' that occurs when the sample size increases.

The tendency of the distribution of sample means to follow a normal distribution, irrespective of the distribution of the raw data, is embodied in the central limit theorem.

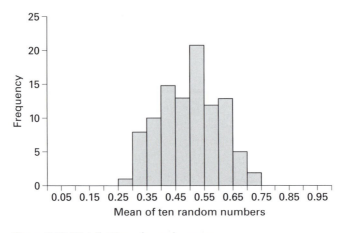

Figure 3.19. Distribution of sample means.

3.8   The central limit theorem

This is an expression of what we have learned in the previous section, namely:

> Suppose a sample consisting of n values is drawn from a population and the mean of that sample is calculated. If this process is repeated many times, the distribution of means tends towards a normal distribution, irrespective of the distribution of the raw data, so long as n is large.[24]

Let the population mean of a distribution of the raw data be μ, and the population standard deviation be σ. The population mean, $\mu_{\bar{x}}$, of the new distribution consisting of sample means is given by $\mu_{\bar{x}} = \mu$. The standard deviation of the new distribution, $\sigma_{\bar{x}}$, is given by

$$\sigma_{\bar{x}} = \frac{\sigma}{\sqrt{n}} \qquad (3.19)$$

The distribution of sample means can be transformed into the standard normal distribution. To do this, replace x by $\bar{x}$ and replace σ by $\sigma_{\bar{x}}$ in equation (3.11), so that z becomes

$$z = \frac{\bar{x} - \mu}{\sigma_{\bar{x}}} \qquad (3.20)$$

---

[24] If the underlying distribution is symmetric with a central peak, such as a triangular distribution, an n of 4 or more will produce means that are normally distributed. If the underlying distribution is extremely asymmetric, n may need to be in excess of 20.

When σ is not known, the best we can do is replace σ in equation (3.19) by s so that

$$\sigma_{\bar{x}} \approx \frac{s}{\sqrt{n}} \qquad (3.21)$$

The population standard deviation, σ, of the raw data is independent of sample size. By contrast, the standard deviation of the distribution of sample means, $\sigma_{\bar{x}}$, *decreases* as n increases as given by equation (3.21). This is consistent with the observation of the 'narrowing' of the distributions of the sample means shown in figures 3.18 and 3.19 for $n=2$ and $n=10$ respectively. The ability to reduce $\sigma_{\bar{x}}$ by increasing n is important as it permits us to quote a confidence interval for the population mean which can be as small as we choose so long as we are able to make sufficient repeat measurements.

3.8.1 Standard error of the sample mean

$\sigma_{\bar{x}}$ given in equation (3.19) is most often referred to as the *standard error of the mean*.[25] We can apply equation (3.19) when quoting confidence limits for the population mean or true value of a quantity.[26] This is extremely valuable since we are often interested in obtaining, through repeat measurements, a better estimate of the true value of a physical quantity. We begin by drawing the probability distribution for the sample means as shown in figure 3.20.

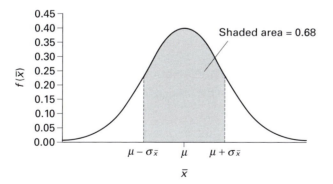

Figure 3.20. Normal distribution of sample means.

---

[25] $\sigma_{\bar{x}}$ is sometimes written as SE $\bar{x}$.
[26] See appendix 4 for a derivation of equation (3.19).

We discovered in section 3.6.1 that the 68% confidence interval for normally distributed data can be written $\mu - \sigma \leq x \leq \mu + \sigma$. Similarly, the 68% confidence interval for the distribution of sample means is $\mu - \sigma_{\bar{x}} \leq \bar{x} \leq \mu + \sigma_{\bar{x}}$ as indicated in Figure 3.20.

The probability that a sample mean, $\bar{x}$, lies between $\mu - \sigma_{\bar{x}}$ and $\mu + \sigma_{\bar{x}}$ may be written

$$P\left(\mu - \frac{\sigma}{\sqrt{n}} \leq \bar{x} \leq \mu + \frac{\sigma}{\sqrt{n}}\right) = 0.68 \tag{3.22}$$

Finding the confidence interval for $\bar{x}$ is less important than finding the confidence interval for the population mean (or true value), μ. By rearranging the terms within the brackets, the 68% confidence limits for the population mean becomes

$$P\left(\bar{x} - \frac{\sigma}{\sqrt{n}} \leq \mu \leq \bar{x} + \frac{\sigma}{\sqrt{n}}\right) = 0.68 \tag{3.23}$$

Equation (3.23) is certainly more useful than equation (3.22), but there still is a problem: the population standard deviation, σ, appears in this equation and we do not know the value of this. The best we can do is replace σ by the estimate of the population standard deviation, s. As discussed in section 3.5.4, this approximation is regarded as reasonable so long as $n \geq 30$. Equation (3.23) becomes

$$P\left(\bar{x} - \frac{s}{\sqrt{n}} \leq \mu \leq \bar{x} + \frac{s}{\sqrt{n}}\right) = 0.68 \tag{3.24}$$

The 68% confidence interval for μ is

$$\bar{x} - \frac{s}{\sqrt{n}} \leq \mu \leq \bar{x} + \frac{s}{\sqrt{n}} \tag{3.25}$$

We use the properties of the normal distribution to give the limits associated with other confidence intervals. The X% confidence interval is written

$$\bar{x} - z_{X\%}\frac{s}{\sqrt{n}} \leq \mu \leq \bar{x} + z_{X\%}\frac{s}{\sqrt{n}} \tag{3.26}$$

The value of $z_{X\%}$ corresponding to the X% confidence interval is shown in table 3.5.

Example 14

Using the heat sink temperature data in table 3.6:

   (i) Calculate the 95% confidence interval for the population mean for these data.

   (ii) How many values would be required to reduce the 95% confidence interval for the population mean to $(\bar{x}-0.1)\,°C$ to $(\bar{x}+0.1)\,°C$?

ANSWER

   (i) The mean, $\bar{x}$, and standard deviation, s, of the data in table 3.6, are[27]

$$\bar{x}=18.41\,°C \qquad s=0.6283\,°C$$

Using equation (3.26), with $z_{95\%}=1.96$,

$$z_{95\%}\frac{s}{\sqrt{n}}=1.96\times\frac{0.6283\,°C}{\sqrt{40}}=0.1947\,°C$$

The 95% confidence interval for the population mean is from $(18.41-0.1947)\,°C$ to $(18.41+0.1947)\,°C$, i.e. from $18.22\,°C$ to $18.60\,°C$.

   (ii) To decrease the interval to $(\bar{x}-0.1)\,°C$ to $(\bar{x}+0.1)\,°C$ we require that

$$z_{95\%}\frac{s}{\sqrt{n}}=0.1\,°C$$

Rearranging this equation gives

$$n=\left(\frac{z_{95\%}s}{0.1\,°C}\right)^{2}$$

Substituting $z_{95\%}=1.96$ and $s=0.6283\,°C$ gives

$$n=\left(\frac{1.96\times0.6283\,°C}{0.1\,°C}\right)^{2}=152$$

Exercise H

Thirty repeat measurements are made of the density of a high quality multigrade motor oil. The mean density of the oil is found to be 905 kg/m^3 and the standard deviation 25 kg/m^3. Use this information to calculate the 99% confidence interval for the population mean of the density of the motor oil.

---

[27] We retain four figures for s to avoid rounding errors in later calculations.

3.8.1.1   APPROXIMATING $\sigma_{\bar{x}}$

In situations in which a few values (say up to 12) are to be considered, there is a rapid method by which the standard error of the mean, $\sigma_{\bar{x}}$, can be approximated which is good enough for most purposes. $\sigma_{\bar{x}}$ can be written as,[28]

$$\sigma_{\bar{x}} \approx \frac{\text{range}}{n} \tag{3.27}$$

where n is number of values and range = (maximum value – minimum value). In particular, equation (3.27) is useful when the standard error is required to be within ±20% of the value that would be obtained using equation (3.21).

Example 15

Table 3.7 shows the input offset voltages of five operational amplifiers. Use equations (3.21) and (3.27) to estimate the standard error of the mean, $\sigma_{\bar{x}}$, of these values to two significant figures.

ANSWER

Using equation (1.16) we obtain $s = 1.795$ mV. Substituting s into equation (3.21) we then obtain $\sigma_{\bar{x}} = s/\sqrt{n} = (1.795 \text{ mV})/\sqrt{6} = 0.73$ mV. Using equation (3.27) we find $\sigma_{\bar{x}} \approx \text{range}/n = (5 \text{ mV})/6 = 0.83$ mV.

Table 3.7. *Input offset voltages for six operational amplifiers.*

Input offset voltage (mV)	4.7	5.5	7.7	3.4	2.7	5.8

Exercise I

Table 3.8 shows eight values of the energy gap of crystalline germanium at room temperature. Use equations (3.21) and (3.27) to estimate the standard error of the mean, $\sigma_{\bar{x}}$, of the values in table 3.8 to two significant figures.

Table 3.8. *Values of the energy gap of germanium, as measured at room temperature.*

Energy gap (eV)	0.67	0.63	0.72	0.66	0.74	0.71	0.66	0.64

---

[28] See Lyon (1980) for a discussion of equation (3.27).

3.8.2 Excel®'s CONFIDENCE() function

The $X\%$ confidence interval of the population mean for any set of data may be found using Excel®'s CONFIDENCE function. The syntax of the function is

CONFIDENCE(α,standard deviation, sample size)

α is sometimes referred to as the 'level of significance'. Its place in data analysis is discussed when we consider hypothesis testing in chapter 8. For the moment we only require the relationship between α and the confidence level, $X\%$. We can write

$$\alpha = \frac{(100\% - X\%)}{100\%} \tag{3.28}$$

As an example, a 95% confidence level corresponds to $\alpha = 0.05$.

Example 16

Fifty measurements are made of the coefficient of static friction between a wooden block and a flat metal table. The data are normally distributed with a mean of 0.340 and a standard deviation of 0.021. Use the CONFIDENCE() function to find the 99% confidence interval for the population mean.

ANSWER
Use equation (3.28) to find α.

$$\alpha = \frac{(100\% - 99\%)}{100\%} = 0.01$$

Sheet 3.8(a) shows the CONFIDENCE() function entered into cell A1. Sheet 3.8(b) shows the value returned in cell A1 after the Enter key is pressed. The confidence interval can be written

Mean $-$ CONFIDENCE(0.01,0.021,50) to Mean $+$ CONFIDENCE(0.01,0.021,50)

i.e. $0.340 - 0.0077$ to $0.340 + 0.0077$. The 99% confidence interval for the population mean of the static friction data is therefore 0.3323 to 0.3477.

Sheet 3.8. *Use of* CONFIDENCE() *function.*
(a) *numbers entered into* CONFIDENCE()
function;   (b) *result of calculation.*

	A	B	C
1	=CONFIDENCE(0.01,0.021,50)		
2			

	A	B	C
1	0.00765		
2			

3.9 The *t* distribution

While the shape of the normal distribution describes well the variability in the mean when sample sizes are large, it describes the variability less well when sample sizes are small. It is important to be aware of this, as many experiments are carried out in which the number of repeat measurements is small (say less than ten). Essentially the difficulty stems from the assumption made that the estimate of the population standard deviation, s, is a good approximation to σ. The variation in values is such that, for small data sets, s is not a good approximation to σ, and the quantity

$$\left(\frac{\bar{x} - \mu}{s/\sqrt{n}} \right)$$

where n is the size of the sample, does not follow the standard normal distribution, but another closely related distribution, referred to as the 't' distribution. If we write

$$t = \left(\frac{\bar{x} - \mu}{s/\sqrt{n}} \right) \tag{3.29}$$

we may study the distribution of t as n increases from $n=2$ to $n=\infty$. The probability density function, $f(t)$, can be written in terms of the variable, t, as[29]

$$f(t) = K(\nu) \left(1 + \frac{t^2}{n-1} \right)^{-n/2} \tag{3.30}$$

where $K(\nu)$ is a constant which depends on the number of degrees of freedom. $K(\nu)$ is chosen so that

$$\int_{-\infty}^{\infty} f(t)\,dt = 1 \tag{3.31}$$

Figure 3.21 shows the general shape of the t probability density function. On the face of it, figure 3.21 has a shape indistinguishable to that of the normal distribution with a characteristic bell shape in evidence. The difference becomes clearer when we compare the t and the standard normal distributions directly. Equation (3.30) predicts a different distribution for each sample size, n, with the t distribution tending to the standard normal distribution as $n \rightarrow \infty$. Figure 3.22 shows a comparison of the t distribution curve with $n=6$ and that of the standard normal probability distribution. An important difference between the family of t and the normal distribu-

[29] See Hoel (1984) for more information on the t probability density function.

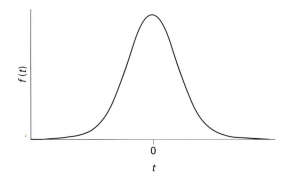

Figure 3.21. *t* distribution curve.

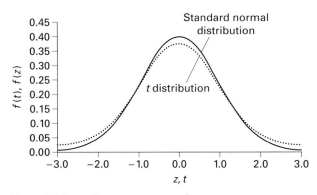

Figure 3.22. *t* and *z* curves compared.

tions is the extra area in the tails of the *t* distributions. As an example, consider the normal distribution in which there is 5% of the area confined to the tails of the distribution. A confidence interval of 95% for the population mean is given by

$$\bar{x} - 1.96 \frac{\sigma}{\sqrt{n}} \quad \text{to} \quad \bar{x} + 1.96 \frac{\sigma}{\sqrt{n}}$$

By comparison, the 95% confidence interval for the population mean, when σ is not known, is given by

$$\bar{x} - t_{95\%,\nu} \frac{s}{\sqrt{n}} \quad \text{to} \quad \bar{x} + t_{95\%,\nu} \frac{s}{\sqrt{n}} \tag{3.32}$$

where ν is the number of degrees of freedom.

Table 3.9 gives the *t* values for the 68%, 90%, 95% and 99% confidence levels of the *t* distribution for various degrees of freedom, $\nu = n - 1$. The table shows $t_{95\%,\nu}$ is greater than 1.96 for $\nu \leq 99$, so that the 95% confidence interval is necessarily wider for the *t* distribution than the 95% confidence interval for the normal distribution that has the same standard deviation.

Table 3.9. *t values for various confidence levels and degrees of freedom.*[30]

Number of values, n	Degrees of freedom, ν	$t_{68\%,\nu}$	$t_{90\%,\nu}$	$t_{95\%,\nu}$	$t_{99\%,\nu}$
2	1	1.82	6.31	12.71	63.66
4	3	1.19	2.35	3.18	5.84
10	9	1.05	1.83	2.26	3.25
20	19	1.02	1.73	2.09	2.86
30	29	1.01	1.70	2.05	2.76
50	49	1.00	1.68	2.01	2.68
100	99	1.00	1.66	1.98	2.63
10000	9999	0.99	1.65	1.96	2.58

Table 3.10. *Diameters of steel balls.*

Diameter (mm)				
4.75	4.65	4.60	4.80	4.70
4.70	4.60	4.65	4.75	4.60

Example 17

The diameter of ten steel balls is shown in table 3.10. Using these data, calculate:

(i) $\bar{x}$;

(ii) s;

(iii) the 95% and 99% confidence interval for the population mean.

ANSWER

(i) The mean of the data in table 3.10 is 4.680 mm.

(ii) The standard deviation of data is 0.07149 mm.

(iii) The 95% confidence interval for the mean is given by expression (3.32) and the 99% confidence interval by

$$\bar{x} - t_{99\%,9} \frac{s}{\sqrt{n}} \quad \text{to} \quad \bar{x} + t_{99\%,9} \frac{s}{\sqrt{n}}$$

Using table 3.10 we have the 95% confidence interval as

$$4.680 \text{ mm} - \frac{2.26 \times 0.07149 \text{ mm}}{\sqrt{10}} \quad \text{to} \quad 4.680 \text{ mm} + \frac{2.26 \times 0.07149 \text{ mm}}{\sqrt{10}}$$

---

[30] A more extensive table can be found in appendix 1.

i.e. 4.629 mm to 4.731 mm. For the 99% confidence interval we have

$$4.680 \text{ mm} - \frac{3.25 \times 0.07149 \text{ mm}}{\sqrt{10}} \quad \text{to} \quad 4.680 \text{ mm} + \frac{3.25 \times 0.07149 \text{ mm}}{\sqrt{10}}$$

i.e. 4.607 mm to 4.753 mm.

Exercise J

Calculate the 90% confidence interval for the population mean of the data in table 3.10.

3.9.1 Excel®'s TDIST() and TINV() functions

For a specified value of t, the TDIST() function gives the probability in the tails of the t distribution, so long as the number of degrees of freedom, ν, is specified. The syntax for the TDIST() function is

TDIST(t, degrees of freedom, tails)

If the tails parameter is set to 2, the function gives the area in both tails of the distribution. If the tails parameter is set to 1, then the area in one tail is given.

Example 18

(i) Calculate the area in the tails of the t distribution when $t = 1.5$ and $\nu = 10$.
(ii) Calculate the area between $t = -1.5$ and $t = 1.5$.

ANSWER

(i) Sheet 3.9(a) shows the formula required to calculate the areas in both tails entered into cell A1. Sheet 3.9(b) shows the area in the tails returned in cell A1 when the Enter key is pressed.
(ii) The area between $t = -1.5$ and $t = 1.5$ is equal to 1 − (area in tails) = $1 - 0.16451 = 0.83549$.

Sheet 3.9. *Use of TDIST() function.*
(a) *numbers entered into* TDIST() *function;*   (b) *result of calculation.*

	A	B	C
1	=TDIST(1.5,10,2)		
2			

	A	B	C
1	0.16451		
2			

If the total area in symmetric tails of the t distribution is p, then the TINV() function gives the value of t corresponding to that area, as long as the number of degrees of freedom, ν, is specified. The syntax for the TINV() function is

TINV(p, degrees of freedom)

Example 19

If the area in the tails of the t distribution is 0.6, calculate the corresponding value of t, assuming that the number of degrees of freedom, $\nu, = 20$.

ANSWER

Sheet 3.10(a) shows the TINV() function entered into cell A1. Sheet 3.10(b) shows the value returned in cell A1 after the Enter key is pressed.

Sheet 3.10. *Use of* TINV() *function*.

(a) *numbers entered into* TINV() *function*;

	A	B	C
1	=TINV(0.6,20)		
2			

(b) *result of calculation*.

	A	B	C
1	0.53286		
2			

3.10 The lognormal distribution

While the normal distribution plays a central role in the analysis of experimental data in the physical sciences, not all data are well described by this distribution. It is important to recognise this as, for example, data rejection techniques can be used (such as that described in section 5.8.2) which assume that a particular distribution (usually the normal) applies to data. When it does not, data may be rejected which legitimately belong to the 'true' distribution. One distribution that describes the data obtained in many experiments in the physical sciences (and is closely related to the normal distribution) is the so called 'lognormal' distribution.

Suppose the continuous random variable, x, follows a lognormal distribution. If $y = \ln(x)$, then the random variable, y, is normally distributed. Examples of the probability density function for the lognormal distribution are shown in figure 3.23. The population mean of the transformed x values is 1.7 for both curves (i.e. μ_y is 1.7). Standard deviations of the transformed x values are 0.2 and 0.5 as indicated in figure 3.23.

Occurrences of lognormally distributed data are quite common in

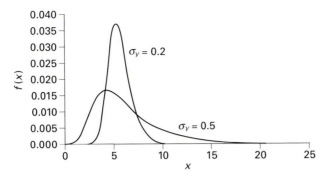

Figure 3.23. The lognormal probability density function, $f(x)$ versus x.

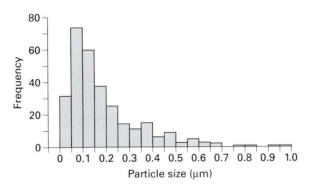

Figure 3.24. Distribution of the size of particles on a film of titanium.

the physical sciences.[31] As an example, consider the histogram shown in figure 3.24 which shows the distribution of particles on the surface of a vacuum deposited film of titanium.[32] The data in figure 3.24 are not normally distributed. However, when a logarithmic transformation is applied to these data and the transformed data are displayed as a histogram, we obtain figure 3.25. The distribution of the transformed data in figure 3.25 is more 'normal-like' than the raw data in figure 3.24. Specifically, the histogram in figure 3.25 has a central peak, is symmetrical about the peak and the frequencies in the tails are small. Though the logarithmic transformation of data has produced a histogram that appears to be 'more nearly' normal, is there a more convincing way to establish whether data follow a normal distribution? The answer is 'yes' and we will consider this next.

[31] Metals (such as copper) in river sediments, sulphate in rainwater and iodine in groundwater specimens have all been found to be approximately lognormally distributed. Crow and Shimizu (1988) offer a detailed discussion of the lognormal distribution.

[32] When atoms of titanium are deposited onto a surface, 'large' particles which adversely affect the quality of the film are also deposited.

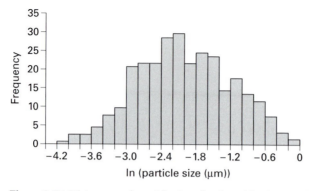

Figure 3.25. Histogram of particle size after logarithmic transformation.

3.11 Assessing the normality of data

How can we be assured that data gathered in an experiment follow a normal distribution? This is an important question, as many of the analysis techniques (such as that of 'least squares' described in chapters 6 and 7) assume that data are normally distributed. While there are statistical tests (such as the chi-squared test described in chapter 8) which can be used to test for 'normality', an alternative approach is to plot the values in such a way that, if the values *are* normally distributed, they lie along a straight line. As it is easy to identify deviations from a straight line 'by eye', this approach is quite appealing. There are a number of similar ways that data may be plotted in order to produce a straight line (so long as the data are normally distributed) including 'normal quantile' and 'normal probability plots'.[33] We focus on the normal quantile plot as it is easy to implement in Excel® (or any other spreadsheet package).

3.11.1 The normal quantile plot

Suppose the fraction, *f*, of a sample is less than or equal to the value $q(f)$. We term $q(f)$ the *quantile* for that fraction. For a particular probability distribution, $q(f)$ can be determined for any value of *f* and any sample size. We expect that the experimental quantile (i.e. the quantile for the data obtained in the experiment) will be very similar to the theoretical quantile

---

[33] See Moore and McCabe (1989) for details of these plots.

if the data follow the theoretical distribution closely. Put another way, if we plot the experimental data against the theoretical quantile we would expect the points to lie along a straight line. A quantile plot may be constructed as follows:[34]

1. Order the experimental data x_i from smallest to largest.
2. Calculate the fraction of values, f_i, less than or equal to the ith value using the relationship

$$f_i = \frac{i - \frac{3}{8}}{n + \frac{1}{4}} \tag{3.33}$$

where n is the number of values in the sample.
3. Calculate the quantile, $q(f_i)$, for all the fractions, f_i ($i = 1$ to n), for the standard normal distribution using

$$q(f_i) = 4.91 [f_i^{0.14} - (1 - f_i)^{0.14}] \tag{3.34}$$

4. Plot a graph of x_i versus $q(f_i)$.

Example 20

Table 3.11 shows the diameter (in μm) of small spherical particles deposited on the surface of a film of titanium. Construct a normal quantile plot for these data.

ANSWER

Table 3.12 shows f_i and $q(f_i)$. The values of particles size are shown ordered from smallest to largest (in the interests of brevity, the table shows the first 10 values in the sample only). Figure 3.26 shows x_i versus $q(f_i)$ for all the data in table 3.11. The extreme non-linearity for $q(f_i) > 1$ indicates that the data are not adequately described by the normal distribution.

Table 3.11. *Diameter of small particles on a film of titanium (in μm).*

0.075	0.110	0.037	0.065	0.147	0.106	0.158	0.163	0.149	0.131	0.136	0.106
0.068	0.206	0.037	0.097	0.123	0.968	0.110	0.147	0.081	0.062	0.421	0.188

---

[34] This approach follows that of Walpole, Myers and Myers (1998).

Table 3.12. f_i and $q(f_i)$ for the data shown in table 3.11.

i	x_i (ordered)	f_i	$q(f_i)$
1	0.037	0.025773	-1.95007
2	0.037	0.06701	-1.49944
3	0.062	0.108247	-1.23521
4	0.065	0.149485	-1.03703
5	0.068	0.190722	-0.8732
6	0.075	0.231959	-0.73025
7	0.081	0.273196	-0.6011
8	0.097	0.314433	-0.48148
9	0.106	0.35567	-0.36854
10	0.106	0.396907	-0.26024

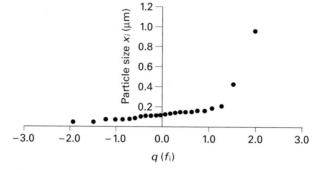

Figure 3.26. Normal quantile plot for data in table 3.11.

Exercise K

Transform the data in table 3.11 by taking the natural logarithms of the values in the table.

  (i) Construct a normal quantile plot of the transformed data.
  (ii) What can be inferred from the shape of this plot?

3.12 Population mean and continuous distributions

When n repeat measurements are made of a single quantity, we express the mean, $\bar{x}$, as

$$\bar{x} = \frac{\sum x_i}{n} \tag{3.35}$$

When n becomes very large, $\bar{x} \to \mu$, where μ is the population mean. We have shown in section 3.5 that the probability of obtaining a value of x can be obtained with the assistance of the probability density function, $f(x)$. Now we go one step further to show how this function is related to the population mean, μ. We begin by considering another equation, equivalent to equation (3.35) which allows us to calculate $\bar{x}$. We can write

$$\bar{x} = \frac{\sum f_i x_i}{\sum f_i} = \frac{f_1 x_1}{\sum f_i} + \frac{f_2 x_2}{\sum f_i} + \frac{f_3 x_3}{\sum f_i} + \cdots \frac{f_n x_n}{\sum f_i} \tag{3.36}$$

where f_i is the frequency of occurrence of the value x_i. As $n \to \infty$, $\bar{x} \to \mu$, and $f_i / \sum f_i$ becomes the *probability*, p_i, of observing the value, x_i. As $n \to \infty$, equation (3.36) becomes

$$\bar{x} \to \mu = p_1 x_1 + p_2 x_2 + p_3 x_3 + \cdots + p_n x_n \tag{3.37}$$

This may be written more compactly as

$$\mu = \sum p_i x_i \tag{3.38}$$

Equation (3.38) is most easily applied when dealing with quantities that take on discrete values, such as occurs when throwing dice or counting particles emitted from a radioactive source. When dealing with continuously varying quantities such as length or mass we can write the probability, p, of observing the value x in the interval Δx as

$$p = f(x)\,\Delta x \tag{3.39}$$

where $f(x)$ is the probability density function introduced in section 3.3.

If we allow Δx to become very small, the equation for the population mean for continuous variables (equivalent to equation (3.38) for discrete variables) may be written[35]

$$\mu = \int_{-\infty}^{\infty} x f(x)\,dx \tag{3.40}$$

---

[35] Here we have shown the limits of the integral extending from $+\infty$ to $-\infty$, indicating that any value of x is possible. In general, the limits of the integral may differ from $+\infty$ to $-\infty$. Most importantly, if the limits are written as a and b, then $\int_a^b f(x)\,dx = 1$.

Example 21

A probability density function, $f(x)$, is written

$$f(x) = \frac{4}{3} - x^2 \text{ for } 0 \leq x \leq 1$$

$f(x) = 0$ for other values of x

Use this information to determine the population mean for x.

ANSWER

Using equation (3.40) we write (with appropriate change of limits in the integration)

$$\mu = \int_0^1 x\left(\frac{4}{3} - x^2\right) dx = \int_0^1 \left(\frac{4x}{3} - x^3\right) dx$$

Performing the integration, we have

$$\mu = \left[\frac{2}{3}x^2 - \frac{x^4}{4}\right]_0^1 = [0.6\dot{6} - 0.25] = 0.41\dot{6}$$

Exercise L

Given that a distribution is described by the probability density function

$$f(x) = 0.2e^{(-0.2x)} \text{ for } 0 \leq x \leq \infty$$

determine the population mean of the distribution.

3.13 Population mean and expectation value

The mean value of x where the probability distribution governing the distribution of x is known is also referred to as the *expectation value* of x and is sometimes written $\langle x \rangle$. For continuous quantities $\langle x \rangle$ is written

$$\langle x \rangle = \int_{-\infty}^{\infty} x f(x)\, dx \tag{3.41}$$

where $f(x)$ is the probability density function.

A comparison of equations (3.41) and (3.40) indicates that $\langle x \rangle \equiv \mu$. The idea of expectation value can be extended beyond finding the mean of any function of x. For example, the expectation value of x^2 is given by

$$\langle x^2 \rangle = \int_{-\infty}^{\infty} x^2 f(x)\, dx \tag{3.42}$$

In general, if the expectation value of a function $g(x)$ is required, we write

$$\langle g(x) \rangle = \int_{-\infty}^{\infty} g(x)f(x)\,dx \qquad (3.43)$$

From the point of view of data analysis, a useful expectation value is that of the square of the deviation of values from the mean, as this is the variance, σ^2, of values. Writing $g(x) = (x-\mu)^2$ and using equation (3.43) we have

$$\sigma^2 = \langle (x-\mu)^2 \rangle = \int_{-\infty}^{\infty} (x-\mu)^2 f(x)\,dx \qquad (3.44)$$

3.14 Review

Data gathered in experiments usually display variability. This could be due to inherent variations in the quantity being measured, the influence of external factors such as electrical noise or even the care, or lack of it, shown by the experimenter in the course of carrying out the experiment. Accepting that variability is a 'fact of life', we need a formal way of expressing variability in data. We discovered in this chapter how some of the basic ideas in probability can be applied to data gathered in experiments. Probability, and its application to the study of the distribution of experimental data gives a concise, numerical way of expressing the variability in quantities. Specifically, with reference to the normal and t distributions, we are able to quote a confidence interval for the mean of a set of data in situations where data sets are large or small. In this chapter we have shown that the calculation of probabilities and confidence intervals can be eased considerably by the use of Excel®'s built in statistical functions.

We will apply what we know of the normal distribution and the t distribution in chapter 5 when we consider the general topic of uncertainty in experiments and how uncertainties combine when measurements are made of many physical quantities. In the next chapter we consider discrete quantities and probability distributions used to describe the variability in such quantities.

Problems

1. A probability density function is written

$$f(x) = A - x \qquad \text{for } 0 \le x \le 1$$
$$f(x) = 0 \qquad \text{for other values of } x.$$

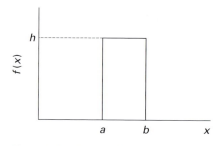

Figure 3.27. Rectangular probability density function.

(i)   Sketch the graph of $f(x)$ versus x.
(ii) Calculate A using the relationship, $\int_0^1 f(x)\,dx = 1$.
(iii) What is the probability that x lies between

(a) 0.0 and 0.1?
(b) 0.9 and 1.0?

2. Consider the rectangular distribution shown in figure 3.27. Given that

$f(x) = h$ for $a \le x \le b$
$f(x) = 0$ for $x < a$ or $x > b$

then

(i)   using $\int_a^b f(x)\,dx = 1$, show that $h = 1/(b-a)$;
(ii) show that the population mean, μ, of the distribution is $\mu = (a+b)/2$.

3. Show that points of inflexion occur in the curve of $f(z)$ versus z at $z = \pm 1$, where $f(z)$ is given by equation (3.12). (Use the result that, at a point of inflexion, $d^2 f(z)/dz^2 = 0$.)

4. An experiment is performed to test the breaking strength of thermoplastic fibres. Out of 1000 fibres tested, 78 fibres have breaking strengths in excess of 945 MPa and 41 fibres have breaking strengths of less than 882 MPa. Assuming the normal distribution to be valid for the data, calculate the mean and standard deviation of the fibre strength.

5. A constant current is supplied to 16 silicon diodes held at constant temperature. The voltage across each diode is recorded. Table 3.13 shows the voltage data. Use these data to estimate the mean and standard deviation of the population from which the diodes were taken. If 200 diodes are tested, how many would you expect to have voltages is excess of 0.6400 V? (Assume the normal distribution to be valid.)

6. The focal lengths of many lenses are measured and found to have a mean of 15.2 cm and a standard deviation of 1.2 cm.

Table 3.13. *Voltage data for silicon diodes.*

Voltage across diodes (V)							
0.6380	0.6421	0.6458	0.6395	0.6389	0.6364	0.6411	0.6395
0.6390	0.6464	0.6420	0.6428	0.6385	0.6401	0.6432	0.6405

Table 3.14. *Pressure data (units µPa).*

147	128	135	137	145
153	129	125	139	142

   (i) Assuming the focal lengths are normally distributed:

      (a) Use the NORMDIST() function on Excel® to calculate the cdf for focal lengths in the range 13.0 cm to 17.0 cm in steps of 0.2 cm.

      (b) Plot a graph of cdf versus focal length.

   (ii) Using your cdf values found in part (i), find the probability that a lens chosen at random has a focal length in the range:

      (a) 13.0 cm to 15.0 cm;

      (b) 16.0 cm to 17.0 cm.

7. Use the Excel® function NORMSDIST() to draw up a standard normal cumulative table for values of z ranging from $z=-2.00$ to $z=2.00$ in steps of 0.01. Plot a graph of NORMSDIST(z) versus z.

8. The resonant frequency in an ac circuit is measured 30 times. The mean of the data is found to be 2106 Hz with a standard deviation of 88 Hz. Assuming the data to be normally distributed, use the CONFIDENCE() function in Excel® to calculate the 90% confidence interval for the population mean.

9. The pressure in a vacuum chamber is measured each day for ten days. The data obtained are given in table 3.14 (units μPa). Using the t distribution, calculate the 95% confidence interval for the mean of the population from which these data are drawn.

10.

   (i) Calculate the area in the one tail of the t distribution between $t=2$ and $t=\infty$ for $\nu=1$.

   (ii) Calculate the area in one tail of the standard normal distribution between $z=2$ and $z=\infty$.

Table 3.15. *Concentration of titanium dioxide in basement rocks.*

Concentration of TiO_2 (wt%)									
0.35	0.37	0.59	0.14	0.55	0.74	1.99	1.81	0.15	1.58
0.63	0.46	0.19	0.79	0.80	0.99	2.34	1.76	1.82	0.82
0.44	0.45	0.20	0.55	0.57	1.96	0.82	2.14	2.22	4.25

Table 3.16. *Radius of microspheres added to a polymer.*

Radius of microspheres (μm)									
16.9	10.3	11.3	9.0	11.3	4.0	11.4	6.3	13.2	15.7
23.5	8.9	8.4	13.1	10.9	11.0	26.8	12.4	11.3	30.4
8.9	23.0	12.1	11.8	27.3	14.6	15.2	10.8	17.1	11.8
8.4	7.0	13.5	7.5	32.1	11.4	5.5	10.3	19.7	10.1
8.8	6.2	14.3	11.2	16.2	8.9	10.2	14.1	10.3	13.5
3.5	11.8	13.8	35.1	5.7	5.6	3.8	21.9	39.9	10.0
7.4	16.0	16.9	8.0	18.3	16.9	11.0	5.6	20.2	17.8
10.2	8.6	18.8	14.5	16.1	34.6	14.0	7.0	15.5	6.6
16.2	11.3	14.1	14.6	13.0	10.6	6.2	35.2	27.1	20.6
5.8	7.0	15.9	12.6	12.1	13.2	4.8	10.7	7.6	7.4

(iii) Repeat part (i) for $\nu=2$ to 100 and plot a graph of the area in the tail versus ν.

(iv) For what value of ν is the area in the tail between $t=2$ and $t=\infty$ equal to the area in the tail of the standard normal distribution between $z=2$ and $z=\infty$ when both areas are expressed to one significant figure?

11. The amount of titanium dioxide in specimens of basement rocks drawn from a geothermal field is shown in table 3.15.

(i) Use a normal quantile plot to help decide whether the values in table 3.15 follow a normal distribution.

(ii) Transform the data in table 3.15 by taking the natural logarithms of the values. Using a normal quantile plot establish whether the transformed values follow a normal distribution. (That is, do the data follow a lognormal distribution?)

12. Microspheres were added to a polymer in order to study the ultrasonic properties of inhomogeneous materials. Values for the radii of 100 microspheres are shown in table 3.16.

(i) Plot a histogram of the values in table 3.16.

(ii) Take the natural logarithm of the values in table 3.16. Plot a histogram of the transformed values.

(iii) Use a normal quantile plot to establish whether the transformed values follow a normal distribution.

Chapter 4
Data distributions II

4.1 Introduction

In chapter 3 we considered the normal distribution largely due to its similarity to the distribution of data observed in many experiments involving repeat measurements of a quantity. In particular, the normal distribution is useful for describing the spread of values when continuous quantities such as temperature or time interval are measured. Another important category of experiment involves counting. For example, we may count the number of defective devices supplied by a manufacturer, the number of electrons scattered by a gas or the number of beta particles emitted by a radioactive source. In these situations, distributions that describe discrete quantities must be considered. In this chapter we consider two such distributions important in science: the binomial and Poisson distributions.

4.2 The binomial distribution

One type of experiment involving discrete variables entails removing an object from a population and classifying that object in one of a finite number of ways. For example, we might test an electrical component and classify it 'within specification' or 'outside specification'. Due to the underlying (and possibly unknown) processes causing components to fail to meet the specification, we can only give a probability that any particular component tested will satisfy the specification. When n objects are removed from a population and tested, or when a coin is tossed n times, we speak of performing *n trials*. The result of a test (e.g. 'pass') or the result of a coin toss (e.g. 'head') is referred to as an *outcome*.

Some experiments consist of trials, in which the outcome of each trial can be classified as a success (S) or a failure (F). If this is the case, and if the probability of a success does not change from trial to trial, we can use the *binomial* distribution to determine the probability of a given number of successes occurring, for a given number of trials.[1] The binomial distribution is useful if we want to know the probability of, say, obtaining one or more defective light emitting diodes (LEDs) when an LED is drawn from a large population of 'identical' LEDs, or the probability of obtaining four 'heads' in six tosses of a coin. In addition, by considering a situation in which the number of trials is very large but the probability of success on a single trial is small (sometimes referred to as a 'rare' outcome), we are able to derive another discrete distribution important in science: the Poisson distribution.

How a 'success' is defined largely depends on the circumstances. For example, if integrated circuits (ICs) are tested, a successful outcome could be defined as a circuit that is fully functional. Let us write the probability of obtaining such a success as *p*. If a circuit is not fully functional, then it is classed as a failure. Denote the probability of a failure by *q*. As success or failure are the only two possibilities, we must have

$$p + q = 1 \tag{4.1}$$

As an example, suppose after testing many ICs, it is found that 20% are defective. If one IC were chosen at random, the probability of a failure is 0.2 and hence the probability of a success is 0.8. If four ICs are drawn from the population, what is the probability that all four are successes? To answer this we apply one of the rules of probability discussed in section 3.2.1. So long as the trials are independent (so that removing any IC from the population does not affect the probability that the next IC drawn from the population is a success) then

$$P(4 \text{ successes}) = p \times p \times p \times p = 0.8 \times 0.8 \times 0.8 \times 0.8 = 0.4096$$

Going a stage further in complexity, if four ICs are removed from the population, what is the probability that *exactly* two of them are fully functional? This is a little more tricky as, given that four ICs are removed and tested, two successes can be obtained in several ways such as two

---

[1] The name binomial distribution derives from the fact that the probability of *r* successful outcomes from *n* trials equates to a term of the binomial expansion $(p + q)^n$, where *p* is the probability of a success in a single trial, and *q* is the probability of a failure.

successes followed by two failures or two failures followed by two successes. The possible combinations of success (S) and failure (F) for two successes from four trials are

SSFF, FFSS, SFFS, FSSF, SFSF, FSFS

The probability of the combination SSFF occurring is

$$P(\text{SSFF}) = p \times p \times q \times q = 0.8 \times 0.8 \times 0.2 \times 0.2 = 0.0256$$

The probability that each of the other five combinations, SFFS, FFSS etc. occurring is also 0.0256, so that the total probability of obtaining exactly two successes from four trials is

$$0.0256 + 0.0256 + 0.0256 + 0.0256 + 0.0256 + 0.0256$$
$$= 6 \times 0.0256 = 0.1536$$

This is an extremely tedious way to determine a probability, but if we generalise what we have done, an equation emerges which greatly assists probability calculations in situations where the binomial distribution is valid.

4.2.1 Calculation of probabilities using the binomial distribution

What is the probability of exactly r successes in n trials? To answer this we proceed as follows:

1. If there are r successes out of n trials, then there must be $n - r$ failures.
2. The probability of r consecutive successes followed by $n - r$ consecutive failures is $p^r q^{n-r}$, where p is the probability of a success for a single trial and q is the probability of a failure for a single trial.
3. The number of combinations of r successes from n trials is given by[2]

$$C_{n,r} = \frac{n!}{(n-r)!r!} \tag{4.2}$$

where $n!$ is 'n factorial', given by

$$n! = n \times (n-1) \times (n-2) \times (n-3) \cdots \times 2 \times 1 \tag{4.3}$$

Multiplying the number of combinations in step 3 by the probability of any one combination in step 2 gives the probability, $P(r)$, of r successes from n trials:

$$P(r) = C_{n,r} p^r q^{n-r} \tag{4.4}$$

[2] There are a variety of symbols used to represent the combinations formula including $\binom{n}{r}$ and $_nC_r$. See Meyer (1975) for more details on combinations.

Example 1

5% of thermocouples in a laboratory need repair. Assume the population of thermocouples to be large enough for the binomial distribution to be valid. If ten thermocouples are withdrawn from the population, what is the probability that:

    (i) Two are in need of repair?

    (ii) Two or fewer are in need of repair?

    (iii) More than two are in need of repair?

ANSWER

We designate a good thermocouple (i.e. one not in need of repair) as a success. Given that 5% of thermocouples need repair, the probability, q, of a failure is 0.05. Therefore the probability of a success, p, is 0.95.

    (i) If two thermocouples are in need of repair (failures) then the other eight must be good (i.e. successes). We require the probability of eight successes ($r=8$) from ten trials ($n=10$). Using equation (4.2),

$$C_{n,r} = \frac{n!}{(n-r)!r!} = \frac{10!}{(10-8)!8!} = 45$$

Equation (4.4) becomes

$$P(8) = C_{10,8}p^8q^2 = 45 \times 0.95^8 \times 0.05^2 = 0.07463$$

    (ii) If two or fewer thermocouples are in need of repair, then eight, nine or ten must be good. Therefore the probability required is

$$P(8 \leq r \leq 10) = P(8) + P(9) + P(10)$$

which may be written

$$\sum_{r=8}^{r=10} P(r)$$

Following the steps in part (i)

$$P(9) = C_{10,9}p^9q^1 = 10 \times 0.95^9 \times 0.05 = 0.31512$$
$$P(10) = C_{10,10}p^{10}q^0 = 1 \times 0.95^{10} = 0.59874$$

$$P(8 \leq r \leq 10) = \sum_{r=8}^{r=10} = 0.07463 + 0.31513 + 0.59874 = 0.9885$$

    (iii) If more than two thermocouples need repair, then zero, one, two, three, four, five, six or seven must be good. We require the probability $P(0 \leq r \leq 7)$ which is given by

$$P(0 \leq r \leq 7) = P(0) + P(1) + P(2) + P(3) + P(4) + P(5) + P(6) + P(7)$$

and which may be more concisely written

$$P(0 \le r \le 7) = \sum_{r=0}^{r=7} P(r)$$

We can use the result that $\sum_{r=0}^{r=10} P(r) = 1$ (i.e. the summation of probability of all possible outcomes is equal to 1) which may be written

$$\sum_{r=0}^{r=7} P(r) + \sum_{r=8}^{r=10} P(r) = 1$$

Therefore

$$\sum_{r=0}^{r=7} P(r) = 1 - \sum_{r=8}^{r=10} P(r)$$

From part (ii) of this question $\sum_{r=8}^{r=10} P(r) = 0.9885$, so that

$$\sum_{r=0}^{r=7} P(r) = 1 - 0.9885 = 0.0115$$

Exercise A

The assembly of a hybrid circuit requires the soldering of 58 electrical connections. If 0.2% of electrical connections are faulty, what is the probability that an assembled circuit will have:

  (i) No faulty connections?
  (ii) One faulty connection?
  (iii) More than one faulty connection?

4.2.2 Probability of a success, p

How do we know the probability of a success, p, in any particular circumstance? Some situations are so familiar and well defined, such as tossing a coin or rolling a die that we regard the probability of a particular outcome occurring as known before any trials are made. For example, owing to the symmetry of a die, there are six equally possible outcomes on a single throw. The probability of a particular outcome, say a six, is therefore $1/6$. In other situations, such as those requiring a knowledge of the probability of finding a defective device, we must rely on past experience where the proportion of defective devices from a large sample has been determined. Unlike the situation with the die, in which the probability of a particular outcome is fixed (so long as the die has not been tampered with), the probability of obtaining a defective device may change, especially if some element of the manufacturing process has been modified.

4.2.3 Excel®'s BINOMDIST() function

Calculating the probability of r successful outcomes in n trials using equation (4.4) is tedious if more than a few outcomes must be considered. For example, if we want to know the probability that 10 or less outcomes are successful from 20 trials, we require the *cumulative* probability given by

$$P(r \le 10) = \sum_{r=0}^{r=10} P(r) \qquad (4.5)$$

where $P(r)$ is given by equation (4.4). Excel®'s BINOMDIST() function allows for the calculation of the cumulative probability, such as that given by equation (4.5), as well as the probability of exactly r successes from n trials as given by equation (4.4). The syntax of the function is

> BINOMDIST(number of successes, number of trials, probability of a success, cumulative)

Cumulative is a parameter which is set to TRUE if the cumulative probability is required and to FALSE if the probability of exactly r successful outcomes is required.

For example, if the probability of a success on a single trial is 0.45 and we require the probability of 10 or less successes from 20 trials, we would enter into a cell =BINOMDIST(10,20,0.45,TRUE) as shown in sheet 4.1. After pressing the Enter key, the number 0.750711 is returned in cell A1. If we require the probability of *exactly* 10 successes from 20 trials, we replace TRUE by FALSE, as shown in sheet 4.2. After pressing the Enter key, the number 0.159349 is returned in cell A1.

Sheet 4.1. *Use of* BINOMDIST() *to calculate cumulative probability.*

	A	B	C
1	=BINOMDIST(10,20,0.45,TRUE)		
2			

Sheet 4.2. *Use of* BINOMDIST() *to calculate probability of r successful outcomes.*

	A	B	C
1	=BINOMDIST(10,20,0.45,FALSE)		
2			

Exercise B

Given the number of trials, $n=60$, and the success on a single trial, $p=0.3$, use the BINOMDIST() function in Excel® to determine:

(i) $P(r=20)$;

(ii) $P(r\leq20)$;

(iii) $P(r<20)$;

(iv) $P(r>20)$;

(v) $P(r\geq20)$.

4.2.4   Mean and standard deviation of binomially distributed data

If the probability of a success is p, and an experiment is performed which consists of n trials, then we expect np outcomes to be successful. It is possible in a real experiment for the actual number of successful outcomes from n trials to vary from 0 to n, but if the experiment is carried out many times, then the mean number of successes will tend to np. If we designate the mean number of successes over many experiments as $\bar{r}$ then, as the number of experiments becomes very large, $\bar{r}$ tends to the population mean, μ.

As an example, consider an experiment in which ten transistors are tested. Previous experience has shown that the probability of obtaining a good transistor on a single trial is 0.7. It follows that the population mean, μ, is

$$\mu = 10 \times 0.7 = 7$$

One hundred samples each consisting of ten transistors are removed from a large population and tested. The number of good transistors in each sample is shown in table 4.1. If we calculate the mean of the values in table 4.1 we find $\bar{r}=6.75$, which is close to the population mean[3] of 7.

Another important parameter representative of a population is its standard deviation. For an experiment consisting of n trials, where the probability of success is p and the probability of failure is q, the population standard deviation, σ, is given by[4]

[3] We cannot really know the population mean except in simple situations such as the population mean for the number of heads that would occur in ten tosses of a coin. In this situation we would make the assumption that the probability of a head is 0.5 so that in ten tosses we would expect five heads to occur.

[4] See Meyer (1975) for a proof of equation (4.6).

Table 4.1. *Number of good transistors in samples each of size n = 10.*

6	10	6	4	9	6	7	5	6	8
7	4	5	5	6	6	5	9	8	8
8	6	10	4	7	6	5	6	6	8
8	8	6	7	6	5	6	7	5	7
6	7	7	7	8	6	8	7	8	8
6	7	6	8	5	7	4	7	7	4
7	5	8	6	5	8	5	6	9	5
7	9	8	9	8	7	9	7	8	8
5	7	8	8	7	6	8	8	2	6
7	6	8	7	5	8	7	6	9	9

$$\sigma = \sqrt{npq} \tag{4.6}$$

For the data in table 4.1, $n = 10$, $p = 0.7$ and $q = 0.3$. Using equation (4.6), $\sigma = 1.449$.

The value for σ can be compared with the estimate of the population standard deviation, s, calculated using equation (1.16). We find that

$$s = 1.5$$

Exercise C

2% of electronic components are known to be defective. If these components are packed in containers each containing 400 components, determine:

(i) The population mean of defective items in each container.
(ii) The population standard deviation of defective items in each container.
(iii) The probability that a container has two or more defective components.

4.2.5 Normal distribution as an approximation to the binomial distribution

The probability distribution for any number, n, and probability of success, p, can be displayed graphically. For example, the probability of r successes, $P(r)$, from four trials, when $p = 0.2$ is shown in figure 4.1. The probability distribution in figure 4.1 has little in common with the normal and t distributions considered in chapter 3. Specifically:

- The distribution is asymmetric.
- There is no well defined peak.

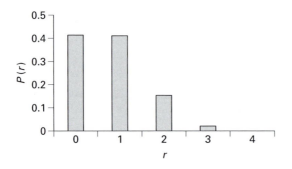

Figure 4.1. Binomial distribution for $n=4$ and $p=0.2$.

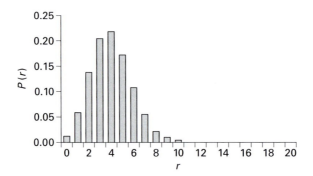

Figure 4.2. Binomial distribution with $n=20$ and $p=0.2$.

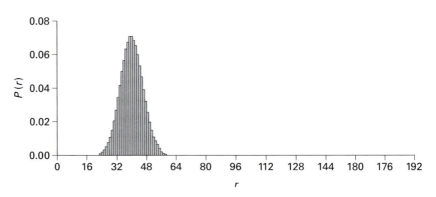

Figure 4.3. Binomial distribution with $n=200$ and $p=0.2$.

However, if the number of trials increases (and therefore the number of possible successes increases) there is a gradual change in the shape of the probability distribution as indicated in figures 4.2 and 4.3.

Figures 4.2 and 4.3 bear a striking resemblance to the normal distribution. The similarity is so close, in fact, that the normal distribution with population mean $\mu=np$, and standard deviation $\sigma=\sqrt{npq}$, is often used as an approximation to the binomial distribution for the purpose of calculating probabilities. This is due to the fact that probabilities may be calculated more easily using the normal rather than the binomial distribution when a cumulative probability is required and the number of trials, n, is large.

As an example of the use of the normal distribution as an approximation to the binomial distribution, suppose we require the probability of three successful outcomes from ten trials when $p=0.4$. Using the normal distribution we find $P(3)=0$, as finite probabilities can only be determined if an *interval* is defined. If we consider the number '3' as a rounded value from a continuous distribution with continuous random variable, x, then we would take $P(3)$ as $P(2.5 \leq x \leq 3.5)$. Comparing probabilities calculated using the binomial and normal distributions we find:

Binomial distribution calculation:

$$P(3) = C_{10,3}\,0.4^3 0.6^7 = 0.2150$$

Normal distribution calculation: Normal distribution approximation with mean $\mu = 10 \times 0.4 = 4$, and standard deviation $\sigma = \sqrt{10 \times 0.4 \times 0.6}$
$= 1.549$

$$P(3) \approx P(2.5 \leq x \leq 3.5) = P(-\infty \leq x \leq 3.5) - P(-\infty \leq x \leq 2.5) \qquad (4.7)$$

The terms on the right hand side of equation (4.7) may be most easily determined using the NORMDIST() function described in section 3.5.1. Specifically, entering =NORMDIST(3.5,4,1.549,TRUE) into a cell in an Excel® spreadsheet returns the number 0.3734. Similarly, entering =NORMDIST(2.5,4,1.549,TRUE) into a cell returns the number 0.1664. It follows that

$$P(2.5 \leq x \leq 3.5) = 0.3734 - 0.1664 = 0.2070$$

The approximate probability as determined by the normal distribution is about 4% less than that as given by the binomial distribution which is good enough for most purposes. In this example there is nothing to be gained in terms of efficiency by using the normal distribution as an approximation to the binomial distribution. By contrast, in situations in which the number of trials, n, is large (say more than 20) and the calculation of cumulative probabilities is required, probability calculations using the normal distribution become attractive. When n is very large and r is not too close to n or zero (say, $n > 2000$ and $1800 > (n-r) > 200$), it is not possible to calculate $C_{n,r}$ as given by equation (4.2) as an 'overflow' error occurs when using most computer based data analysis packages such as Excel®.[5] In this situation there is little alternative but to use the normal distribution as an approximation to the binomial distribution in order to calculate probabilities.

[5] If overflow is to be avoided, the range of values for n and $(n-r)$ is even more restricted when using a pocket calculator to determine $C_{n,r}$.

Example 2

Given the number of trials, $n = 100$ and the probability of a success on a single trial, $p = 0.4$, determine the probability of between 38 and 52 successes (inclusive) occurring using:

(i) the binomial distribution;
(ii) the normal distribution.

ANSWER

(i) Using the binomial distribution we require the probability, P, given by

$$P(38 \leq r \leq 52) = P(r \leq 52) - P(r < 38) \tag{4.8}$$

The cumulative probabilities on the right hand side of equation (4.8) are most conveniently found using the BINOMDIST() function on Excel®. Entering the formula =BINOMDIST(52,100,0.40,TRUE) into a cell in an Excel® spreadsheet returns the value 0.9942.

Entering =BINOMDIST(37,100,0.40,TRUE) into a cell[6] returns the value 0.3068. It follows that

$$P(38 \leq r \leq 52) = P(r \leq 52) - P(r < 38) = 0.9942 - 0.3068 = 0.6874$$

(ii) Using the normal distribution to solve this problem requires that we calculate the area under a normal distribution curve between 37.5 and 52.5. Using x to represent the random variable, we write

$$P(37.5 \leq x \leq 52.5) = P(-\infty \leq x \leq 52.5) - P(-\infty \leq x \leq 37.5)$$

The population mean, μ, is given by $\mu = np = 100 \times 0.4 = 40$ and the population standard deviation, σ, is given by $\sigma = \sqrt{npq} = \sqrt{100 \times 0.4 \times 0.6} = 4.899$.

$P(37.5 \leq r \leq 52.5)$ may be found using Excel®'s NORMDIST() function.[7] Specifically, entering =NORMDIST(52.5,40,4.899,TRUE) into a spreadsheet cell returns the number 0.9946. Similarly, entering =NORMDIST(37.5,40,4.899,TRUE) into a cell returns the value 0.3049. It follows that

$$P(-\infty \leq x \leq 52.5) = 0.9946 \quad \text{and} \quad P(-\infty \leq x \leq 37.5) = 0.3049$$

so that

$$P(-\infty \leq x \leq 52.5) - P(-\infty \leq x \leq 37.5) = 0.9946 - 0.3049 = 0.6897$$

---

[6] Note that $P(r \leq 37) = P(r < 38)$.
[7] The NORMDIST() function is discussed in section 3.5.1.

Exercise D

If the probability of a success $p = 0.3$ and the number of trials $n = 1000$, use the binomial distribution and normal approximation to the binomial distribution to determine the probability of:

  (i) between 290 and 320 successes (inclusive) occurring;

 (ii) more than 320 successes occurring.

4.3 The Poisson distribution

When introducing the binomial distribution we used the term 'trial' to describe the process of sampling a population. There are important physical processes in which outcomes may be counted, but where the term 'trial' is less appropriate. An example is the process of radioactive decay. A radioactive source emits particles due to the instability of the nuclei within the source. The emission process is random so it is not possible to say when a particular nucleus is going to decay through the emission of a particle. Nevertheless, even for a small radioactive source, there are so many nuclei that in a given time interval (say 1 minute) it is very likely that one or more nuclei will decay, ejecting a particle in the process. A system incorporating a detector and some electronics can be used to count particles emitted from the radioactive source. Characteristics of the radioactive counting experiment are:

- An interval of time is chosen and the number of events (in this example an 'event' is the detection of a particle) occurring in that time interval is counted.
- The probability of an event occurring in an interval of time does not change.
- The mean number of events in a chosen time interval is a constant for the system being considered.
- The occurrence of an event can be counted but the non-occurrence cannot. So, for example, we might count five particles emitted from a source in 1 minute, but the question 'how many particles were *not* emitted in that minute?' has no meaning.

If we regard the occurrence of an event, such as detecting a high energy particle, as a 'success' then the binomial distribution can be used as the starting point for the derivation of an equation for the probability of an event occurring in a specified interval of time.

If the probability of success on a single trial, p, is very small, but the number of trials, n, is very large, then the expression for the probability of r successes, $P(r)$, as given by equation (4.4) may be approximated by[8]

$$P(r) = \frac{\mu^r e^{-\mu}}{r!} \tag{4.9}$$

where μ is the mean number of successes in the interval chosen. Equation (4.9) may be regarded as a new discrete probability distribution, called the Poisson distribution. The distribution is valid when n is large and p is small and p does not vary from one interval to the next.

Figure 4.4 shows how the probability given by equation (4.9) depends on the value of μ. For $\mu = 1$ the probability distribution is clearly asymmetric, but as μ increases we again find that the characteristic shape becomes more and more 'normal-like'. In situations where $\mu \geq 5$, the normal distribution can be used as an excellent approximation to the Poisson distribution and this facilitates computation of probabilities which would be very difficult or impossible to determine using equation (4.9).

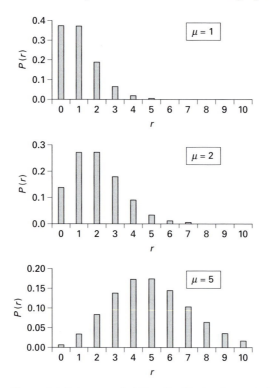

Figure 4.4. Poisson probability distribution for $\mu = 1, 2$ and 5.

[8] See Scheaffer and McClave (1995) for a discussion of equation (4.9).

Example 3

Given that $\mu = 0.8$, calculate $P(r)$ when:

(i) $r = 0$;
(ii) $r = 1$;
(iii) $r \geq 2$.

ANSWER

(i)   Using equation (4.9), $P(0) = 0.8^0 e^{-0.8}/0! = 0.4493$ (note that $0! = 1$).
(ii) Similarly, $P(1) = 0.8^1 e^{-0.8}/1! = 0.3595$.
(iii) $P(r \geq 2) = P(2) + P(3) + P(4)\cdots + P(\infty)$. A convenient way to calculate this sum is to recognise that the sum of the probabilities for all possible outcomes $= 1$, i.e. $P(0) + P(1) + P(2) + P(3) + P(4)\cdots + P(\infty) = 1$. It follows that

$$P(2) + P(3) + P(4)\cdots + P(\infty) = 1 - (P(0) + P(1)) = 1 - (0.4493 + 0.3595) = 0.1912$$

Exercise E

Given that $\mu = 0.5$, use equation (4.9) to determine:

(i) $P(r = 0)$;
(ii) $P(r \leq 3)$;
(iii) $P(2 \leq r \leq 4)$.

4.3.1 Applications of the Poisson distribution

As well as nuclear counting experiments, there are other situations in which the Poisson distribution is valid. These include counting the number of:

- X-rays scattered by air;
- dust particles detected per unit volume in a clean room;
- electrons scattered as they accelerate towards a specimen in an electron microscope;
- pinholes per unit area in a thermally evaporated thin metal film;
- particles emitted from a radioactive source.

Strictly, radioactive decay does not satisfy the requirement of the Poisson distribution, as the probability that an event occurs is not constant but decreases with time. The reason for this is that the decay rate depends on how many undecayed nuclei remain at any particular instant. As that number is continually decreasing *due* to radioactive decay, the probability of detecting a decay also decreases. However, so long as the fraction of the total number of nuclei that decay over the period of the experiment is small (which requires that the half life of the decay is much longer than the duration of the experiment), then the Poisson distribution can be applied with some confidence.

4.3.2 Standard deviation of the Poisson distribution

To determine the population standard deviation for the Poisson distribution, we begin with the binomial distribution and determine the consequences of allowing the number of trials, n, to become very large while at the same time letting the probability of success on a single trial, p, become very small.

The standard deviation, σ, for the binomial distribution is written

$$\sigma = \sqrt{npq} = \sqrt{np(1-p)}$$

If p is very small, then $1 - p \approx 1$, so that

$$\sigma \approx \sqrt{np}$$

but $np = \mu$, so that

$$\sigma \approx \sqrt{\mu} \tag{4.10}$$

A useful indicator of whether data are 'Poisson distributed' is to calculate the mean and standard deviation of the number of events counted in a particular time interval or interval of area, volume etc. If the condition $\sigma \approx \sqrt{\mu}$ does not hold then the data are not Poisson distributed. As usual, we are not able to know σ and μ, and so we must be satisfied with their best estimates, s and $\bar{x}$, respectively.

Example 4

Table 4.2 shows the number of cosmic rays detected in time intervals of 10 minutes at the surface of the earth. Use the data to estimate:

   (i) the population mean;
  (ii) the population standard deviation.

Calculate the probability of detecting in any 1 minute:

  (iii) zero cosmic rays;
  (iv) more than two cosmic rays.

The experiment is continued so that the number of counts occurring in 1200 successive time intervals of 1 minute are recorded.

(v) In how many intervals would you expect the number of counts to exceed 2?

ANSWER

(i) The mean of the values in table 4.2 is 6.48. This is the best estimate of the population mean.

(ii) Using equation (4.10), the population standard deviation $\sigma \approx \sqrt{6.48} = 2.55$.

(iii) Using equation (4.9), $P(0) = 6.48^0 e^{-6.48}/0! = 1.534 \times 10^{-3}$.

(iv) $P(r>2) = P(3) + P(4) + P(5) + \cdots + P(\infty) = 1 - P(0) + P(1) + P(2)$

Now $P(1) = 9.939 \times 10^{-3}$ and $P(2) = 3.220 \times 10^{-2}$ so that

$$P(3) + P(4) + P(5) + \cdots + P(\infty) = 1 - (1.534 \times 10^{-3} + 9.939 \times 10^{-3} + 3.220 \times 10^{-2}) = 0.9563$$

(v) The expected number of 1 minute intervals in which more than two counts occur $= NP(r>2)$, where N is the total number of intervals:

$NP(r>2) = 1200 \times 0.9563 = 1148$ intervals

Table 4.2. *Number of cosmic rays detected at the earth's surface.*

5	3	2	7	10	4	5	7	7	7
10	7	6	10	9	5	9	7	9	6
8	5	9	9	3	6	6	8	2	11
7	6	4	9	4	5	1	4	6	6
7	4	10	3	9	6	9	5	10	7

Exercise F

Small spots of contamination appear on the surface of a ceramic conductor when it is exposed to a humid atmosphere. The contamination degrades the quality of electrical contacts made to the surface of the ceramic conductor. Table 4.3 shows the number of spots identified in 50 non-overlapping regions of area 100 μm^2 at the surface of a particular sample exposed to high humidity. Use the data in table 4.3 to determine the mean number of spots per 100 μm^2. A silver electrode of area 100 μm^2 is deposited on the surface of the conductor. Assuming the positioning of the electrode is random, calculate the probability that the silver electrode will:

(i)   cover no spots of contamination;
(ii) cover one spot;
(iii) cover more than two spots.

Table 4.3. *Number of contamination spots on the surface of a ceramic conductor.*

2	1	1	0	0	1	0	2	2	2
1	2	1	0	1	0	1	0	0	1
0	1	0	0	3	1	1	2	4	0
1	2	1	0	0	0	0	2	2	1
0	1	0	3	2	1	2	2	0	2

4.3.3   Excel®'s POISSON() function

Excel®'s POISSON() function allows for the determination of probability based on equation (4.9), as well as the cumulative probability given by $P(r \leq R)$, where

$$P(r \leq R) = \sum_{r=0}^{r=R} P(r) \qquad (4.11)$$

The syntax of the function is

POISSON(number of events, mean, cumulative)

Cumulative is a parameter which is set to TRUE for a cumulative probability given by equation (4.11). The cumulative parameter is set to FALSE if the probability of exactly r successful outcomes is required.

As an example, if the mean number of cosmic rays detected in a time interval of 1 minute is 8, and we require the probability of detecting 4 or less cosmic rays in any 1 minute interval, we would type =POISSON(4,8,TRUE) into a cell of an Excel® Worksheet, as shown in sheet 4.3. After pressing the Enter key, the number 0.099632 is returned in cell A1. On the other hand, if we require the probability of *exactly* 4 cosmic rays being detected in a one minute time interval, we would type =POISSON(4,8,FALSE) into a cell in Excel® as shown in sheet 4.4. After pressing the Enter key, the number 0.057252 is returned in cell A1.

Sheet 4.3. *Using the* POISSON() *function to calculate cumulative probability.*

	A	B	C
1	=POISSON(4,8,TRUE)		
2			

Sheet 4.4. *Using the* POISSON() *function to calculate the probability of r events occurring.*

	A	B	C
1	=POISSON(4,8,FALSE)		
2			

Exercise G

Table 4.4 shows the number of X-rays detected in 100 time intervals where each interval was of 1 second duration. Use the data in the table to determine the mean number of counts per second. Assuming the Poisson distribution to be applicable, use Excel® to determine the probability of observing in a 1 second time interval:

  (i)   zero counts;
 (ii)   one count;
(iii)   three counts;
 (iv)   between two and four counts (inclusive);
  (v)   more than six counts.

Table 4.4. *Data from an X-ray counting experiment.*

1	1	1	1	1	0	0	6	1	2
0	4	1	1	2	3	1	0	3	3
4	1	1	0	0	2	4	1	3	5
6	0	1	1	4	6	0	0	0	1
1	1	2	2	1	1	0	2	3	1
1	4	0	2	0	0	3	3	2	4
0	2	2	1	1	2	2	0	0	1
1	1	3	2	2	0	0	2	0	1
1	1	2	2	1	0	1	4	0	1
1	1	0	1	0	2	2	0	0	3

4.3.4 Normal distribution as an approximation to the Poisson distribution

Figure 4.4 indicates that when the mean, μ, equals 5, the shape of the Poisson distribution is very similar to that of the normal distribution. When μ equals or exceeds 5, the normal distribution is preferred to the Poisson distribution when the calculation of probabilities is required. This is due to the fact that summing a large series is tedious, unless computational aids are available. Even with aids like Excel®, some summations cannot be determined. When r, μ^r and $r!$ are large, the result of a calculation using equation (4.9) can exceed the numerical range of the computer causing an 'overflow' error to occur.

Example 5

In an X-ray counting experiment, the mean number of counts in a period of 1 minute is 200. Use the Poisson and normal distributions to calculate the probability that in any other 1 minute period the number of counts occurring would be exactly 200.

ANSWER

To find the probability using the Poisson distribution we use the equation (4.9), with $\mu = r = 200$, i.e.

$$P(r) = \frac{\mu^r e^{-\mu}}{r!}$$

so that

$$P(200) = \frac{200^{200} e^{-200}}{200!}$$

This is where we must stop, as few calculators or computer programs can cope with numbers as large as $200!$ or 200^{200}.

Using the normal distribution we use the approximation that

$$P(r = 200) \cong P(199.5 \leq x \leq 200.5)$$

where x is a continuous random variable. Take the population standard deviation of the normal distribution to be

$$\sqrt{\mu} = \sqrt{200} = 14.142$$

Now

$$P(199.5 \leq x \leq 200.5) = P(-\infty \leq x \leq 200.5) - P(-\infty \leq x \leq 199.5) \tag{4.12}$$

The two terms on the right hand side of equation (4.12) may be determined using the NORMDIST() function on Excel®.[9] Entering NORMDIST(200.5,200,14.142,TRUE) into a cell returns the number 0.5141 and entering NORMDIST (199.5,200,14.142,TRUE) into another cell returns the number 0.4859. It follows that

$$P(r=200)=0.5141-0.4859=0.0282$$

Exercise H

Using the information in example 5, determine the probability that in any 1 minute time period the number of counts lies between the inclusive limits of 180 and 230.

4.4 Review

When random processes, such as coin tossing, radioactive decay or the scattering of particles, lead to outcomes which may be counted, we turn to discrete probability distributions to assist in determining the probability that certain outcomes will occur. The starting point for discussing discrete distributions is to consider the binomial distribution. In particular, we found that the binomial distribution is useful for determining probabilities when there are only two possible outcomes from a single trial and the probability of those outcomes is fixed from trial to trial. This requires that we know (or are able to determine experimentally) the probability of a successful outcome occurring in a single trial. In addition, the binomial distribution is helpful in deriving another important distribution, the Poisson distribution. This distribution is appropriate when the frequency of occurrence of 'rare' events is required, such as the decay of a radioactive nucleus. Both the binomial and the Poisson distributions become more 'normal-like' as the number of trials increases. Owing to the comparative ease with which calculations may be performed using the normal distribution, it is used extensively as an approximation to the binomial and Poisson distributions when determining probabilities or frequencies.

Whether an experiment deals with discrete or continuous quantities, the outcome of a measurement or a trial cannot be predicted with certainty. We are therefore forced to acknowledge that 'uncertainty' is an inherent and very important part of all experimental work. With the results

[9] See section 3.5.1 for details of the NORMDIST() function.

obtained in chapters 3 and 4 we are in a position to discuss the important topic of uncertainty in experimental measurements and this allows the quantifying of uncertainties when random effects exist.

Problems

1. Determine,

$$C_{10,5}, \; C_{15,2}, \; C_{42,24}, \; C_{580,290}$$

(suggestion: use the COMBIN() function in Excel®).

2. An active matrix screen for a laptop computer is controlled by 480 000 transistors. If the probability of a faulty transistor occurring during manufacture of the screen is 2×10^{-7}, determine the probability that a screen will have:

   (i) one faulty transistor;
  (ii) more than one faulty transistor.

If 5000 screens are supplied to a computer manufacturer, how many screens would you expect to have more than one faulty transistor?

3. 10% of ammeters in a laboratory have blown fuses and so are unable to measure electric currents. An electronics experiment requires the use of three ammeters. If 25 students are each given three ammeters, how many students would you expect to have:

   (i) Three functioning ammeters?
  (ii) Two functioning ammeters?
 (iii) Less than two functioning ammeters?

4. An electroencephalograph (EEG) is an instrument which uses small electrodes pressed against the scalp to detect the electrical activity of the human brain. When an electrode is pressed against the scalp there is a probability of 0.89 that it will make good enough electrical contact to the skin to allow faithful recording of brain activity. In a study, 24 electrodes are used and of those at least 20 must make good electrical contact with the scalp for an acceptable assessment of brain activity to be made.

   (i) What is the probability that at least 20 electrodes will make good contact to the scalp?

A new type of electrode is trialled which has a probability of 0.72 of making good contact with the scalp.

(ii) How many electrodes must be pressed against the scalp so that the probability that at least 20 of them will make good contact is the same as in part (i)?

5. When a specimen is bombarded with high energy electrons, (such as occurs in an electron microscope) it emits X-rays. In an experiment the number of 14.4 keV X-rays emerging from a specimen was counted in ten successive 100 second time intervals. The data obtained are shown in table 4.5. Use the data to estimate:

Table 4.5. *Number of X-rays counted in ten periods each of duration 100 s.*

118	131	136	119	131
122	119	129	124	98

(i) the population mean for the number of counts in 100 s;
(ii) the population standard deviation.

6. A thin tape of superconductor is inspected for flaws that would affect its capacity to carry a high electrical current. Sixty strips of tape, each of length 1 m, are inspected and the number of flaws in each strip is recorded. These are shown in table 4.6.

Table 4.6. *Number of flaws in superconducting tapes.*

Number of flaws, N	Number of 1 m strips of tape with N flaws
0	18
1	19
2	12
3	9
4	1
5	1

(i) Calculate the mean number of flaws per metre.
(ii) Assuming the Poisson distribution to be valid, determine the expected frequency for zero, one, two, three, four, and five flaws (hint: first calculate $P(0)$, $P(1)$ etc. using equation (4.9)).

7. An environmental scanning electron microscope (ESEM) is used to image specimens in a 'poor' vacuum environment. Owing to the high

concentration of air molecules in the microscope, electrons emitted from the electron gun in an ESEM are scattered by the molecules as they travel towards a specimen. An acceptable image of the specimen is produced as long as 20% or more of the electrons reach the specimen unscattered.

Assuming the Poisson distribution may be used to describe the probability of electron scattering in an ESEM and that 20% of the electrons are unscattered, determine:

(i) the number of times, on average, that an electron is scattered as it travels from electron gun to specimen;

(ii) the probability that an electron will undergo more than four scattering 'events'.

8. The mean number of cyclones to strike a region in Western Australia per year is 0.2. What is the probability that over a period of 20 years, one or more cyclones will strike that region?

Chapter 5

Measurement, error and uncertainty

5.1 Introduction

Physicists, chemists and other physical scientists are proud of the quantitative nature of their disciplines. By subjecting nature to ever closer scrutiny, new relationships between quantities are discovered and established relationships are pushed to the limits of their applicability. When 'numbers' emerge from an experiment, they can be subjected to quantitative analysis, compared to the 'numbers' obtained by other experimenters and be expressed in a clear and concise manner using tables and graphs. If an unfamiliar experiment is planned, an experimenter usually executes a pilot experiment. The purpose of such an experiment might be to assess the effectiveness of the experimental methods being used, or to offer a preliminary examination of a theoretical prediction. An experimenter then typically moves to the next stage in which a more thorough experiment is performed and where there is increased emphasis on the quality of the data gathered. The analysis of these data often provides crucial and defensible evidence sought by the experimenter to support (or refute) a particular theory or idea.

The goal of an experiment might be to determine the value of a single quantity, for example the value of the charge on an electron. Experimenters are aware that influences exist, some controllable and others less so, that conspire to adversely affect the values they obtain. Despite an experimenter's best efforts, some uncertainty in an experimentally determined value remains. In the case of the charge on the electron, its value is recognised to be of such importance in science that considerable effort has gone into establishing an accurate value for it. Currently the 'best' value for the charge on the electron is $(1.60217733 \pm 0.00000049) \times 10^{-19}$ C. A very important part of the expression for the charge is the number following the $\pm$ sign. This is the *uncertainty* in the value for the electronic charge and though the uncertainty is rather small compared to the size of the charge, it is not zero. In general, every value obtained through measurement has some uncertainty and though the uncertainty may be minimised by thorough planning, prudent choice of measuring instrument and careful execution of the experiment, it cannot be eliminated entirely.

In this chapter we consider uncertainties, how they arise, may be identified and categorised, strategies for their minimisation and how they propagate when several quantities each with its own uncertainty are brought together in a calculation. We will use many of the ideas describing variability in data discussed in chapters 3 and 4 and show how they can be adapted to provide quite general methods for combining uncertainties.

5.2 The process of measurement

Looking for patterns in data which may reveal an important relationship between quantities demands confidence in the data gathered, since theories supported by dubious data mislead and divert attention from better theories or explanations. Reliable data are a consequence of reliable measurement and therefore it is important to appreciate the factors that can affect data gathered during an experiment.

Measurement in science can involve something as simple as using an alcohol-in-glass thermometer to measure the temperature of a room, to scanning probe microscopy incorporating state of the art technology which permits features to be measured that are of the order of nanometres in size. The process of measurement in general requires:

- Choice of quantity or quantities to be measured. The measured quantity is sometimes referred to as the *measurand*.
- A sensor to be chosen which is responsive to changes in the measurand. The sensor is situated so that accurate sensing occurs.
- Transformation of the output of the sensor to a form which allows easy display and/or recording of the output.
- Recording or logging the transformed sensor output.

Factors that might affect the accuracy of experimental data are best illustrated by example. We consider temperature measurement, as many experiments involve the study of the variation of a quantity with temperature, such as the pressure of a gas, electrical conductivity of an alloy, or rate of chemical reaction. Figure 5.1 shows the block diagram of a system used

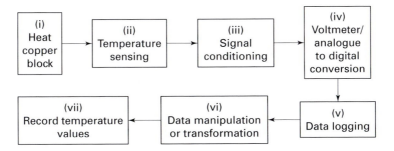

Figure 5.1. System for determining the temperature of a copper block.

to measure the temperature of a copper block. The process of temperature measurement indicated in figure 5.1 involves:

(i) Heat transfer from the copper block to the temperature sensor as the temperature of the block rises above that of the sensor.

(ii) A change of a physical attribute of the temperature sensor (such as its electrical resistance) in response to the temperature change of the sensor. The sensor then supplies an output signal, often in the form of a voltage.

(iii) Modification of the output signal of the sensor. This is often referred to as 'signal conditioning'. For example, the signal may be amplified, or perhaps a filter used to attenuate ac electrical noise at 50 Hz or 60 Hz.

(iv) Measurement of the output of the signal conditioner using a voltmeter or analogue to digital converter on a data acquisition card.

(v) Recording the voltage displayed by the voltmeter 'by hand' in a notebook, or automatically on a computer disc.

(vi) Transformation of voltage to temperature[1] using a calibration curve or 'look-up' table.[2]

(vii) Recording values of temperature in a notebook or a computer.

Though we have considered temperature measurement in this example, a system that measures pressure, flow rate, strain, light intensity or any other physical quantity would contain similar elements.

Processes (i) to (vii) described above each influence the values of temperature recorded as part of an experiment. A complete analysis of all the influences is quite challenging, but table 5.1 describes some of the

---

[1] If a resistance thermometer is used, it is likely that voltages will be first transformed into resistances before a temperature versus resistance calibration curve is used to determine the value of temperature.

[2] If a simple mathematical relationship between the output of a transducer and the measured quantity cannot be found, a look-up table can be used in which each 'output' value from a transducer has a corresponding 'input' value of the measured quantity associated with it.

Table 5.1. *Factors influencing values of temperature obtained through measurement.*

Factor	Explanation
Thermal resistance between temperature sensor and copper block	No matter how intimate the contact between temperature sensor and copper block, there is some resistance to the flow of heat between the two. This affects the response time of the sensor so that, in dynamic experiments where temperatures are changing with time, the sensor temperature lags behind the temperature of the block.
Size and constitution of temperature sensor	A temperature sensor with a large heat capacity responds slowly to temperature changes. If the temperature of the copper block is higher than that of the sensor then when the sensor is brought in contact with the block it will lower the block's temperature.
Lead wires to temperature sensor	In the case of thermocouples, thermistors and resistance thermometers, the leads attached to these sensors form a conduction pathway for heat. If the temperature of the sensor is below the surroundings, heat is conducted to the sensor along the leads thus raising its temperature. Conversely, if the temperature of the sensor is higher than that of the surroundings, heat is conducted away from the sensor thus reducing its temperature.
Gain stability of signal conditioning amplifier	Changing the electrical properties of the signal conditioning system can adversely affect the measurement of temperature. For example, the ageing of components can cause the gain of an amplifier to change over time.
Sampling time and voltmeter resolution	The time to complete the measurement of voltage (i.e. the sampling time) using a digital voltmeter can vary from typically 1 ms to 1 s. Any variation in the voltage that occurs over the period of the measurement cannot be known as the voltmeter presents an average value. (A typical hand held digital voltmeter makes two to three measurements per second). The resolution of a digital voltmeter puts a limit on the change in voltage that can be detected. For 3½ digit voltmeters on the 2 V range, the resolution is 1 mV.
Accuracy of calibration curve	A calibration curve is used to transform one quantity (say resistance or voltage) into temperature. A generic curve may be provided by the sensor manufacturer which relates the sensor output to temperature. Deviation from the curve for a particular sensor due to manufacturing variability is a source of uncertainty.

factors that affect the value(s) of temperature recorded along with a brief explanation.[3] All of the factors described in table 5.1 introduce experimental *error* into values obtained through measurement. By error we are not referring to a mistake such as recording incorrectly a value that appears on the display of a meter, but a deviation from the 'true value' of a quantity being measured. Such a deviation might be caused, for example, by the limit of resolution of an instrument or the finite response time of a sensor. Experimental errors prevent us from being able to quote an 'exact value' for any quantity determined through experiment. By careful consideration of all sources of error in an experiment we can quote an *uncertainty* in the measured value. A detailed consideration of the sources of uncertainties in an experiment might suggest ways by which some or all of those uncertainties can be minimised. While identifying and quantifying uncertainties is important, it is equally important to use any insight gained to reduce the uncertainties, if possible, through better experimental design.

In addition, quoting uncertainties allows for values to be compared. For example, the density of a particular alloy obtained by two separate experimenters might be 6255 kg/m^3 and 6138.2 kg/m^3. If we have no knowledge of the uncertainty in each value, we cannot know if these values are consistent with each other. By contrast, if we are told that the values are (6255 ± 10) kg/m^3 and (6138.2 ± 2.9) kg/m^3 we are in a much better position to argue that the values are *not* consistent and that further investigation is required. When quoting results we are obliged to communicate all relevant details including the uncertainty in the value as well as the 'best' value.

5.3   True value, error and uncertainty

It is often helpful to imagine that a quantity, such as the mass of a body, has a 'true' value and it is the true value that we seek through measurement. If measurement procedures and instruments are perfect and no outside influences conspire to affect the value, then we should be able to determine the true value of a quantity to arbitrary precision. Recognising that neither experimental methods nor instruments are perfect and that although outside influences can be minimised they can never be completely eliminated, the best we can do is obtain a 'best' estimate of the true

---

[3] Table 5.1 is not exhaustive and for a fuller discussion of the measurement of temperature and its challenges see Tompkins and Webster (1988) and Nicholas and White (1982).

value. If we *could* know the true value of a quantity then, in an experiment to determine the temperature, *t*, of a body we could perhaps write

$$t = 45.1783128212\,°C$$

The reality is that, for example, offset drift in the measuring instrument, electrical noise and resolution limits cause measured values to vary from the true value by such an amount as to make the figures after the decimal point in *t* above quite meaningless in most situations. We refer to the difference between the measured value and the true value as the experimental *error*. Representing the true value of a quantity by the symbol μ, the error in the *i*th measured value, δx_i, is given by

$$\delta x_i = x_i - \mu \tag{5.1}$$

where x_i is the *i*th value. If the errors are random then we expect both positive and negative δx_i, as x_i will take on values greater and less than μ. If we could sum the errors due to many repeat measurements then the errors would have a 'cancelling effect' such that

$$\sum_{i=1}^{i=n} \delta x_i \approx 0 \tag{5.2}$$

where *n* is the number of measurements made.

Substituting for δx_i from equation (5.1) gives[4]

$$\sum (x_i - \mu) \approx 0 \tag{5.3}$$

It follows that

$$\sum x_i - n\mu \approx 0$$

or

$$\mu \approx \frac{\sum x_i}{n} = \bar{x} \tag{5.4}$$

Equation (5.4) supports what we have argued already in chapters 1 and 3, namely that the best estimate of the true value of a quantity is the sample mean and as $n \to \infty$, $\bar{x} \to \mu$.

The 'true value' of a quantity discussed in this chapter is a very close relation to the 'population mean' discussed in chapter 3. The sample mean, $\bar{x}$, tends to the population mean when the number of values used to calcu-

---

[4] All summations are carried out between $i = 1$ and $i = n$.

late $\bar{x}$ tends to infinity. However, the terms 'true value' and 'population mean' are not always interchangeable. As an example, ten repeat measurements of the breakdown voltage, V_B, of a *single* zener diode might be made to estimate the true value of V_B for that diode. In this case we could replace the term 'true value' by 'population mean' to describe the value being estimated. However, if the breakdown voltage of each of ten *different* zener diodes is measured then we can estimate the mean of the breakdown voltages for the population from which the zener diodes were drawn, but the term 'true value' has little meaning in this context.

When communicating a value obtained through measurement, we should quote a confidence interval for the true value (or the population mean) being sought. If we use μ to represent the true value, then we can say that the confidence interval is

$$\bar{x} - u \leq \mu \leq \bar{x} + u \tag{5.5}$$

where u is the *uncertainty* in $\bar{x}$ which is used to define the limits between which μ is expected to lie. A more compact way to write the confidence interval is

$$\mu = \bar{x} \pm u \tag{5.6}$$

5.3.1 Calculation of uncertainty, u

The value assigned to u depends upon how confident we want to be that μ lies between the limits $\bar{x} - u$ and $\bar{x} + u$. If we want to be 99.9% confident that μ lies between the specified limits then those limits are necessarily wider than if we are content with a confidence level of 95%. The confidence interval can be expressed as a multiple of the standard error of the mean, $\sigma_{\bar{x}}$. For example, when the number of values used in the calculation of the mean is large (>30), the 68% confidence interval for μ occurs when $u = \sigma_{\bar{x}}$, so that

$$\mu = \bar{x} \pm \sigma_{\bar{x}} \tag{5.7}$$

In general, u in equation (5.6) may be replaced by $t_{X\%, \nu} \sigma_{\bar{x}}$, where $t_{X\%, \nu}$ is the critical t value[5] corresponding to the $X\%$ confidence level and ν is the number of degrees of freedom.[6] Equation (5.6) becomes

$$\mu = \bar{x} \pm t_{X\%, \nu} \sigma_{\bar{x}} \tag{5.8}$$

---

[5] See section 3.9.

[6] $\nu = n - 1$, where n is the numbers of values – see section 3.9.

Table 5.2. *Critical t values for various confidence levels and degrees of freedom, ν.*

Degrees of freedom, ν	$t_{68\%,\nu}$	$t_{90\%,\nu}$	$t_{95\%,\nu}$	$t_{99\%,\nu}$
1	1.82	6.31	12.71	63.66
3	1.19	2.35	3.18	5.84
9	1.05	1.83	2.26	3.25
19	1.02	1.73	2.09	2.86
29	1.01	1.70	2.05	2.76
49	1.00	1.68	2.01	2.68
99	1.00	1.66	1.98	2.63

Table 5.2 contains critical t values for various degrees of freedom and confidence levels (a more complete table appears in appendix 1).

To avoid confusion, it is important to indicate what confidence level has been adopted when quoting an uncertainty. In problems and exercises in this chapter we will assume (unless stated otherwise) that the 95% confidence interval applies when expressing an uncertainty.

Example 1
Ten repeat measurements of the breakdown voltage of a zener diode are shown in table 5.3. Calculate the 95% confidence interval for the true value of the breakdown voltage.

ANSWER
Using equation (5.8) we write

$$\mu = \bar{x} \pm t_{X\%, \nu} \sigma_{\bar{x}}$$

Now $\sigma_{\bar{x}} \approx s/\sqrt{n}$ (see section 3.8), where s is the estimate of the population standard deviation, given by

$$s = \left[\frac{\sum (x_i - \bar{x})^2}{n-1} \right]^{\frac{1}{2}}$$

Using the data in table 5.3

$$n = 10, \ \nu = 9, \ \bar{x} = 5.624 \text{ V}$$

this gives $s = 0.07834$ V, so that

$$\sigma_{\bar{x}} \approx \frac{s}{\sqrt{n}} = \frac{0.07834 \text{ V}}{\sqrt{10}} = 0.02477 \text{ V}$$

Using table 5.2, $t_{95\%,9} = 2.26$, so that

$$\mu = (5.624 \pm 2.26 \times 0.02477) \text{ V}$$
$$= (5.624 \pm 0.056) \text{ V}$$

In this example we have quoted the uncertainty to two significant figures. We adopt this convention throughout this text.

Table 5.3. *Ten values for the breakdown voltage, V_B, of a zener diode.*

V_B (V)	5.62	5.49	5.55	5.61	5.72	5.54	5.63	5.67	5.70	5.71

Exercise A

Table 5.4 contains 20 values of light intensity, I, (in lux). Use these values to determine:

  (i) the mean light intensity;
  (ii) the estimate of the population standard deviation of the light intensity;
  (iii) the 99% confidence interval for the true value of the light intensity.

Table 5.4. *Light intensity values.*

I (lx)	348	328	359	380	378	389	310	349	376	332	340	320	317	334	339	312	343	346	357	347

As long as values are affected only by random errors, equation (5.8) is an excellent way of expressing the confidence interval for the true value of a quantity or the population mean of a set of data. However, it is unwise to take for granted that random effects dominate when quantities are measured. Other sources of error exist to catch the unwary and so it is appropriate that we look in more detail at the causes of error and how we may identify and account for these causes. The influence of errors is to impair the precision and accuracy of values obtained through experiment. It is important that we clarify the scientific meaning of precision and accuracy as they are often used interchangeably in everyday language.

5.4 Precision and accuracy

Repeat measurements of a quantity reveal underlying scatter or variability in that quantity. A *precise* experiment is one in which the scatter of the experimental data is small. As an example, consider values of the acceleration of a free-falling body measured at sea level close to the equator as shown in table 5.5. The mean of these values is 9.605 m/s². The values are clustered close to the mean, suggesting that effects that would act to scatter the data are small. Representing the spread of the values by the standard deviation calculated using equation (1.16) gives $s = 0.0307$ m/s². While s is taken as a measure of the precision of the set of values, the precision of the mean of the values is expressed by the standard error of the mean.

Table 5.5. *Values for acceleration caused by gravity, g, measured at the equator.*

g (m/s²)	9.62	9.57	9.64	9.64	9.58	9.59	9.57	9.63

In the above discussion of precision there is no mention of how close the measured values are to the true value. In fact, values may be precise (i.e. show little scatter) but be far from the true value. A value (or the mean of many values) is regarded as *accurate* if it is close to the true value. The difficulty with this description is that the true value is rarely if ever known, so how is it possible to say something about the accuracy of a particular value? This is not an easy or trivial question to answer, but where measurements are designed to establish the value of a quantity or constant of wide interest to the scientific community, such as the energy gap of a semiconductor, the boiling point of ethanol or the acceleration caused by gravity, it is possible to compare values with those published by other experimenters who have devoted themselves to the accurate determination of those quantities.

We can compare the values of the acceleration due to gravity in table 5.5 with the value acknowledged as being accurate for the acceleration at sea level at the equator[7] which is 9.780 m/s². None of the values in table 5.5 is as large as 9.780 m/s², suggesting that these values are *not* accurate. It appears that, although the scatter is quite small, each value is less than the true value by about 0.2 m/s². Some influence is at play which is consistently causing all the values to be too low. Such consistent or system-

---

[7] See Young and Freedman (1996) for values of the acceleration due to gravity.

atic influences have no effect on the precision of measurements but they do dramatically affect the accuracy of a measurement, since the mean of many values will never tend to the true value so long as systematic effects prevail.

We can summarise precision and accuracy as follows:

- Values are *precise* when the scatter in the values about the mean is small *but* this does not imply that the values are close to the true value.
- Values are *accurate* when they are close to the true value.
- The mean of *n* values is accurate when the mean tends to the true value as *n* becomes large *but* the deviation of individual values from the mean may be large.

The goal in experimental science is for values obtained through experiment to be both accurate and precise (close to the true value with little scatter). Experiments in which the precision is high are sometimes shown later to have produced data of poor accuracy.

5.5 Random and systematic errors

Random errors cause values to lie above and below the true value and, due to the scatter they create, they are usually easy to recognise. So long as only random errors exist, the mean of the values tends towards the true value as the number of repeat measurements increases.

Another type of error is that which causes the measured values to be consistently above or consistently below the true value. This is termed a *systematic* error, also sometimes referred to as a *bias* error. An example of this is the zero error of a spring balance. Suppose with no mass attached to the spring, the balance reads 0.01 N. We can 'correct' for the zero error by:

- if the facility is available, adjusting the balance to read zero; or
- subtracting 0.01 N from every value obtained using the balance.

No value, including an offset, can be known exactly so the correction applied to values will have its own error. As a consequence we cannot assume that a correction applied to a value has completely eliminated the systematic error.

Some systematic errors are difficult to identify and have more to do with the way a measurement is made than with any obvious inadequacy or limitation in the measuring instrument. Section 5.9 considers the causes and treatment of systematic errors. Systematic errors have no effect on the precision of values, but rather act to impair measurement accuracy.

5.6 Random errors

Many effects can cause values obtained through experiment to be scattered above and below the true value. Those effects may be traced to limitations in a measuring instrument or perhaps be related to a human trait, such as reaction time. In some experiments, values may show scatter due to the occurrence of physical processes that are inherently random. Table 5.6 describes three examples of sources of random error.

5.6.1 Common sources of error

Classifying the sources of error in table 5.6 as random is clear-cut as the physical processes causing the error are well characterised. We next consider several other sources of error normally classified as random but indicate that in some cases the same source may be responsible for both random and systematic error.

Table 5.6. *Examples of sources of random error.*

Description	Explanation
Johnson noise	A voltage that varies randomly with time appears across any component or material that has electrical resistance. The random voltage is referred to as Johnson noise. The cause of the noise is the random thermal motion of electrons in the material and so one way the noise can be reduced is to lower the temperature of the material.[8]
Radioactive decay	The nuclei of some chemical elements are unstable and become more stable by emitting a particle. The process is random and so it is not possible to predict in advance at what instant a particular nucleus will emit a particle. As a consequence, the number of particles detected in a given time interval by a nuclear counter varies in a random manner from one interval to the next.[9]
Molecular motion	Forces are exerted on bodies due to the random motion of molecules in gases and liquids. Though the forces are very small, in highly sensitive pressure measuring systems, or those incorporating fine torsion fibres, the forces are sufficient to be detected.

[8] See Simpson (1987) for a discussion of noise due to random thermal motion.
[9] Chapter 4 deals with the statistics of radioactive decay.

Resolution error

Every instrument has a limit of resolution and any small changes in a quantity less than that limit go undetected. The uncertainty due to the limit of resolution is routinely taken as one half of the smallest division that appears on the scale of the instrument. So, for example, for a metre rule with a smallest scale division of 1 mm, the uncertainty due to the limit of resolution is taken as ± 0.5 mm. For a micrometer with a smallest division of 0.01 mm, the uncertainty is taken as ± 0.005 mm. Assuming that the rounding of values that occurs when making a measurement is such that no bias is introduced, i.e. sometimes a value is rounded up to the nearest half division and other times it is rounded down, then we can regard the resolution error as a random error. Similarly, the least significant digit appearing on an instrument with a digital display is the limit of resolution of the instrument. For example, if the voltage indicated by a $3\frac{1}{2}$ digit voltmeter is 1.43 V, then the voltage lies between 1.425 V and 1.435 V. It is therefore reasonable to quote the uncertainty due to the resolution limit of the instrument as 0.005 V, so that the voltage can be expressed as (1.430 ± 0.005) V. The uncertainty due to the limit of resolution of an instrument represents the smallest uncertainty that may be quoted in a value obtained through a single measurement. Other sources of uncertainty, such as those due to offset or calibration errors or variability in the quantity being measured, very often exceed that due to the instrument resolution and so they also need to be quantified.

Parallax error

If you move your head when viewing the position of a pointer on a stop watch or the top of a column of alcohol in an alcohol-in-glass thermometer, the value (as read from the scale of the instrument) changes. The position of the viewer with respect to the scale introduces *parallax* error. In the case of the alcohol-in-glass thermometer, the parallax error is a consequence of the displacement of the scale from the top of the alcohol column, as shown in figure 5.2.

The parallax error may be random if the eye moves with the top of the alcohol column such that the eye is positioned sometimes above or below the 'best viewing' position shown by figure 5.2(c). However, it is possible that the experimenter either consistently views the alcohol column with respect to the scale from a position below the column (figure 5.2(a)) or above the column (figure 5.2 (b)). In such situations the parallax error would be systematic, i.e. values obtained would be consistently below or above the true value.

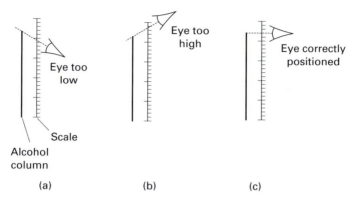

Figure 5.2. Errors introduced by parallax.

Reaction time error

Clocks and watches are available that routinely measure the time between events to a precision of milliseconds or better. If the starting and stopping of a watch is synchronised with events 'by hand' then the error introduced is often considerably larger than the resolution of the watch. Events that are hand timed introduce errors which are typically 200 ms to 300 ms and therefore time intervals of the order of seconds are not really suited to hand timing. Though the timing error *may* be random, it is possible that the experimenter consistently starts or stops the watch prematurely so that the systematic component of the total error is larger than the random component.

These examples indicate that it is not always obvious how a particular error should be categorised, since the way an instrument is used has an important bearing on whether an error should be regarded as random or systematic.

5.7 Absolute, fractional and percentage uncertainties

The uncertainty in a value may be expressed in the unit in which the value is measured. Such an uncertainty is referred to as the *absolute* uncertainty.[10] As an example, if the mass, m, of a body is written

$$m = (45.32 \pm 0.15) \text{ g}$$

then 0.15 g is the absolute uncertainty in the mass.

---

[10] For brevity, the word 'absolute' is usually omitted.

In some situations we may prefer the uncertainty to be expressed as a fraction of the best estimate of the true value of the quantity. This is referred to as the *fractional* uncertainty and is written

$$\text{fractional uncertainty} = \frac{u}{|\bar{x}|} \qquad (5.9)$$

where $\bar{x}$ is the mean of values.[11] In the case of the mass of the body discussed above, we have

$$\text{fractional uncertainty} = \frac{0.15\,\text{g}}{45.32\,\text{g}} = 3.3 \times 10^{-3}$$

Note that the fractional uncertainty has no units, as any units appearing in the numerator and denominator of equation (5.9) 'cancel'.

The uncertainty in a value can be expressed as a percentage of the best estimate of the value by multiplying the fractional uncertainty by 100%, so that

$$\text{percentage uncertainty} = \frac{u}{|\bar{x}|} \times 100\% \qquad (5.10)$$

Example 2

In a report on density measurements made of sea water, the density, ρ, is given as $(1.05 \pm 0.02)\,\text{g/cm}^3$. Express the uncertainty as both fractional and percentage uncertainties.

ANSWER

Using equation (5.9), the fractional uncertainty is given by

$$\frac{u}{|\bar{x}|} = \frac{0.02\,\text{g/cm}^3}{1.05\,\text{g/cm}^3} = 0.019$$

Using equation (5.10), the percentage uncertainty becomes

$$\text{fractional uncertainty} \times 100\% = 0.019 \times 100\% = 1.9\%$$

Exercise B

The temperature, t, of a furnace is given as $t = 1450\,°\text{C} \pm 5\%$. Express the uncertainty as both fractional and absolute uncertainties.

---

[11] The modulus of the mean is used to avoid the fractional uncertainty assuming a negative value.

5.7.1 Combining uncertainties caused by random errors

Often, the value of a quantity obtained through measurement is entered into an equation, possibly along with other values determined by experiment. Assuming that each value has some uncertainty, the value emerging from the equation must also have some uncertainty. To calculate the 'combined' uncertainty we apply some results from differential calculus.

5.7.2 Equations containing a single variable

If y is a function of x we can write[12]

$$\frac{\delta y}{\delta x} \approx \frac{dy}{dx} \tag{5.11}$$

where δy is a small change in y, δx is a small change in x and dy/dx is the derivative of y with respect to x. Equation (5.11) can be rewritten

$$\delta y \approx \left(\frac{dy}{dx}\right)\delta x \tag{5.12}$$

The approximation indicated by equation (5.12) is sufficiently good for our purposes when δx is much less than x (which is usually the case) that we usually replace $\approx$ by $=$. The symbol δx may also be used to represent the error in x, in which case δy is the error in y. As the errors in x and y are unknown, there is merit in recasting equation (5.12) in terms of the quantities we *are* able to determine, namely the uncertainties in x and y which we write as u_x and u_y respectively. Equation (5.12) becomes

$$u_y = \left|\frac{dy}{dx}\right| u_x \tag{5.13}$$

where the derivative is evaluated at the 'best' estimate of x, which is usually the mean, $\bar{x}$.

The modulus of the derivative is used to prevent u_y from assuming a negative value when dy/dx is negative (uncertainties always have a positive sign). In situations in which we take the uncertainty to be the standard error of the mean of repeat measurements, equation (5.13) is modified to give

$$\sigma_{\bar{y}} = \left|\frac{dy}{dx}\right| \sigma_{\bar{x}} \tag{5.14}$$

---

[12] See Meadows (1981) as a useful reference for this section.

where $\sigma_{\bar{y}}$ is the standard error in the mean of y and $\sigma_{\bar{x}}$ is the standard error in the mean of x.

Example 3

The radius of a metal sphere is (2.10 ± 0.15) mm. Calculate the surface area of the sphere and the uncertainty in the surface area.

ANSWER

The surface area of a sphere, A, of radius, r, is given by

$$A = 4\pi r^2 \tag{5.15}$$

Substituting $r = 2.1$ mm into equation (5.15) gives $A = 55.418$ mm^2. In order to find the uncertainty in A, we rewrite equation (5.13) in terms of A and r, i.e.

$$u_A = \left| \frac{dA}{dr} \right| u_r \tag{5.16}$$

If $A = 4\pi r^2$, then $dA/dr = 8\pi r = 52.779$ mm. Using equation (5.16) gives

$$u_A = \left| \frac{dA}{dr} \right| u_r = 52.779 \text{ mm} \times 0.15 \text{ mm} = 7.917 \text{ mm}^2$$

Rounding the uncertainty to two significant figures, gives $A = (55.4 \pm 7.9)$ mm^2.

Exercise C

1. An oscilloscope is used to view an electrical signal which varies sinusoidally with time. Given that the period of the signal, T, is (0.552 ± 0.012) ms, calculate the frequency, f, of the signal and the uncertainty in the frequency, where

$$f = \frac{1}{T}$$

2. The volume, V, of a sphere of radius, r, is given by

$$V = \frac{4}{3}\pi r^3$$

Using the value for r and the uncertainty in r given in example 3, calculate V and the uncertainty in V.

5.7.3 Equations containing more than one variable

Many equations contain two or more variables. If the values of the variables have some uncertainty, how does the uncertainty in each combine to give an uncertainty in the final 'result'? To answer this, let y depend on x and z. Write the change in y, δy, as[13]

$$\delta y \approx \left(\frac{\partial y}{\partial x}\right)\delta x + \left(\frac{\partial y}{\partial z}\right)\delta z \tag{5.17}$$

where δx and δz are small changes in x and z respectively, $\partial y/\partial x$ is the partial derivative of y with respect to x, and $\partial y/\partial z$ is the partial derivative of y with respect to z. Writing equation (5.17) in terms of uncertainties gives

$$u_y = \left|\frac{\partial y}{\partial x}\right|u_x + \left|\frac{\partial y}{\partial z}\right|u_z \tag{5.18}$$

where u_y is the uncertainty in y, u_x is the uncertainty in x, and u_z is the uncertainty in z. The moduli of $\partial y/\partial x$ and $\partial y/\partial z$ are employed in equation (5.18) to ensure that u_y is always positive. The derivatives are evaluated at $x = \bar{x}$ and $z = \bar{z}$.

If u_x and u_z are replaced by the standard errors in the means of x and z, then u_y becomes the standard error in the mean of y, so that

$$\sigma_{\bar{y}} = \left|\frac{\partial y}{\partial x}\right|\sigma_{\bar{x}} + \left|\frac{\partial y}{\partial z}\right|\sigma_{\bar{z}} \tag{5.19}$$

Example 4

The velocity, v, of a wave travelling along a stretched string is given by

$$v = \left(\frac{T}{\mu}\right)^{\frac{1}{2}} \tag{5.20}$$

where T is the tension in the string and μ is the mass per unit length of the string. If $T = (3.2 \pm 0.2)$ N and $\mu = (1.24 \pm 0.05) \times 10^{-3}$ kg/m, calculate v and the uncertainty in v.

ANSWER

Using equation (5.20),

$$v = \left(\frac{3.2 \text{ N}}{1.24 \times 10^{-3} \text{ kg/m}}\right)^{\frac{1}{2}} = 50.8 \text{ m/s}$$

[13] For convenience we consider only two variables, x and z. However, this approach can be extended to any number of variables.

To find the uncertainty in v, u_v, we rewrite equation (5.18) as

$$u_v = \left| \frac{\partial v}{\partial T} \right| u_T + \left| \frac{\partial v}{\partial \mu} \right| u_\mu \tag{5.21}$$

where u_T is the uncertainty in the tension and u_μ is the uncertainty in the mass per unit length. With reference to equation (5.20)

$$\frac{\partial v}{\partial T} = \frac{1}{2} \left(\frac{1}{\mu T} \right)^{\frac{1}{2}} = \frac{1}{2} \left(\frac{1}{1.24 \times 10^{-3}\,\text{kg/m} \times 3.2\,\text{N}} \right)^{\frac{1}{2}} = 7.938\,\text{s/kg}$$

$$\frac{\partial v}{\partial \mu} = -\frac{1}{2} \left(\frac{T}{\mu^3} \right)^{\frac{1}{2}} = \frac{1}{2} \left(\frac{3.2\,\text{N}}{1.24 \times 10^{-3}\,\text{kg/m}} \right)^{\frac{1}{2}} = -2.048 \times 10^4\,\text{m}^2/(\text{kg·s})$$

Substituting values into equation (5.21) gives

$$u_v = 7.938\,\text{s/kg} \times 0.2\,\text{N} + 2.048 \times 10^4\,\text{m}^2/(\text{kg·s}) \times 0.05 \times 10^{-3}\,\text{kg/m} = 2.61\,\text{m/s}$$

We now combine v and u_v to give $v = (50.8 \pm 2.6)$ m/s.

Exercise D

The pressure difference, p, between two points in a flowing fluid is related to the density of the fluid, ρ, and the speed of the fluid, v, by the equation

$$p = \tfrac{1}{2}\rho v^2 \tag{5.22}$$

Given that $\rho = (0.986 \pm 0.013) \times 10^3$ kg/m³ and $v = (1.840 \pm 0.023)$ m/s, calculate p and the uncertainty in p.

Equation (5.18) is extremely useful when combining uncertainties, but it tends to overestimate the size of the uncertainty in y. This is because if random errors in x and z are independent, there is a probability that the errors will partially cancel, resulting in a smaller uncertainty in y than that given by equation (5.18). If we cannot be assured that errors are independent then it is reasonable to use equation (5.18), as this represents the *maximum* uncertainty in y for given uncertainties in x and z. The consequence of independent errors is considered next.

5.7.4   **Most probable uncertainty**

Consider the situation in which the relationship between y, x and z is

$$y = x + z \qquad (5.23)$$

so that the error in y, δy, can be written[14]

$$\delta y = \delta x + \delta z \qquad (5.24)$$

where δx and δz are the errors in x and z respectively.

We cannot know the errors δx and δz but we *do* know that if the errors are random, δx and δz can take on positive and negative values. In situations in which the errors on the right hand side of equation (5.24) have opposite signs, there will be partial cancellation, such that δy is smaller than either δx or δz. At other times, δx and δz have the same sign so that no cancellation occurs and the magnitude of δy is larger than either δx or δz. Equations (5.18) and (5.19) are overly pessimistic expressions of uncertainty as they do not allow for 'partial cancellation' in errors. If random errors in x and z are independent (i.e. the error in x is not related to the error in y), then when y is a function of x and z, the *most probable uncertainty* in y, u_y, is given by

$$u_y = \left[\left(\frac{\partial y}{\partial x} u_x \right)^2 + \left(\frac{\partial y}{\partial z} u_z \right)^2 \right]^{\frac{1}{2}} \qquad (5.25)$$

where u_x and u_z are the uncertainties in x and z respectively.

If the uncertainty is taken to be the standard error in the mean of each quantity, we have

$$\sigma_{\bar{y}} = \left[\left(\frac{\partial y}{\partial x} \sigma_{\bar{x}} \right)^2 + \left(\frac{\partial y}{\partial z} \sigma_{\bar{z}} \right)^2 \right]^{\frac{1}{2}} \qquad (5.26)$$

where $\sigma_{\bar{y}}$ is the standard error in the mean of y, $\sigma_{\bar{x}}$ is the standard error in the mean of x and $\sigma_{\bar{z}}$ is the standard error in the mean of z.

Appendix 2 gives a brief derivation of equations (5.25) and (5.26). We now consider an example which illustrates the superiority of equation (5.25) over equation (5.19) when errors in values are independent.

Table 5.7 contains repeat measured values of the (dc) voltage, V, across a resistor and current, I, through the resistor. The power, P, dissipated in the resistor, given by $P = VI$, is also shown in the table. The As and Bs preceding the values in the table indicate whether the value is above (A)

---

[14] This approach to calculating the uncertainty in y can be extended to any number of variables used to represent values with uncertainty.

Table 5.7. *Values of voltage and current for a resistor.*

Voltage, V (mV) (mean, $\bar{V}=14.78$ mV)	Current, I, (mA) (mean, $\bar{I}=0.251$ mA)	Power, $P=VI$, (μW) (mean, $\bar{P}=3.71$ μW)
A15.16	A0.266	A4.033
A15.66	B0.224	B3.508
B14.50	A0.280	A4.060
B14.23	A0.271	A3.856
B14.68	A0.253	A3.714
A15.51	B0.235	B3.645
A14.90	B0.230	B3.427
B14.21	B0.228	B3.240
A15.44	A0.273	A4.215
B13.53	B0.250	B3.383

or below (B) the mean of values in that column. Examining the rows seems to indicate no correlation between the deviation of the values in columns 1 and 2. For example, a value of voltage above the mean in the first column does not coincide consistently with a value of current above the mean in the second column. Turning now to the standard errors of the values in each column, we find[15]

$$\sigma_{\bar{V}}=0.216\text{ mV} \qquad \sigma_{\bar{I}}=0.00659\text{ mA} \qquad \sigma_{\bar{P}}=0.103\ \mu\text{W}$$

The value of $\sigma_{\bar{P}}$ of 0.103 μW found using table 5.7 may be compared with those determined using equations (5.19) (maximum uncertainty) and (5.26) (most probable uncertainty).

As $P=VI$, it follows that

$$\frac{\partial P}{\partial V}=I \qquad \frac{\partial P}{\partial I}=V$$

Using equation (5.19), and evaluating the derivatives at $I=\bar{I}$ and $V=\bar{V}$ gives

$$\sigma_{\bar{P}}=\left|\frac{\partial P}{\partial V}\right|\sigma_{\bar{V}}+\left|\frac{\partial P}{\partial I}\right|\sigma_{\bar{I}}$$

$$=0.251\text{ mA}\times0.216\text{ mV}+14.78\text{ mV}\times0.00659\text{ mA}$$

$$=0.152\ \mu\text{W}$$

---

[15] See section 3.8 for details of calculating standard errors.

Using equation (5.26), we write

$$\sigma_{\bar{P}} = \left[\left(\frac{\partial P}{\partial V} \sigma_{\bar{V}} \right)^2 + \left(\frac{\partial P}{\partial I} \sigma_{\bar{I}} \right)^2 \right]^{\frac{1}{2}}$$

$$= [(0.251 \text{ mA} \times 0.216 \text{ mV})^2 + (14.78 \text{ mV} \times 0.00659 \text{ mA})^2]^{\frac{1}{2}}$$

$$= 0.111 \ \mu\text{W}$$

Comparing the standard errors calculated using equations (5.19) and (5.26) with the standard error in the mean power, $\sigma_{\bar{P}} = 0.103 \ \mu\text{W}$, as determined using the values in the third column of table 5.7, reveals that equation (5.26) gives an estimate for $\sigma_{\bar{P}}$ (0.111 μW) which is superior to that of equation (5.19) (0.152 μW). In short, when random errors in measured values are independent, equation (5.26) is preferred to equation (5.19) when combining standard errors.

Exercise E

1. Light incident at an angle, i, on a plane glass surface is refracted and enters the glass. The refractive index of the glass, n, can be calculated using

$$n = \frac{\sin i}{\sin r}$$

where r is the angle of refraction. Assuming errors in i and r are independent, calculate n and the uncertainty in n, given that $i = (52 \pm 1)°$ and $r = (32 \pm 1)°$. Note, for $y = \sin x$, the approximation $\delta y / \delta x \approx dy / dx$ is valid for x expressed in radians.

2. In an optics experiment, the distance, u, from an object to a convex lens is (125.5 ± 1.5) mm. The distance, v, from the lens to the image is (628.0 ± 1.5) mm. Assuming that the errors in u and v are independent:

   (i) Calculate the focal length, f, of the convex lens and the uncertainty in f, given that f is related to u and v by the equation

$$\frac{1}{f} = \frac{1}{u} + \frac{1}{v}$$

   (ii) Calculate the linear magnification, m, of the lens and the uncertainty in m given that

$$m = \frac{v}{u}$$

Review of combining uncertainties

When a quantity y depends on x and z we have considered two approaches to evaluating the uncertainty in y given the uncertainties in x and z. When random errors in x and z are independent, equation (5.25) is preferred when calculating uncertainties, otherwise equation (5.18) should be used. Rather than assuming the independence of errors, it is preferable for an experimenter to consider this matter carefully and ask the question: 'is the error in x dependent on, or correlated with, the error in z?' If the answer is no then equation (5.25) should be used.

5.8 Coping with extremes in data variability

When repeat measurements are made of a particular quantity, it is reasonable to anticipate that values will vary due to random errors. However, occasionally we encounter situations in which most of the values are 'well behaved' in that all but one or two values appear to be consistent with the others. The 'one or two values' far from the others are usually referred to as 'outliers'. The question is, what should be done with outliers? Should they be included in the analysis or should they be ignored?

At the other extreme, occasionally a situation occurs in which repeat measurements of a quantity produce values which do not differ one from the next. In this situation the standard deviation as given by equation (1.16) is zero and so the standard error in the mean is also zero. Can we really assert that the uncertainty in the mean is zero, i.e. that we have succeeded in determining the true value of the quantity?

In the next two sections we deal with the issue of outliers and situations in which the standard deviation in measured values is zero.

5.8.1 Outliers

Identification of outliers is important as they may indicate the existence of a novel effect or that some intermittent occurrence, perhaps due to electrical interference, is adversely influencing the measurement process. Another, perhaps more common, cause of outliers is due to the incorrect recording of the measured value. Trying to identify the cause of an outlier is very important, and should precede any effort to eliminate it, as systems that automatically eliminate outliers may expel the most important data before anyone has had the opportunity of assessing the cause of the outlier or its significance. Where possible, it is preferable to repeat the experiment in an effort to establish if an outlier is, in fact, a 'mistake' or an occasional, but regular occurrence.

It is not always possible to repeat an experiment due to time, cost or other reasons and we must do the best we can with the data available. However, caution is always warranted when deciding to omit an outlier based on statistical grounds only, as the process often requires the assumption that the data follow a particular distribution (most often the normal distribution). If the assumption is not valid then removing an outlier may not be justified. Where possible, the actual distribution of data should be compared with the assumed distribution (for example by assessing the normality of data as described in section 3.11).

5.8.2 Chauvenet's criterion

One approach to deciding whether an outlier should be removed from a group of values is to begin by asking the question:

> Given the scatter displayed by the data as a whole, what is the probability of obtaining a value at least as far from the mean as the outlier?

If this probability is very low then we have good statistical grounds for omitting the outlier from any further analysis. We assume that the scatter data being considered is well described by the normal distribution.

Let us consider a specific example. Table 5.8 shows mass loss from nine samples of a ceramic when they are heated in an oxygen free environment. An inspection of the values in table 5.8 indicates that the value $x=6.6$ mg seems to differ considerably from the others. Should this value be eliminated, and if so on what grounds? To assess the scatter in the values we first determine the mean mass loss, $\bar{x}$, and the standard deviation, s, using all the values in table 5.8. This gives

$$\bar{x} = 5.711 \text{ mg} \qquad s = 0.3551 \text{ mg}$$

The characteristic spread of the values is represented by the standard deviation, s, so it is sensible to express the 'separation' between the mean and the outlier in terms of the standard deviation. To do this we calculate z given by[16]

---

[16] Note we should really use the population standard deviation, σ, in the denominator of equation (5.27). As we cannot know the value of σ, we must replace it by the estimate of its value, s.

Table 5.8. *Mass loss from the ceramic YBa$_2$Cu$_3$O$_7$ heated in an oxygen free environment.*

Mass loss, x (mg)	5.6	5.4	5.7	5.5	5.6	6.6	5.8	5.7	5.5

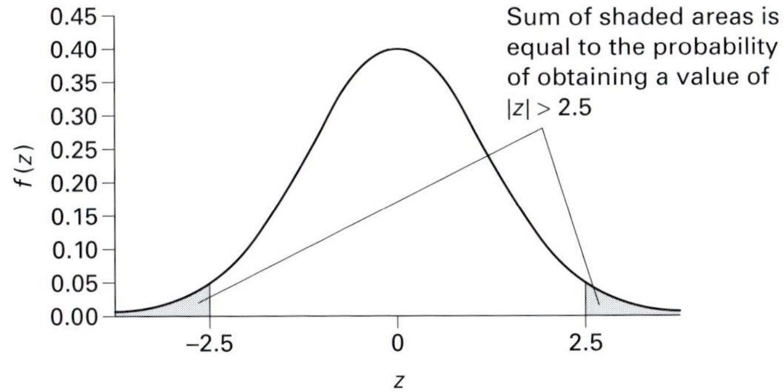

Figure 5.3. Standard normal probability distribution. The sum of the areas in the tails is equal to the probability that a value is more than 2.5 standard deviations from the mean.

$$z = \frac{x_{OUT} - \bar{x}}{s} \qquad (5.27)$$

where x_{OUT} is the value of the outlier. Inserting values from this example gives

$$z = \frac{6.6 \text{ mg} - 5.711 \text{ mg}}{0.3551 \text{ mg}} = 2.50$$

To find the probability that a value differs from the mean by a least 2.5 standard deviations, we assume that the values are drawn from a normal distribution. Figure 5.3 shows the standard normal distribution. $P(-\infty \le z \le -2.5)$ obtained using table 1 in appendix 1 is equal to 0.006 21. As the tails of the distribution are symmetric, we double this value to find the probability represented by the total shaded area in figure 5.3, i.e. the probability that a value lies at least as far from the mean as the outlier is $2 \times 0.006 21 = 0.012 42$. Put another way, if we obtain, for example, 100 values, we would expect $100 \times 0.012 42 \approx 1$ to be at least as far from the mean as the outlier. As only 9 measurements have been made, the number of values we would expect to be at least as far from the mean as the outlier is $9 \times 0.012 42 \approx 0.1$.

We can use the number of values expected to be at least as far from the mean as the outlier as the basis of a criterion upon which to accept or reject an outlier. If the expected number of values at least as far from the mean as the outlier is ≤0.5 then the outlier should be eliminated and the mean and standard deviation of the remaining values recalculated. This criterion in referred to as Chauvenet's criterion and it should be applied only once to a set of values, i.e. if a value is rejected and the mean and standard deviation is recalculated, the criterion should not be applied to any other value.

In summary, the steps in applying Chauvenet's criterion are:

(i)   Calculate the mean, $\bar{x}$, and the standard deviation, s, of the values being considered.

(ii) Identify a possible outlying value, x_{OUT}. Calculate z, where $z = (x_{OUT} - \bar{x})/s$.

(iii) Use the normal probability tables to determine the probability, P, that a value will occur at least as far from the mean as x_{OUT}.

(iv) Calculate the number, N, expected to be at least as far from the mean as x_{OUT}, $N = nP$, where n is the number of values and P is the probability that a value is at least as far from the mean as x_{OUT}.

(v)   If $N \leq 0.5$, reject the outlier.

(vi) Recalculate $\bar{x}$ and s but do not reapply the criterion to the remaining data.

Example 5

Table 5.9 shows values for the 'hand timing' of the free fall of an object through a vertical distance of 5 m. Use Chauvenet's criterion to decide if any value should be rejected.

ANSWER

The mean, $\bar{x}$, and standard deviation, s, of the values in table 5.9 are 1.084 s and 0.099 91 s respectively. The value furthest from the mean is 0.90 s, i.e. $x_{OUT} = 0.90$ s. Calculating z using equation (5.27) gives

$$z = \frac{x_{OUT} - \bar{x}}{s} = \frac{0.90\,\text{s} - 1.084\,\text{s}}{0.099\,91\,\text{s}} = -1.84$$

Using table 1 in appendix 1 gives $P(-\infty \leq z \leq -1.84) = 0.032\,88$. The total probability that a value lies at least as far from the mean as the outlier is $P = 2 \times 0.032\,88 = 0.065\,76$. The number of values, N, expected to be at least as far from the mean as the outlier is $N = nP \approx 10 \times 0.066 = 0.66$. As this number is greater than 0.5, do *not* reject the value $x = 0.90$ s.

Table 5.9. *Values of time for an object to fall freely through a distance of 5 m.*

Time, x (s)	1.11	0.90	1.13	1.13	1.16	1.23	0.94	1.08	1.12	1.04

Exercise F

In a fluid flow experiment, the time, t, for a fixed volume of water to flow from a vessel is measured. Table 5.10 shows five values for t, obtained through repeat measurements.

   (i)   Calculate the mean and standard deviation of these values.

  (ii)   Identify the value in table 5.10 furthest from the mean.

 (iii)   Apply Chauvenet's criterion to the outlier to decide whether the outlier should be eliminated.

 (iv)   If the outlier *is* eliminated, recalculate the mean and standard deviation of the remaining values.

Table 5.10. *Value of time for liquid to flow from a vessel.*

t (s)	215	220	217	235	222

5.8.3 Dealing with values that show no variability

Every value differs from the true value by an amount equal to the experimental error. It is possible for the random error to be so small that it is less than the resolution of the instrument. In this case every measurement will yield the same value. Take as an example the values of voltage measured across a 1 kΩ resistor using a $3\frac{1}{2}$ digital multimeter as shown in table 5.11. The mean of these values is 3.2 mV and the standard deviation is zero. As the standard deviation is zero, it follows that the standard error in the mean is also zero, so we write (for any level of confidence)

$$V = (3.2 \pm 0) \text{ mV}$$

Table 5.11. *Values of voltage across a 1 kΩ resistor.*

Voltage, V (mV)	3.2	3.2	3.2	3.2	3.2	3.2	3.2	3.2

Is such a statement reasonable? No, it is not. Irrespective of the size of the uncertainty in a value due to random errors, there will always be uncertainty due to systematic errors, such as the imperfection in the calibration of an instrument. We consider systematic errors in section 5.9.

If, for the moment, we assume that the systematic errors can be considered as negligible, we are still able to quote an uncertainty due to random errors even when all values obtained through measurement are the same. When measured values show no variation, it is possible to determine an upper limit for the standard deviation, and therefore an upper limit to the standard error in the mean so long as the following assumptions are valid:

(i) Random errors, though small, still exist but are masked due to the resolution limit of the instrument.
(ii) Random errors are normally distributed.

We begin by considering the resolution of the instrument. In the case of the digital multimeter mentioned above, we can say that if the instrument indicates 3.2 mV, then the voltage cannot exceed 3.25 mV (otherwise the instrument would indicate 3.3 mV), nor can the voltage be less than 3.15 mV (otherwise the instrument would indicate 3.1 mV). Next we ask, if we obtain eight values through repeat measurements (as shown in table 5.11), what is the probability, P, that none of the values will lie outside the range 3.15 mV to 3.25 mV?

To estimate P we suppose that, given n values, the expected number, N, to lie outside the range 3.15 mV to 3.25 mV is 0.5. We choose $N=0.5$ so that a reasonable upper limit for the probability, P, may be established. If N were chosen to be greater than 0.5 then we could expect at least one value to differ from the others. We can write

$$N = nP \tag{5.28}$$

If $N=0.5$ and $n=8$, then $P=\frac{0.5}{8}=0.0625$.

We can use the standard normal probability distribution to find the value of z which corresponds to the total area in the tails of the distribution of 0.0625. Figure 5.4 illustrates the situation when $P=0.0625$ so that the area in each tail in 0.03125. Using table 1 in appendix 1 gives the value of z when $P=0.03125$ as $z=-1.86$. We now write z as

$$z = \frac{x_{MIN} - \bar{x}}{s} \tag{5.29}$$

where x_{MIN} is the smallest value of x that will cause the value indicated by the instrument to remain unchanged and s is the estimate of the popula-

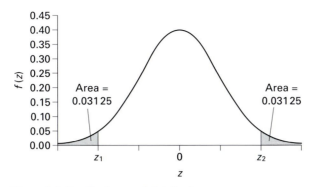

Figure 5.4. Standard normal distribution.

tion standard deviation. In this example, the smallest value of voltage sensed by the meter such that the meter indicates 3.2 mV is 3.15 mV. Rearranging equation (5.29) gives

$$s = \frac{x_{MIN} - \bar{x}}{z} \tag{5.30}$$

Substituting $x_{MIN} = 3.15$ mV, $\bar{x} = 3.2$ mV and $z = -1.86$, gives $s = 0.0269$ mV.

s could be smaller than 0.0269 mV, but if s were larger than 0.0269 mV we would expect to observe at least one value to be $\neq 3.2$ mV. Regarding 0.0269 mV as the upper limit for s, we can express the upper limit for the standard error in the mean as

$$\sigma_{\bar{x}} \approx \frac{s}{\sqrt{n}} = \frac{0.0269 \, \text{mV}}{\sqrt{8}} = 0.0095 \, \text{mV}$$

Summarising the steps to determining upper limits for the standard deviation and standard error when all values are the same, we have:

   (i) If n is the number of repeat measurements, calculate the area in one of the tails of the normal distribution equivalent to the probability, P, where $P = 1/4n$.
   (ii) Use table 1 in appendix 1 to determine the value of z corresponding to P.
   (iii) Determine the limit, Δx, due to the resolution of the instrument (for example, for a micrometer with smallest scale divisions of 0.01 mm, $\Delta x = 0.01$ mm$/2 = 0.005$ mm).
   (iv) As $|z| = \Delta x / s$, use values of z and Δx from parts (ii) and (iii) to determine s.
   (v) Calculate the standard error of the mean, $\sigma_{\bar{x}}$, using $\sigma_{\bar{x}} \approx s/\sqrt{n}$.

It is strongly emphasised that, if values due to repeat measurements are constant, then the random errors are so small that they are very likely to be insignificant and that the uncertainties will be dominated by systematic errors to such an extent that random errors can be ignored. The main message of this section is that even if values do not change when we make repeat measurements, we can quote an upper limit to the standard deviation and standard error if we know the resolution of the instrument and this avoids the difficulties arising from a statement that the uncertainty in measured values is zero.

Exercise G

On measuring the diameter, d, of a copper wire with a micrometer with a smallest scale division of 0.01 mm, it is found that $d = 1.22$ mm. The measurement is repeated a further four times at other positions along the wire. No change in d is observed.

Assuming that measurements are affected only by random errors:

(i)   Determine an upper limit for the standard deviation of the values and the standard error of the mean.
(ii) Calculate the 95% confidence interval for the true value of the diameter.

5.9   Uncertainty due to systematic errors

Systematic errors, like random errors, cause measured values to differ from the true value. In contrast to random errors, which cause measured values to be scattered both above and below the true value, systematic errors introduce a bias so that measured values are consistently above or consistently below the true value. Random errors 'betray' themselves by apparent variability in measured values but systematic errors are characterised by a lack of variability and this makes them very difficult to detect (or, put another way, very easy to overlook). Experienced experimenters are always on the 'look out' for sources of systematic error and devote time to identifying and where possible compensating for such errors. It is not uncommon for systematic errors to be several times larger than random errors and this is why we should consider carefully causes of systematic error and where possible include their effect in an expression of the uncertainty in the best estimate of the true value.

5.9.1 Calibration errors and specifications

Measuring instruments are sources of systematic error. An expensive instrument such as a microvoltmeter is often supplied with a calibration certificate which details the performance of the instrument as established by comparison with another, more accurate, instrument.[17] Less expensive voltmeters are supplied with a maker's specification which indicates the limits of accuracy of an instrument. The specification incorporates the expected uncertainty due to the calibration process as well as other factors such as the likely drift with time of the performance of the instrument.

As an example of systematic error due to an instrument, consider a commercially available multimeter. The specifications for two ranges of a $3\frac{1}{2}$ digit multimeter are given in table 5.12. Note accuracies are specified by the manufacturer as $\pm$ (%meter reading + number of digits[18]). The accuracies are valid for temperatures in the range 18°C to 28°C and relative humidity 0% to 70%.

Example 6

A $3\frac{1}{2}$ digit multimeter is used to measure a steady voltage across a resistor. The measurement is made on the 2 V range at 22°C and relative humidity of 50%. The value appearing on the display of the meter is 1.898 V. Use the information in table 5.12 to determine the uncertainty in the voltage.

ANSWER

Referring to table 5.12, the accuracy of the meter is $\pm$ (0.5% of reading + one digit). 0.5% of 1.898 V is

$$\frac{1.898\,\text{V}}{100\%} \times 0.5\% = 0.0095\,\text{V}$$

'One digit' on the 2 V range of the meter corresponds to 0.001 V. So that the total systematic uncertainty is 0.0105 V which we would normally round to 0.011 V and we can write $V = (1.898 \pm 0.011)$ V.

---

[17] Eventually the calibration should be traceable back to a standard held by a national measurement laboratory, though for much scientific work such traceability is not pursued.

[18] The digits referred to are the least significant digits that appear on the display.

Table 5.12. *Resolution and accuracy specification for the dc voltage and capacitance ranges of a 3½ digit multimeter.*

	Range	Resolution	Accuracy
dc voltage	200 mV	100 μV	± (0.5% of reading + 1 digit)
	2 V	1 mV	
	20 V	10 mV	
	200 V	100 mV	
	1000 V	1 V	
Capacitance	2 nF	1 pF	± (2% of reading + 4 digits)
	200 nF	100 pF	
	2 μF	1 nF	
	20 μF	10 nF	

Checking the calibration of an instrument is important before measurement begins as all values obtained using the instrument may be viewed with some suspicion until there is some satisfactory evidence that the values are trustworthy. Gross errors may be easy to identify. For example, if the voltage across a 9 V battery is measured with a voltmeter and the voltmeter indicates 15 V then it is likely that something is amiss. It is possible that the meter has malfunctioned and requires attention. Smaller discrepancies may be more difficult to detect. There may not be the time or facilities to fully assess the calibration of an instrument. A useful check that can be done quite quickly is to replace an instrument with another and assess how well the two instruments 'agree'. Though this approach is no substitute for a proper calibration, it can reveal inconsistencies which must be investigated further.

Exercise H

A $3\frac{1}{2}$ digit meter is used to measure the capacitance of a capacitor. The room temperature is 25 °C and the relative humidity is 60%. The display on the meter indicates a capacitance of 156.4 nF. Use the information in table 5.12 to determine the uncertainty in the capacitance due to systematic error in the instrument.

5.9.2 Offset and gain errors

Offset errors are common systematic errors and, in some cases, may be reduced. If the error remains fixed over time, then the measured value can be corrected by an amount so as to compensate for the offset error. As an example, consider a micrometer which is capable of measuring lengths in the range 0 mm to 25.40 mm. The resolution of the instrument (i.e. the separation of the smallest scale division) is 0.01 mm. When the jaws of a particular micrometer are brought into contact, the scale of the micrometer indicates 0.01 mm. We conclude that any measurement made with this instrument will produce a value which is too large by an amount 0.01 mm. To minimise the systematic error we would subtract 0.01 mm from any value obtained using the instrument. If we do this we are making the assumption that the offset error will remain fixed and that may or may not be a reasonable assumption. Also, we must acknowledge that any correction we apply will itself have some uncertainty, so we cannot hope to 'cancel' the offset error exactly.

Another important characteristic of a measuring system is its gain. For example, a system used to measure light intensity might consist of a light detector[19] and an amplifier. Focussing for a moment upon the amplifier, its purpose is to increase the small voltage generated by the light detector. If the gain of the amplifier decreases (for example due to a temperature rise in the amplifier) then the output voltage of the amplifier for a given input voltage will similarly decrease. This will cause a systematic error in the measurement of the light intensity. Let us consider this in more detail. For a particular instrumentation amplifier[20] the gain is fixed using an external 'gain setting' resistor, R_g. The voltage gain, G, is given by

$$G = 1 + \frac{40000\ \Omega}{R_g} \tag{5.31}$$

R_g is temperature dependent and so G must also be temperature dependent. For many applications a small change in G with temperature will be unimportant. However, for precision applications the systematic variation of G with temperature must be included in the assessment of the uncertainty in values obtained using the amplifier.[21]

---

[19] A typical light detector would be a photodiode.

[20] Instrumentation amplifiers are used in precision applications when small signals, often from sensors, are amplified. Usual voltage gains for such amplifiers are in the range 10 to 1000. In the example discussed here we consider the INA101 instrumentation amplifier manufactured by Burr-Brown.

[21] There are other parameters within the amplifier that are influenced by temperature variations, such as the voltage offset of the device. Here we will focus solely on the influence of temperature on the gain setting resistor and ignore other sources of systematic error.

Example 7

An instrumentation amplifier is used to amplify a small voltage generated by a photodiode. The gain of the amplifier is given by equation (5.31). The relationship between the gain setting resistor, $R_g(t)$, and temperature of the resistor, t, is

$$R_g(t) = R_g(0)(1 + \alpha t) \tag{5.32}$$

where $R_g(0)$ is the resistance at $0°C$, t is the temperature in $°C$ and α is the temperature coefficient of resistance in $°C^{-1}$. If a wire wound resistor made from nichrome is used as a gain setting resistor with $R_g(0) = 318.3\ \Omega$ and $\alpha = 4.1 \times 10^{-4}\ °C^{-1}$, determine:

(i) The value of $R_g(t)$ when $t = 20°C$ and $t = 40°C$.

(ii) The gain of the amplifier at $0°C$, $20°C$ and $40°C$.

ANSWER

(i) Using equation (5.32):

$$R_g(20) = 318.3(1 + 4.1 \times 10^{-4} \times 20)\Omega = 320.9\ \Omega$$
$$R_g(40) = 318.3(1 + 4.1 \times 10^{-4} \times 40)\Omega = 323.5\ \Omega$$

(ii) Using equation (5.31) to find the gain at $0°C$, $20°C$ and $40°C$:

$$G(0°C) = 1 + \frac{40000\ \Omega}{318.3\ \Omega} = 126.7$$

$$G(20°C) = 1 + \frac{40000\ \Omega}{320.9\ \Omega} = 125.6$$

$$G(40°C) = 1 + \frac{40000\ \Omega}{323.5\ \Omega} = 124.6$$

Exercise I

During an experiment, the temperature of the room in which the experiment is carried out is nominally $23°C$. Due to the effect of an air-conditioning system, the room temperature varies between $21°C$ and $25°C$. The gain of an amplifier used in the experiment is given by equation (5.31).

(i) A gain setting resistor with the following characteristics is used:

$$R_g(0) = 21.23\ \Omega \qquad \alpha = 3.9 \times 10^{-3}°C^{-1}$$

Find the gain of the amplifier and estimate the uncertainty due to systematic error in the gain of the amplifier caused by the temperature dependence of the gain setting resistor.

(ii) What assumptions did you make in order to carry out part (i)?

5.9.3 Loading errors

Instruments influence the quantities they are used to measure. Take as an example a Pitot tube used to measure the speed of a fluid.[22] By introducing the tube into the stream there is a slight effect on the flow of the fluid. Similarly, an ammeter measuring the electrical current in a circuit influences the value obtained because the measurement process requires that a reference resistor is placed in the path of the current. Placing a reference resistor into the circuit reduces the current slightly. The error introduced by the instrument in these circumstances is referred to as a *loading* error.

Another example of a loading error can be seen when the temperature of a body is measured using a thermometer. As the thermometer is brought into contact with the body, heat is transferred between the body and the thermometer unless both are at the same temperature. If the temperature of the thermometer is less than the body it is contact with, then the thermometer will act to lower the temperature of the body. The effect of this type of loading can be reduced by choosing a thermometer with a thermal mass much less than the thermal mass of the body whose temperature is to be measured.

Though it is often difficult to determine the effect of loading errors, there are situations in which the loading error can be quantified, especially when electrical measurements are made. Consider the situation in which a voltage across a resistor is measured using a voltmeter. Figure 5.5 shows the basic set up. With the values of resistance shown in the figure, we would expect a little more than half the voltage between the terminals of the battery to appear across the 5.6 MΩ resistor. Calculation predicts that a $3\frac{1}{2}$ digit voltmeter should indicate 4.67 V. In practice, when a particular voltmeter is used, it is found to indicate a value of 3.81 V. While the discrepancy could be due to the two resistors, which may have resistances far from their nominal values, the actual cause of the 'discrepancy' is the voltmeter. All

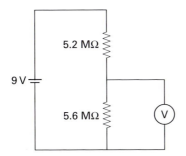

Figure 5.5. Voltmeter used to measure the voltage across a 5.6 MΩ resistor.

[22] See Khazan (1994) for a discussion of transducers used to measure fluid flow.

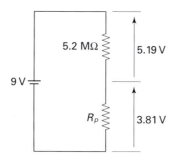

Figure 5.6. Circuit with resistor and voltmeter replaced by an equivalent resistor, R_p.

voltmeters have a large, but not infinite, internal resistance, R_{int}. As R_{int} is in parallel with the 5.6 MΩ resistor, some of the current that would have gone through the 5.6 MΩ resistor is diverted through the voltmeter. The loading effect of the voltmeter causes the voltage across the 5.6 MΩ resistor to be reduced in comparison to the value it would take if the voltmeter were removed. We can calculate the value of the internal resistance of the voltmeter by first redrawing figure 5.5 to include the loading effect of the voltmeter. This is shown in figure 5.6, where R_p is the effective resistance of the 5.6 MΩ in parallel with the internal resistance of the voltmeter. If the voltage across R_p is 3.81 V, then the voltage across the 5.2 MΩ resistor must be $(9-3.81)\,V=5.19\,V$. The current through the 5.2 MΩ resistor and R_p must be the same as they are in series, so that

$$\frac{5.19\ V}{5.2\ M\Omega}=\frac{3.81\ V}{R_p}$$

It follows that

$$R_p=\frac{3.81\ V\times5.2\ M\Omega}{5.19\ V}=3.82\ M\Omega$$

The effective resistance, R_p, for two resistors, R and R_{int}, connected in parallel is given by

$$\frac{1}{R_p}=\frac{1}{R}+\frac{1}{R_{int}} \tag{5.33}$$

If $R_p=3.82$ MΩ and $R=5.6$ MΩ, then using equation (5.33) we obtain $R_{int}=12.0$ MΩ.

We can see that when R_{int} is comparable in size to R the loading effect is considerable. To reduce the loading effect, R_{int} should be much greater than R.

Exercise J

An experimenter replaces the voltmeter in figure 5.5 by another with a higher internal resistance, R_{int}. What must be the minimum value for R_{int} if the experimenter requires the voltage indicated by the meter to differ by no more than 1% from the voltage that would appear across the 5.6 MΩ in the absence of the meter? (In this problem consider only the effects on the measured values due to loading.)

Quantifying the loading error can be difficult and in many cases we rely on experience to suggest whether a loading effect can regarded as negligible.

5.9.4 Dynamic effects

In situations in which calibration, offset, gain and loading errors have been minimised or accounted for, there is still another source of error that can cause measured values to be above or below the true value and that error is due to the temporal response of the measuring system. Static, or at least slowly varying, quantities are easy to 'track' with a measuring system. However, in situations in which a quantity varies rapidly with time, the response of the measuring system influences the values obtained. Elements within a measuring system which may be time dependent include:

- the transducer used to sense the change in a physical quantity;
- the electronics used in the signal conditioning of the output from the transducer;
- the measuring device such as a voltmeter or analogue to digital converter used to measure the signal after conditioning has occurred;
- the recording component of the system, which could be the experimenter or an automatic data logging system.

When dealing with the temporal response of a measuring system as a whole it is useful to categorise systems as zero, first or second order. The order of the response relates to whether the input and output of the system can be adequately described by a zero, first or second order ordinary differential equation. For example, a first order system could be described by the equation

$$a_1 \frac{dy}{dt} + a_0 y = bx \tag{5.34}$$

where a_1, a_0 are constants which depend on the characteristics of the measurement system, b is a proportionality constant, y is the output signal from the measurement system and x is the input signal. Both x and y are functions of time. We will consider briefly zero and first order systems.[23]

5.9.5 Zero order system

An 'ideal' measuring system is zero order, as the output of the system follows the input of the system at all times. There is no settling time with a zero order system and the output is in phase with the input. Figure 5.7 shows the response of a zero order measuring system to a sudden or 'step' change at the input to the system at time $t = t^*$.

  Though figure 5.7 shows the response of a zero order measuring system to a step input, the system would respond similarly to any variation of the input signal, i.e. the shape of the output waveform would mimic that of the input waveform. In practice, a zero order measuring system is an ideal never found as no system truly responds 'instantly' to large step changes at the input to the system.

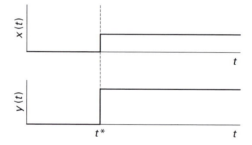

Figure 5.7. Input to $x(t)$, and output from $y(t)$, a zero order measuring system.

5.9.6 First order system

While a zero order measuring system is often a good approximation to a 'real' system in situations where measured quantities vary slowly, we must accept that all measuring systems have an inherent time dependence. The time dependence can be illustrated by considering an alcohol-in-glass thermometer used to measure the temperature of a hot water bath. If the thermometer is immersed in the water at $t = t^*$ then figure 5.8 shows the response of the thermometer as well as the response that would have

---

[23] Bentley (1988) deals with second order measurement systems.

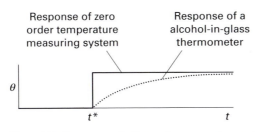

Figure 5.8. Response of a first order measuring system to a step change at the input of the system.

occurred had the thermometer responded instantly to the change in temperature. The shape of the temperature, θ, versus time, t, graph in figure 5.8 arises because heat takes a finite time to be transferred from the water to the thermometer and only when that transfer has ceased will both the thermometer and the water be at the same temperature. By looking at the equation which describes the time dependence of the temperature indicated by the thermometer we are able to explain the detailed form of the temperature variation with time and reveal what factors affect the response time of the thermometer. If the final temperature indicated by the thermometer is θ_f and the temperature indicated at any instant is θ, then the differential equation representing the system can be written[24]

$$\tau\frac{d\theta}{dt} + \theta = \theta_f \tag{5.35}$$

where τ is referred to as the time constant of the system. Equation (5.35) is an example of an equation which describes the response of a first order system. A detailed analysis shows that if we can approximate the bulb of the thermometer to a sphere filled with liquid, then

$$\tau = \frac{\rho r c}{3h} \tag{5.36}$$

where ρ is the density of the liquid in the thermometer, c is the specific heat of the liquid in the thermometer, h is the heat transfer coefficient for heat passing across the wall of the bulb of the thermometer and r is the radius of the thermometer bulb.

Solving equation (5.35) gives

$$\theta = \theta_f(1 - e^{-t/\tau}) \tag{5.37}$$

---

[24] See Doebelin (1995) for more detail on first and higher order instruments.

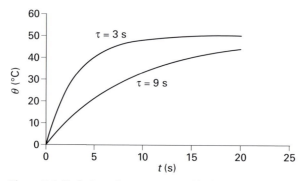

Figure 5.9. Variation of temperature with time as indicated by two thermometers with different time constants.

Figure 5.9 shows a graphical representation of equation (5.37) for $\theta_f = 50\,°C$ and two time constants, $\tau = 3$ s and $\tau = 9$ s. Equation (5.37) and figure 5.9 indicate that for the time response of a first order system to be improved, i.e. approach that of a zero order system, τ should be made as small as possible. In the case of an alcohol-in-glass thermometer we have little control over the quantities appearing in equation (5.36) that affect the value of τ. However, if we have a number of liquid-in-glass thermometers to choose from, then for a given size of thermometer, we would expect those containing low density liquid with a small specific heat to have the smaller value of τ.

Example 8

A particular thermometer has a time constant of 5 s. If the thermometer is immersed in hot water, how much time must elapse before the value indicated by the thermometer is within 5% of the final value?

ANSWER

We begin by rearranging equation (5.37) to give

$$\frac{\theta}{\theta_f} = (1 - e^{-t/\tau}) \tag{5.38}$$

If the temperature indicated by the thermometer is within 5% of the final value, then $\theta/\theta_f = 0.95 = (1 - e^{-t/\tau})$. It follows that

$$e^{-t/\tau} = 0.05 \quad \text{or} \quad \frac{-t}{\tau} = \ln(0.05) = -2.996$$

therefore

$$t = 2.996\tau = 2.996 \times 5 \text{ s} = 14.98 \text{ s}$$

For a step change in temperature, the thermometer will approach within 5% of the final value after the time elapsed from the step change is greater than or equal to about 3τ. This can be generalised to any first order measuring system, i.e. after a step change at the input of the system, a time interval of at least 3τ must elapse before the output of the system is within 5% of the final value.

Exercise K

An experimenter has determined the time constant of a first order measuring system to be 2.5 s. If there is a step change at the input of this system at $t=0$, how long must the experimenter wait before the value indicated by the system is within 0.1% of the final value (assume random errors are negligible).

5.10 Combining uncertainties caused by systematic errors

An important difference between systematic and random errors is that, for random errors, repeat measurements assist in determining the underlying distribution of values. Repeating measurements does not help to determine the distribution of systematic errors. If the uncertainty due to systematic errors is estimated to be u_s, where u_s is effectively an upper limit for the uncertainty due to systematic errors, then we can assume that the true value, μ, we seek (in the absence of any random errors) lies between limits given by

$$\mu = x \pm u_s \qquad (5.39)$$

where x is the measured value.[25] It is possible that u_s has been chosen such that the probability of the true value lying between $x \pm u_s$ is 1. This will cause difficulties later when we wish to sum uncertainties due to random and systematic errors so we will be slightly conservative and assume that the probability of the true value lying between $x \pm u_s$ is 0.95.

As with random errors, systematic errors may or may not be independent. As an example suppose a micrometer is used to measure the dimensions of a rectangular block of metal. If the micrometer has an (uncorrected) offset error, then that error will affect the measurement of each dimension of the metal block in the same way, i.e. errors in each value obtained through measurement are dependent or correlated.

---

[25] Dietrich (1991) discusses the distribution of systematic errors in some detail.

If a measured quantity, y, depends on x and z, then we can write the uncertainty in y, u_{sy}, due to dependent systematic errors in x and z as

$$u_{sy} = \left| \frac{\partial y}{\partial x} \right| u_{sx} + \left| \frac{\partial y}{\partial z} \right| u_{sz} \tag{5.40}$$

where u_{sx} is the uncertainty in x due to systematic errors and u_{sz} is the uncertainty in z due to systematic errors.

In many situations systematic errors are likely to be independent. For example, consider an experiment to determine the resistivity of a material. The experiment requires that the electrical resistance and dimensions of a wire made from the material be measured. Specifically the resistivity, ρ, can be determined from

$$\rho = \frac{Rl}{A} \tag{5.41}$$

where R is the electrical resistance measured between the ends of the wire, l is the length of the wire and A is the cross-sectional area of the wire. As the instruments used to measure R, l and A differ there is unlikely to be any dependence between the systematic errors in these quantities. It follows that applying equation (5.40) is likely to overestimate the uncertainty in ρ, and adding uncertainties in the manner we considered when dealing with independent random errors in section 5.7.4 is more appropriate.

In general, if a measured quantity, y, depends on x and z, then we can write the systematic uncertainty in y, u_{sy}, when uncertainties in x and z are *independent* as

$$u_{sy} = \left[\left(\frac{\partial y}{\partial x} u_{sx} \right)^2 + \left(\frac{\partial y}{\partial z} u_{sz} \right)^2 \right]^{\frac{1}{2}} \tag{5.42}$$

where u_{sx} is the uncertainty in x due to systematic errors and u_{sz} is the uncertainty in z due to systematic errors.

5.11 Combining uncertainties due to random and systematic errors

If we write the uncertainty in a value due to random errors as u_r and the uncertainty in a value due to systematic errors as u_s, then, as it is unlikely that the errors are correlated, we can write the combined uncertainty, u, in a value due to both random and systematic errors as

$$u = \sqrt{u_r^2 + u_s^2} \tag{5.43}$$

Equation (5.43) can be applied on the condition that the confidence intervals associated with u_r and u_s are the same. As stated in section 5.10, u_s may be such that, in the absence of random error, the true value definitely lies between $\pm u_s$ of the mean. That is u_s really defines the 100% confidence interval. By committing the 'sin' of taking $x \pm u_s$ to be the 95% confidence interval, we are slightly overestimating the uncertainty due to systematic errors. If we were to insist on using the 100% confidence interval for the systematic errors, then we would have to do the same for the random errors. The difficulty with this is that if we use the normal distribution to describe the random errors, then the 100% confidence interval for the true value of a quantity lies between $\pm\infty$. To say that the true value of a quantity lies between $\pm\infty$ is not very discriminating.

Example 9

A $3\frac{1}{2}$ digit voltmeter is used to measure the output voltage of a transistor. The mean of the ten values of voltage is found to be 5.34 V with a standard deviation of 0.11 V. Given that the voltmeter is operating on the 20 V range and that the accuracy of the voltmeter is 0.5% of the reading + 1 figure, determine:

  (i)   the standard error of the mean;
  (ii) the 95% confidence interval for the true value of the mean, assuming systematic errors are negligible;
  (iii) the 95% confidence interval for the true value of the mean, assuming random errors are negligible;
  (iv) the 95% confidence interval for the true value of the mean, when uncertainties due to random and systematic errors are combined.

ANSWER

  (i) The standard error in the mean, $\sigma_{\bar{x}}$, is found using the equation $\sigma_{\bar{x}} = \sigma/\sqrt{n}$. In this example, $\sigma = 0.11$ V and $n = 10$, so $\sigma_{\bar{x}} = 0.11\,\text{V}/\sqrt{10} = 0.03479$ V.
  (ii) The 95% confidence interval for the true value of the mean, in the absence of systematic error is given by equation (5.8) as $\bar{x} \pm t_{X\%,\nu}\sigma_{\bar{x}}$, where $t_{X\%,\nu}$ is the t value for ν degrees of freedom and $X\%$ is the confidence level. Here $X\% = 95\%$ and $\nu = 9$. From table 2 in appendix 1, $t_{95\%,9} = 2.262$. It follows that the 95% confidence interval for the true voltage (in the absence of systematic error) is

$$(5.34 \pm 2.262 \times 0.03479)\,\text{V, i.e. } (5.340 \pm 0.079)\,\text{V.}$$

Writing the true value of the voltage as $\bar{x} \pm u_r$, we have $u_r = 0.079$ V.

(iii) The systematic uncertainty found from the instrument specification is 0.5% of the reading + one figure. 0.5% of 5.34 V is 0.027 V and the least significant figure for a $3\frac{1}{2}$ digit voltmeter on the 20 V range is 0.01 V (see table 5.12). The uncertainty due to systematic error, u_s, is therefore

$$u_s = 0.027\,\text{V} + 0.01\,\text{V} = 0.037\,\text{V}$$

It follows that the 95% confidence interval for the true voltage (in the absence of random error) is

$$(5.340 \pm 0.037)\,\text{V}$$

(iv) Now we use equation (5.43) to combine u_r and u_s:

$$u = \sqrt{u_r^2 + u_s^2} = \sqrt{(0.079\,\text{V})^2 + (0.037\,\text{V})^2} = 0.087\,\text{V}$$

Finally, we write the value of the voltage as $(5.340 \pm 0.087)\,\text{V}$.

Exercise L

Wooden metre rules can be used to measure lengths to a precision of better than 1 mm. However, prolonged use exposes the rules to a variety of atmospheric conditions (such as variations in temperature and humidity) which can cause the wood to shrink or expand. Such effects introduce a systematic error into length measurement, so that the uncertainty in a length measured as 1 m using the wooden rule can be as much as 1 mm.

In an optics experiment, a wooden metre rule was used to measure the position of an image produced by a lens. Eight repeat measurements of the image position were obtained and are shown in table 5.13. Use the information in the table to find:

(i) the mean image distance;
(ii) the estimate of the standard deviation of the values in table 5.13;
(iii) the standard error in the mean;
(iv) the 95% confidence interval for the true image distance in the absence of systematic errors;
(v) the 95% confidence interval for the true image distance in the absence of random errors. Assume that the uncertainty due to systematic error is 1 mm when a length of 1 m is measured with the rule;
(vi) the 95% confidence interval for the true image distance when random and systematic errors are accounted for.

Table 5.13. *Value of distance of image from a lens.*

Image distance (mm)	855.5	851.5	855.0	852.5	855.5	851.0	853.0	856.0

5.11.1 Type A and Type B categorisation of uncertainties

We have dealt with uncertainties due to random errors and systematic errors, eventually summing the contributions to the total uncertainty using equation (5.43). It is possible to categorise uncertainties another way. The 'Type A' category includes components of uncertainty that are evaluated using standard statistical techniques and the 'Type B' category includes uncertainties evaluated by other methods such as using tables of published data. While such an approach is arguably less intuitive than that based directly on the random/systematic categorisation of errors, the Type A/Type B categorisation is recognised internationally and is recommended by the International Committee for Weights and Measures. A description of the Type A/Type B method of categorisation can be found in the NIST technical note 1297, published by the National Institute of Standards and Technology, 1994 (Washington).

5.12 Weighted mean

When calculating the mean of values obtained through experiment or observation, the assumption is usually made that all values should be weighted equally. While this is reasonable in most circumstances, there are situations in which weighting should be considered. Weighting recognises that some values are known more precisely than others and so the calculation of the mean is modified to account for this. Specifically, those values that are known more precisely are weighted more heavily than those values that are known to lesser precision.

The standard deviation in each value appears explicitly in the expression for the weighted mean. The weighted mean of several values is written[26]

$$\bar{x}_w = \frac{\sum \dfrac{x_i}{\sigma_i^2}}{\sum \dfrac{1}{\sigma_i^2}} \qquad (5.44)$$

where x_i is the ith value and σ_i is the standard deviation of the population of which the value x_i is a member. Though equation (5.44) is useful, it is usual to combine several means, where each mean has a different standard error. Equation (5.44) is rewritten

---

[26] See appendix 3 for a derivation of equation (5.44).

$$\bar{x}_w = \frac{\sum \dfrac{\bar{x}_i}{\sigma_{\bar{x}_i}^2}}{\sum \dfrac{1}{\sigma_{\bar{x}_i}^2}} \tag{5.45}$$

where $\bar{x}_i$ is the ith mean and $\sigma_{\bar{x}_i}$ is the standard error of the ith mean.

Example 10

The thickness of a platinum film is measured using a profilometer (PF), a scanning electron microscope (SEM) and an atomic force microscope (AFM). Table 5.14 shows the mean of the values obtained by each method and standard error in the mean. Calculate the weighted mean thickness of the film.

ANSWER

Using equation (5.45), we have

$$\bar{x}_w = \frac{\sum \dfrac{\bar{x}_i}{\sigma_{\bar{x}_i}^2}}{\sum \dfrac{1}{\sigma_{\bar{x}_i}^2}} = \frac{\dfrac{325 \text{ nm}}{(7.5 \text{ nm})^2} + \dfrac{330 \text{ nm}}{(5.5 \text{ nm})^2} + \dfrac{329 \text{ nm}}{(4.0 \text{ nm})^2}}{\dfrac{1}{(7.5 \text{ nm})^2} + \dfrac{1}{(5.5 \text{ nm})^2} + \dfrac{1}{(4.0 \text{ nm})^2}} = \frac{37.25 \text{ nm}}{0.1133} = 328.7 \text{ nm}$$

Table 5.14. *Thickness of thin film of platinum determined using three techniques.*

Instrument	Mean thickness (nm)	Standard error in mean (nm)
PF	325.0	7.5
SEM	330.0	5.5
AFM	329.0	4.0

Exercise M

The mean time of a ball to fall a fixed distance under gravity is found by three experimenters to be

(1.25 ± 0.21) s (0.98 ± 0.15) s (1.32 ± 0.32) s

Determine the weighted mean of these times.

5.12.1 Standard error in the weighted mean

If the standard deviation in values varies, the standard error in the weighted mean must take this variation into account. In appendix 4 we show that the weighted standard error, $\sigma_{\bar{x}_w}$, may be expressed as

$$\sigma_{\bar{x}_w} = \left(\frac{1}{\sum \frac{1}{\sigma_i^2}} \right)^{\frac{1}{2}} \tag{5.46}$$

where σ_i is the standard deviation in the ith value. If means are combined, where each mean $\bar{x}_i$ has standard error $\sigma_{\bar{x}_i}$, then the standard error in the weighted mean is

$$\sigma_{\bar{x}_w} = \left(\frac{1}{\sum \frac{1}{\sigma_{\bar{x}_i}^2}} \right)^{\frac{1}{2}} \tag{5.47}$$

Example 11

Calculate the standard error of the weighted mean for the mean thicknesses appearing in table 5.14.

ANSWER

Using equation (5.47),

$$\sigma_{\bar{x}_w} = \left(\frac{1}{\sum \frac{1}{\sigma_{\bar{x}_i}^2}} \right)^{\frac{1}{2}} = \left[\frac{1}{\frac{1}{(7.5\ \mathrm{nm})^2} + \frac{1}{(5.5\ \mathrm{nm})^2} + \frac{1}{(4.0\ \mathrm{nm})^2}} \right]^{\frac{1}{2}} = [8.823(\mathrm{nm})^2]^{\frac{1}{2}}$$

$$= 3.0\ \mathrm{nm}$$

Exercise N

Calculate the standard error of the weighted mean for the time of fall of a ball using information given in exercise M.

5.12.2 Should means be combined?

Though we have considered how a weighted mean may be calculated, we should pause to ask whether it is reasonable to combine means if those means are considerably different. At the heart of the matter is the assumption, or hypothesis, that the mean of values obtained by whatever method is used would tend to the same true value as the number of measurements increases. Suppose, for example, the velocity of sound in air is determined. Method A gives the velocity as (338 ± 12) m/s and method B as (344.5 ± 6.2) m/s. As the difference between the two means is less than the standard error of the means it is reasonable to assert that the same true value underlies values obtained using both method A and method B. By contrast if the velocity of sound in air determined using method A is (338.2 ± 1.2) m/s and by method B the velocity is (344.5 ± 1.1) m/s, we should consider carefully before combining the means as the difference between the means is larger than the standard errors. There may be many reasons for inconsistency between the means, for example:

- The conditions in which each experiment was performed are different (for example the ambient temperature might have changed between measurements carried out by method A and method B). In this situation, the true value of the velocity of sound is not the same in each experiment.
- One or more instruments used in method A or method B may no longer be operating within specification, i.e. a large instrument based systematic error may account for the difference between values obtained.
- There could be systematic error introduced if hand timing were carried out by different experimenters as part of methods A and B.

In chapter 8 we consider more fully whether two means obtained from data are consistent with those data being drawn from the same population.

5.13 Review

It is hard to overstate the importance of careful measurement in science. Whether an experiment involves establishing the effectiveness of an anti-reflective coating on heat transfer into a car, the effect of temperature on the electrical conductive properties of a ceramic material or the determination of the energy output of a reaction occurring within a vessel, the 'values' that emerge from the experiment are subjected to extremely close scrutiny. Careful measurements can reveal an effect that has been overlooked by a less careful worker and lead to an important discovery. But no matter how carefully any experiment is performed, there is always some uncertainty in values established through measurement and so it is very important that we know how to determine, combine and interpret uncertainties. In this chapter we have considered measurements and some of the factors that affect the precision and accuracy of values obtained through experiment. In particular, we have seen that random and systematic errors introduce uncertainty into any measured value, how uncertainty can be quantified, and in some cases how it can be minimised. Our focus in the chapter has been on the factors that affect repeat measurements of a quantity when the conditions in which the measurements are made are essentially unchanged. An important area we consider next is the study of the relationship between physical variables, and how the relationship can be quantified.

Problems

1. The pH of a sample of river water is 5.8. If the uncertainty in the pH is 0.2 determine to two significant figures:

   (i) the fractional uncertainty in the pH;
  (ii) the percentage uncertainty in the pH.

2. Four repeat measurements are made of the height, h, to which a steel ball rebounds after striking a flat surface. The values obtained are shown in table 5.15. Use these values to determine:

   (i) the mean rebound height;
  (ii) the standard error in the rebound height;
 (iii) the 95% confidence interval for the true value of the rebound height.

Table 5.15. *Rebound heights for a steel ball.*

h (mm)	189	186	183	186

3. A profilometer is an instrument used to measure the thickness of thin films. Table 5.16 shows values of thickness of a film of aluminium on glass obtained using a profilometer. Using these values find:

   (i) the mean film thickness;
  (ii) the standard error in the mean thickness;
 (iii) the 99% confidence interval for the population mean thickness.

Table 5.16. *Thickness of aluminium film determined using a profilometer.*

Thickness (nm)	340	310	300	360	300	360

4. The relationship between the critical angle, θ_c, and the refractive index, n, for light travelling from glass into air is

$$n = \frac{1}{\sin \theta_c} \tag{5.48}$$

For a particular glass, $\theta_c = (43 \pm 1)°$. Use this information to determine n and the uncertainty in n.

5. The porosity, r, of a material can be determined using

$$r = \left(1 - \frac{\rho_s}{\rho_t}\right) \tag{5.49}$$

where ρ_s is the density of a specimen of the material as determined by experiment, and ρ_t is the theoretical density of the material. Given that, for a ceramic, $\rho_t = 6.35$ g/cm^3 and $\rho_s = (3.9 \pm 0.2)$ g/cm^3, determine r and the uncertainty in r.

6. X-rays incident on a thin film are no longer reflected when the angle of incidence exceeds a critical angle, θ_c, given by

$$\theta_c = K\lambda \sqrt{\rho} \tag{5.50}$$

where K is a constant for the film, λ is the wavelength of the incident X-rays and ρ is the density of the film. For a gallium arsenide film, $K = 4.44 \times 10^5$ Rad·m$^{\frac{1}{2}}$·kg$^{-\frac{1}{2}}$. Given that $\lambda = 1.5406 \times 10^{-10}$ m, determine θ_c and the uncertainty in θ_c, if $\rho = (6.52 \pm 0.24)$ g/cm^3.

7. The electrical resistance, R, of a wire at temperature, θ, is written

$$R = R_0 (1 + \alpha\theta + \beta\theta^2) \tag{5.51}$$

where α and β are constants with negligible uncertainty given by $\alpha = 3.91 \times 10^{-3}°C^{-1}$ and $\beta = 1.73 \times 10^{-7}°C^{-2}$. R_0 is the resistance at 0°C. Given that $R_0 = 100 \ \Omega$ and that it too has negligible uncertainty, calculate R and the uncertainty in R when $\theta = (65.2 \pm 1.5)°C$.

8. When a ball rebounds from a flat surface, the coefficient of restitution, c, for the collision is given by

$$c = \sqrt{\frac{h_2}{h_1}} \tag{5.52}$$

where h_1 is the initial height of the ball and h_2 is its rebound height. Given that

$$h_1 = (52.2 \pm 0.2) \text{ cm} \qquad h_2 = (23.5 \pm 0.5) \text{ cm}$$

determine c and the uncertainty in c assuming:

(i) errors in h_1 and h_2 are correlated;

(ii) errors in h_1 and h_2 are independent.

9. The rate of heat flow, H, across a material of thickness, l, can be written

$$H = -kA\left(\frac{t_2 - t_1}{l}\right) \qquad (5.53)$$

where k is the thermal conductivity of the material, A is its cross-sectional area and t_2 and t_1 are the temperatures at opposite surfaces of the material $(t_2 > t_1)$. Given that $k = 109$ W/(m·°C), $A = (0.0015 \pm 0.0001)$ m^2, $t_2 = (62.5 \pm 0.5)$ °C, $t_1 = (56.5 \pm 0.5)$ °C and $l = (0.15 \pm 0.01)$ m, determine H and the uncertainty in H, assuming the errors in the values of A, t_1, t_2 and l are independent.

10. The velocity of sound may be determined by measuring the length of a resonating column of air. In an experiment, the length, l, of the resonating column is measured ten times. The values obtained are shown in table 5.17.

(i) Determine the mean of the values in table 5.17.

(ii) Identify the value which lies furthest from the mean.

(iii) Use Chauvenet's criterion to decide whether the outlier identified in part (ii) should be removed.

(iv) If the outlier *is* removed, recalculate the mean.

Table 5.17. *Values for the length of a resonating column of air.*

l (cm)	49.2	48.6	47.8	48.5	42.7	47.7	49.0	48.8	48.3	47.7

11. A $3\frac{1}{2}$ digit voltmeter set on its 200 mV range is used to measure the output of a pressure transducer. Values of voltage obtained are shown in table 5.18.

(i) Assuming that random errors dominate, use the data in table 5.18 to determine the 95% confidence interval for the true voltage.

(ii) If the resolution and accuracy of the meter are given by table 5.12, determine the 95% confidence interval for the true voltage assuming that systematic errors dominate.

(iii) Combine the confidence intervals in parts (i) and (ii) of this question to give a 95% confidence interval which accounts for uncertainty due to both random and systematic errors.

Table 5.18. *Voltages from a pressure transducer determined using a 3½ digit voltmeter.*

Voltage (mV)	167.0	167.8	165.2	163.5	167.5

12. In the process of calibrating a 1 mL bulb pipette, the mass of water dispensed by the pipette was measured using an electronic balance. The process was repeated with the same pipette until ten values were obtained. These values are shown in table 5.19. Using the values in table 5.19 determine:

    (i) the mean mass of water dispensed;
    (ii) the standard error of the mean;
    (iii) the 95% confidence interval for the true value of the mass dispensed by the pipette.

Table 5.19. *Mass of water dispensed by a 1 mL bulb pipette.*

Mass (g)	0.9737	0.9752	0.9825	0.9569	0.9516	0.9617	0.9684	0.9585	0.9558	0.9718

13. The reducing agent sodium thiosulphate is added to an analyte during a titration. The amount of sodium thiosulphate required to bring the reaction to its end point for six repeat titrations is shown in table 5.20.

    (i) Calculate the sample mean and the estimate of the population standard deviation.
    (ii) Identify a possible outlier and apply Chauvenet's criterion to determine whether the outlier should be removed.
    (iii) If the outlier *is* removed, calculate the new mean and standard deviation.

Table 5.20. *Volume of sodium thiosulphate added to bring reaction to end point.*

Volume (mL)	33.2	33.4	33.4	33.9	33.3	33.1

14. The density of an aqueous solution is determined by two experimenters. Experimenter A obtains a density of 1.052 g/cm^3 for the solution with a standard error of 0.023 g/cm^3. Experimenter B obtains a density of 1.068 g/cm^3 for the solution with a standard error of 0.011 g/cm^3. Determine the weighted mean of these densities and the standard error in the weighted mean.

Chapter 6

Least squares I

6.1 Introduction

Establishing and understanding the relationship between quantities measured during an experiment are principal goals in the physical sciences. As examples, we might be keen to know how:

- the size of a crystal depends on the growth time of the crystal;
- the output intensity of a laser varies with the emission wavelength;
- the amount of light absorbed by a chemical species depends on the species concentration;
- the electrical power supplied by a solar cell varies with incident light intensity;
- the viscosity of an engine oil depends upon temperature;
- the hardness of paint varies with curing time.

Once an experiment is completed and the data presented in the form of an x–y graph, an inspection of the data often assists in answering important qualitative questions such as: Is there evidence of a clear trend in the data, if so is that trend linear, and do any of the data conflict with the general trend? Such qualitative considerations are essential and should precede any quantitative analysis.

There are many situations in the physical sciences in which prior knowledge or experience suggests a relationship between measured quantities. Perhaps we are already aware of an equation which predicts how one quantity depends on another. Our goal in this situation might be to discover how well the equation can be made to 'fit' the data.

As an example, in an experiment to study the variation of the viscosity of engine oil between room temperature and 100 °C, we observe that the viscosity of the oil decreases with increasing temperature, but we would like to know more:

- What is the quantitative relationship between viscosity and temperature? Does viscosity decrease linearly with increasing temperature, or in some other way?
- Can we find an equation that represents the variation of viscosity with temperature? Perhaps this would allow us to predict values of viscosity at any given temperature, i.e. permit interpolation between measured temperatures.
- Is it possible to relate the data to a theory of how viscosity depends on temperature?
- Can a useful parameter be estimated from viscosity–temperature data, such as the change in viscosity for a temperature rise of 1 °C?

A powerful and widely used method for establishing a quantitative relationship between quantities is that of *least squares*, also known as *regression*, and it is this method that we will focus upon in this chapter and chapter 7.

The method of least squares is computationally demanding, especially if there are many data to be considered. We will use Excel® to assist with least squares analysis at various points in this chapter after the basic principles have been considered.

6.2 The equation of a straight line

It is quite common for there to be a linear relationship between two quantities measured in an experiment. By fitting a straight line to the data, a quantitative expression may be found that relates the two quantities. What we would really like to do is find the equation that describes the *best* line that passes through (or at least close to) the data. We assume that the equation of the 'best line' is the closest we can come to finding the relationship between the quantities with the data available.

Figure 6.1 shows a straight line drawn on an x–y graph. The y value, y_i, of any point on the line is related to the corresponding x value, x_i, by the equation[1]

$$y_i = a + bx_i$$

[1] It is quite common to find the equation of a straight line written in other ways, such as $y_i = mx_i + c$, $y_i = mx_i + b$ or $y_i = b_0 + b_1 x_i$.

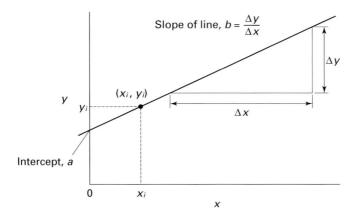

Figure 6.1. Straight line relationship between x and y.

where a is the intercept and b is the slope of the line. If all the x–y data gathered in an experiment were to lie along a straight line, there would be no difficulty in determining a and b and our discussion would end here. We would simply use a rule to draw a line through the points. Where the line intersects the y axis at $x = 0$ gives a. The slope, b, is found by dividing Δy by Δx, as indicated in figure 6.1.

In situations in which 'real' data are considered, even if the underlying relationship between x and y is linear, it is highly unlikely that all the points will lie on a straight line, since sources of error act to scatter the data. So how do we find the best line through the points?

6.2.1   The 'best' straight line through x–y data

When dealing with experimental data, we commonly plot the quantity that we are able to control on the x axis. This quantity is referred to as the *independent* (or the 'predictor') variable. A quantity that changes in response to changes in the independent variable is referred to as the *dependent* (or the 'response') variable, and is plotted on the y axis.

As an example, consider an experiment in which the velocity of sound in air is measured at different temperatures. Here temperature is the independent variable and velocity is the dependent variable. Table 6.1 shows temperature–velocity data for sound travelling through dry air. The data in table 6.1 are plotted in figure 6.2. Error bars are attached to each point to indicate the uncertainty in the values of velocity. From an inspection of figure 6.2, it appears reasonable to propose that there is a linear

Table 6.1. *Temperature–velocity data for sound travelling through dry air.*

Temperature, θ (°C)	Velocity, v (m/s) (± 5 m/s)
−13	322
0	335
9	337
20	346
33	352
50	365

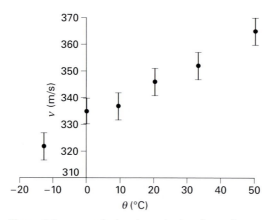

Figure 6.2. *x–y* graph showing velocity of sound versus temperature.

relationship between velocity, v, and temperature, θ. This relationship can be written as

$$v = A + B\theta \tag{6.1}$$

where A and B are constants.

Finding the best straight line through the points *means* finding the best values for A and B, as it is these two parameters that describe the relationship between v and θ. We *can* draw a line 'by eye' through the points using a transparent plastic rule and from that line estimate A and B. The difficulty with this approach is that, when data are scattered, it is difficult to find the position of the best line through the points.

Figure 6.3 shows two attempts at drawing a line through the velocity versus temperature data along with the equation that describes each line.

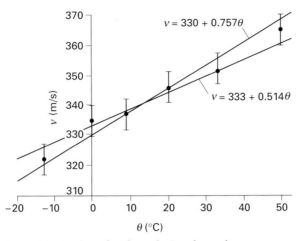

Figure 6.3. Two lines fitted to velocity of sound versus temperature data. The equation describing each line is shown.

Can either of the two lines be regarded as the best line through the points? If the answer is no, then how do we find the best line? The guesswork associated with drawing a line through data by eye can be eliminated by applying the technique of least squares.

6.2.2 Unweighted least squares

To find the best line through x–y data, we need to decide upon a numerical measure of the 'goodness of fit' of the line to the data. One approach is to take that measure to be the 'sum of squares of residuals', which we will discuss for the case where there is a linear relationship between x and y. The least squares method discussed in this section rests on the assumptions described in table 6.2.

Figure 6.4 shows a line drawn through the x–y data. The vertical distances from each point to the line are labelled Δy_1, Δy_2, Δy_3, etc. and are referred to as the *residuals* (or *deviations*). A residual is defined as the difference between the observed y value and the y value on the line for the same x value. Referring to the ith observed y value as y_i, and the ith predicted value found using the equation of the line as $\hat{y}_i$, the residual, Δy_i, is written

$$\Delta y_i = y_i - \hat{y}_i \tag{6.2}$$

Table 6.2. *Assumptions upon which the unweighted least squares method is based.*

Assumption	Comment
y values are influenced by random errors only.	Any measurement is affected by errors introduced by such sources as noise and instrument resolution (see chapter 5). Systematic errors cannot be accounted for using least squares.
y values measured at a particular value of x have a normal distribution.	If errors in measured values are normally distributed, then measured values will exhibit the characteristics of the normal distribution (see chapter 3).
The standard deviations of y values at all values of x are the same.	If this assumption is true then every point on an x–y graph is as reliable as any other and, in using the least squares method to fit a line to data, one point must not be 'weighted' more than any other point. This is referred to as *unweighted* least squares.
There are no errors in the x values.	This is the most troublesome assumption. In most circumstances the x quantity is controlled in an experiment. It is therefore likely to be known to higher precision than the y quantity. But this is not always true. We will consider one situation in which the assumption is not valid, i.e. where the errors in the x values are much larger than the errors in y values.

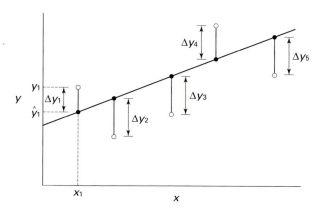

Figure 6.4. x–y graph showing residuals.

where

$$\hat{y}_i = a + bx_i \tag{6.3}$$

We propose that the underlying or 'true' relationship between x and y is given by

$$y = \alpha + \beta x \tag{6.4}$$

where α is the population intercept and β is the population slope. We can never know α and β exactly, but best estimates of α and β, written as a and b respectively, are obtained by applying the 'Principle of Maximum Likelihood' as discussed in appendix 3. The outcome of applying this principle is that the best line is given by values of a and b which minimise the *sum of the square of the residuals, SSR. SSR* is given by

$$SSR = \sum (y_i - \hat{y}_i)^2 \tag{6.5}$$

where $\hat{y}_i$ is given by equation (6.3).

In principle, values of a and b that minimise SSR could be found by 'trial and error', or by a systematic numerical search using a computer. However, when a straight line is fitted to data, an equation for the best line can be found analytically. If the assumptions in table 6.2 are valid, a and b are given by[2]

$$a = \frac{\sum x_i^2 \sum y_i - \sum x_i \sum x_i y_i}{n \sum x_i^2 - \left(\sum x_i\right)^2} \tag{6.6}$$

and

$$b = \frac{n \sum x_i y_i - \sum x_i \sum y_i}{n \sum x_i^2 - \left(\sum x_i\right)^2} \tag{6.7}$$

where n is the number of data points and each summation is carried out between $i = 1$ and $i = n$.

Example 1

Table 6.3 contains x–y data which are shown plotted in figure 6.5. Using these data:

  (i) Find the value for the intercept and slope of the best line through the points.
 (ii) Draw the line of best fit through the points.
(iii) Calculate the sum of squares of residuals, SSR.

---

[2] For derivations of equations (6.6) and (6.7) see section A3.2 in Appendix 3.

ANSWER

To calculate a and b we need the sums appearing in equations (6.6) and (6.7), namely Σx_i, Σy_i, and $\Sigma x_i y_i$, and Σx_i^2. Many pocket calculators are able to calculate these quantities[3] (in fact, some are able to perform unweighted least squares fitting to give a and b directly).

   We offer a word of caution here: As there are many steps in the calculations of a and b, it is advisable *not* to round numbers in the intermediate calculations,[4] as rounding can significantly influence the values of a and b.

(i)   Using the data in table 6.3 we find that $\Sigma x_i = 30$, $\Sigma y_i = 284$, $\Sigma x_i y_i = 1840$ and $\Sigma x_i^2 = 220$. Substituting these values into equations (6.6) and (6.7) (and noting that the number of points, $n = 5$) gives

$$a = \frac{220 \times 284 - 30 \times 1840}{5 \times 220 - (30)^2} = 36.4$$

$$b = \frac{5 \times 1840 - 30 \times 284}{5 \times 220 - (30)^2} = 3.4$$

(ii)   The line of best fit through the data in table 6.3 is shown in figure 6.6.
(iii)   The squares of residuals and their sum, SSR, are shown in table 6.4.

Table 6.3. *Linearly related x–y data.*

x	y
2	43
4	49
6	59
8	63
10	70

Exercise A

Use least squares to fit a straight line to the velocity versus temperature data in table 6.1. Calculate the intercept, a, the slope, b, of the line and the sum of squares of the residuals, SSR.

[3] If there are many data, a spreadsheet is preferred to a pocket calculator for calculating the sums.

[4] This problem is acute when a calculation involves subtracting two numbers that are almost equal, as can happen in the denominator of equations (6.6) and(6.7). Premature rounding can cause the denominator to tend to zero, resulting in very large (and very likely incorrect) values for a and b.

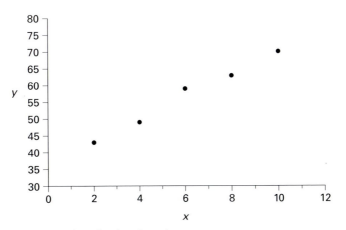

Figure 6.5. Linearly related x–y data.

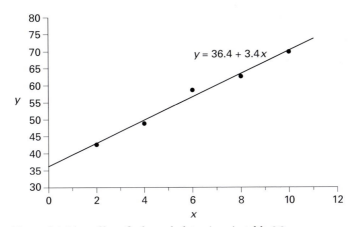

Figure 6.6. Line of best fit through data given in table 6.3.

Table 6.4. *Calculation of sum of squares of residuals.*

x_i	y_i	$\hat{y}_i = 36.4 + 3.4x_i$	$(y_i - \hat{y}_i)^2$
2	43	43.2	0.04
4	49	50.0	1.00
6	59	56.8	4.84
8	63	63.6	0.36
10	70	70.4	0.16
			$SSR = 6.4$

6.2.3 Trendline in Excel®

Excel® may be used to add the best straight line to an x–y graph using the Add Trendline option. Excel® uses equations (6.6) and (6.7) to determine the intercept and slope. Add Trendline can be found in the Chart option on the Menu bar. (Excel®'s Trendline is described in detail in section 2.7.1.)

Advantages of the Trendline option are:

(i) It requires that data be plotted first so that we are encouraged to consider whether it is *reasonable* to draw a straight line through the points.

(ii) The best line is added automatically to the graph.

(iii) The equation of the best line can be displayed if required.

(iv) Excel® is capable of drawing the best line through points for relationships between x and y other than linear, such as a logarithmic or power relationship.

(v) If data are changed then the points on the graph are updated and displayed immediately, as the graph and the line of best fit are linked 'dynamically' to the data.

Though excellent for viewing data and determining the best line through points, Trendline does not:

(i) Give access to the size of the standard errors in a and b. This information is necessary if we wish to quote confidence intervals for intercept and slope.

(ii) Allow 'weighted fitting'. This is required when there is evidence to suggest that some x–y values are more reliable than others. In this situation the best line should be 'forced' to pass close to the more reliable points. Weighted fitting is dealt with in section 6.10.

(iii) Plot residuals. Residuals are extremely helpful for assessing whether it is appropriate to fit a straight line to data in the first place. Residuals are dealt with in section 6.7.

Despite the usefulness of the Trendline option in Excel®, there are often situations in which we need to extract more from the data than a and b. In particular, the uncertainties in a and b, expressed as the standard errors in a and b, are very important as they indicate the number of significant figures to which to quote intercept and slope. We consider uncertainties in intercept and slope next.

6.2.4 Uncertainty in a and b

One of the basic assumptions made when fitting a line to data using least squares is that the dependent variable is subject to random error. It is reasonable to expect therefore that a and b are themselves influenced by the errors in the dependent variable. A preferred way of expressing the uncertainties in a and b is in terms of their respective standard errors, as this permits us to calculate confidence limits for the intercept and slope. Standard errors in a and b may be found using the ideas of propagation of uncertainties discussed in chapter 5. Provided the uncertainty in each y value is the same,[5] the standard errors in a and b are given by σ_a and σ_b, where[6]

$$\sigma_a = \frac{\sigma\left(\sum x_i^2\right)^{\frac{1}{2}}}{\left[n\sum x_i^2 - \left(\sum x_i\right)^2\right]^{\frac{1}{2}}} \tag{6.8}$$

$$\sigma_b = \frac{\sigma n^{\frac{1}{2}}}{\left[n\sum x_i^2 - \left(\sum x_i\right)^2\right]^{\frac{1}{2}}} \tag{6.9}$$

σ is the standard deviation of the observed y values about the fitted line. The calculation of σ is similar to the calculation of the estimate of the population standard deviation, s, of univariate data given by equation (1.16). σ is given by[7]

$$\sigma = \left[\frac{1}{n-2}\sum (y_i - \hat{y}_i)^2\right]^{\frac{1}{2}} \tag{6.10}$$

Example 2

The absorbance–concentration data shown in table 6.5 were obtained during an experiment in which standard silver solutions were analysed by flame atomic absorption spectrometry. Assuming that absorbance is linearly related to the concentration of the silver solution, use unweighted least squares to find the intercept and slope of the best line through the points and the standard errors in these quantities.

---

[5] If repeat measurements are made of y at a particular value of x, the standard deviation in the y values remains the same regardless of the value of x chosen.

[6] See appendix 4 for derivations of equations (6.8) and (6.9).

[7] Strictly, the right hand side of equation (6.10) is an *estimate* of σ, usually written as s, so we should write

$$\sigma \approx s = \left[\frac{1}{n-2}\sum (y_i - \hat{y}_i)^2\right]^{\frac{1}{2}}$$

Table 6.5. *Data obtained in a flame atomic absorption experiment.*

Concentration (ng/mL)	Absorbance
0	0.002
5	0.131
10	0.255
15	0.392
20	0.500
25	0.622
30	0.765

ANSWER

Regarding the concentration as the independent variable, x, and the absorbance as the dependent variable, y, we write $y = a + bx$

Using the data in table 6.5 we find, $\Sigma x_i = 105$ (ng/mL), $\Sigma y_i = 2.667$, $\Sigma x_i y_i = 57.585$ (ng/mL) and $\Sigma x_i^2 = 2275$ (ng/mL)^2. Using equations (6.6) and (6.7),

$$a = \frac{2275 \ (\text{ng/mL})^2 \times 2.667 - 105 (\text{ng/mL}) \times 57.585 (\text{ng/mL})}{7 \times 2275 \ (\text{ng/mL})^2 - (105 \ (\text{ng/mL}))^2} = 4.286 \times 10^{-3}$$

$$b = \frac{7 \times 57.585 \ (\text{ng/mL}) - 105 \ (\text{ng/mL}) \times 2.667}{7 \times 2275 \ (\text{ng/mL})^2 - (105 \ (\text{ng/mL}))^2} = 2.511 \times 10^{-2} \ \text{mL/ng}$$

In this example units have been included explicitly[8] in the calculation of a and b to emphasise that, in most situations in the physical sciences, we deal with quantities that have units and these units must be carried through to the 'final answers' for a and b.

In order to use equations (6.8) and (6.9), first calculate σ as given by equation (6.10). Table 6.6 has been constructed to assist in the calculation of σ. Summing the values in the last column of the table gives

$$\sum (y_i - \hat{y}_i)^2 = 3.2686 \times 10^{-4}$$

Using equation (6.10),

$$\sigma = \left[\frac{1}{(7-2)} \times 3.2686 \times 10^{-4} \right]^{\frac{1}{2}} = 0.008\,085$$

Now using equations (6.8) and (6.9) to find σ_a and σ_b:

[8] In other examples in this chapter, units do not appear (for the sake of brevity) in the intermediate calculations of a and b or σ_a and σ_b.

$$\sigma_a = \frac{\sigma\left(\sum x_i^2\right)^{\frac{1}{2}}}{\left[n\sum x_i^2 - \left(\sum x_i\right)^2\right]^{\frac{1}{2}}} = \frac{0.008085 \times (2275)^{\frac{1}{2}}}{[7 \times 2275 - (105)^2]^{\frac{1}{2}}} = 5.509 \times 10^{-3}$$

$$\sigma_b = \frac{\sigma n^{\frac{1}{2}}}{\left[n\sum x_i^2 - \left(\sum x_i\right)^2\right]^{\frac{1}{2}}} = \frac{0.008085 \times (7)^{\frac{1}{2}}}{[7 \times 2275 - (105)^2]^{\frac{1}{2}}} = 3.056 \times 10^{-4} \text{ mL/ng}$$

Combining a and b with their respective standard errors gives

$$\alpha = (4.3 \pm 5.5) \times 10^{-3} \text{ and } \beta = (2.511 \pm 0.031) \times 10^{-2} \text{ mL/ng}$$

Using the properties of the normal distribution,[9] we can say that the true value for the intercept has a probability of approximately 0.7 of lying between $(4.3 - 5.5) \times 10^{-3}$ and $(4.3 + 5.5) \times 10^{-3}$, i.e. the 70% confidence interval for α is between approximately -1.2×10^{-3} and 9.8×10^{-3}. Similarly, the 70% confidence interval for β is between 2.480×10^{-2} mL/ng and 2.542×10^{-2} mL/ng.

We must admit to a misdemeanour in applying the normal distribution here: In this example we are dealing with a small number of values (seven only), so we should use the t distribution rather than the normal distribution when calculating confidence intervals for α and β. We discuss this further in section 6.2.6.

Table 6.6. *Calculation of squares of residuals.*

x_i (ng/mL)	y_i	$\hat{y}_i = 4.286 \times 10^{-3} + 2.511 \times 10^{-2} x_i$	$(y_i - \hat{y}_i)^2$
0	0.002	0.004286	5.2245×10^{-6}
5	0.131	0.129857	1.3061×10^{-6}
10	0.255	0.255429	1.8367×10^{-7}
15	0.392	0.381000	1.2100×10^{-4}
20	0.500	0.506571	4.3184×10^{-5}
25	0.622	0.632143	1.0288×10^{-4}
30	0.765	0.757714	5.3082×10^{-5}

Exercise B

Calculate σ_a and σ_b for the data in table 6.3.

[9] See section 3.5.

6.2.5 Least squares, intermediate calculations and significant figures

In this text we adopt the convention that uncertainties are rounded to two significant figures. a and b are then presented to the number of figures consistent with the magnitude of the uncertainties. Where a pocket calculator or a spreadsheet has been used to determine a and b, all intermediate calculations are held to the full internal precision of the calculator, rounding only occurring in the presentation of the final parameter estimates.

6.2.6 Confidence intervals for α and β

We discovered in section 3.6 that, if univariate data are scattered normally, a confidence interval may be quoted for the population mean, μ, of those data. For example, for a 95% confidence interval, the true mean of the data has a probability of 0.95 of lying between specified limits.

We are able to determine the confidence interval for the population intercept, α, and that of the population slope, β, of a straight line fitted to data. As is usual, we use the 'rule of thumb' that whenever there are fewer than 30 data points, it is appropriate to use the t distribution rather than the normal distribution when quoting confidence intervals. The 95% confidence interval for α is written as

$$\alpha = a \pm t_{95\%,\nu}\sigma_a \tag{6.11}$$

where ν is the number of degrees of freedom. When fitting a straight line, the experimental data are used to calculate a and b. By using the data to estimate two parameters, the number of degrees of freedom is reduced by 2, so that ν is given by[10]

$$\nu = n - 2$$

where n is the number of data. Similarly, the 95% confidence interval for β is written

$$\beta = b \pm t_{95\%,\nu}\sigma_b \tag{6.12}$$

a and b are calculated using equations (6.6) and (6.7) respectively, σ_a and σ_b are calculated using equations (6.8) and (6.9). $t_{95\%,\nu}$ is the critical t value corresponding to the 95% confidence level.

When the $X\%$ confidence interval is required, $t_{95\%,\nu}$ is replaced in equations (6.11) and (6.12) by $t_{X\%,\nu}$. Table 2 in appendix 1 gives the values of t for various confidence levels, $X\%$, and degrees of freedom, ν.

---

[10] See Devore (1991) for more information on degrees of freedom.

Example 3

Using information given in example 2, calculate the 95% confidence interval for α and β.

ANSWER

This question requires we apply equations (6.11) and (6.12). The relevant information contained in example 2 (retaining extra figures to avoid rounding errors in the final answers) is as follows:

$$a = 4.286 \times 10^{-3}, \sigma_a = 5.509 \times 10^{-3}$$

$$b = 2.511 \times 10^{-2} \text{ mL/ng}, \sigma_b = 3.056 \times 10^{-4} \text{ mL/ng}$$

$$\nu = n - 2 = 7 - 2 = 5$$

We now use table 2 in appendix 1, to find $t_{95\%,5}$:

$$t_{95\%,5} = 2.571$$

From equation (6.11), the 95% confidence interval for α is

$$4.286 \times 10^{-3} \pm 2.571 \times 5.509 \times 10^{-3}$$

i.e.,

$$\alpha = (4 \pm 14) \times 10^{-3}$$

Using equation (6.12) we obtain

$$\beta = (2.511 \times 10^{-2} \pm 2.571 \times 3.056 \times 10^{-4}) \text{ mL/ng}$$

i.e.

$$\beta = (2.511 \pm 0.079) \times 10^{-2} \text{ mL/ng}$$

Exercise C

1. Calculate the 99% confidence intervals for α and β in example 2.
2. The data in table 6.7 were obtained in an experiment to study the variation of the electrical resistance, R, with temperature, θ, of a tungsten wire. Assuming the relationship between R and θ can be written $R = A + B\theta$, calculate:

   (i)   the values of the intercept, a, and slope, b, of the best line through the resistance–temperature data;
   (ii) the standard errors in a and b;
   (iii) the 95% confidence intervals for A and B.

Table 6.7. *Variation of resistance of a tungsten wire with temperature.*

θ (°C)	1	4	10	19	23	28	34	40	47	60	66	78	82
R (Ω)	10.2	10.3	10.7	11.0	11.2	11.4	11.8	12.2	12.5	12.8	13.2	13.5	13.6

6.3 Excel®'s LINEST() function

Calculating a and b and their standard errors using equations (6.6) to (6.10) is tedious, especially if there are many x–y values to consider. It is possible to use a spreadsheet to perform the calculations and this lessens the effort considerably. An even quicker method of calculating a and b is to use the LINEST() function in Excel®. This function estimates the parameters of the line of best fit and returns those estimates into an array of cells on the spreadsheet. The LINEST() function is versatile and we will consider it again in chapter 7. For the moment we use it to calculate a, b, σ_a and σ_b. The syntax of the function is

LINEST(y values, x values, constant, statistics)

y values:	Give range of cells containing the y values.
x values:	Give range of cells containing the x values.
constant:	Set to True to fit the equation $y = a + bx$ to the data. If we want to force the line through the origin (0,0) (i.e. fit the equation $y = bx$ to data) we set the constant to False. See problem 10 at the end of the chapter for a brief consideration of fitting the equation $y = bx$ to data.
statistics:	Set to True to calculate the standard errors[11] in a and b.

Example 4
Consider the x–y data shown in sheet 6.1. Use the LINEST() function to find the parameters a, b, σ_a and σ_b for the best line through these data.

Answer
Data are entered into columns A and B of the Excel® spreadsheet as shown in sheet 6.1. We require Excel® to return a, b, σ_a and σ_b. To do this:

---

[11] In fact, the LINEST() function is able to return other statistics, but we will focus on the standard errors for the moment.

1. Move the cursor to cell D3. With the left hand mouse button held down, pull down and across to cell E4. Release the mouse button. Values returned by the LINEST() function will appear in the four cells, D3 to E4.
2. Type =**LINEST(B2:B9,A2:A9,TRUE,TRUE).**
3. Hold down the Ctrl and Shift keys then press the Enter key.

Sheet 6.1. *x–y data for example 4.*

	A	B
1	x	y
2	1	-2.3
3	2	-8.3
4	3	-11.8
5	4	-15.7
6	5	-20.8
7	6	-25.3
8	7	-34.2
9	8	-37.2

Figure 6.7 shows part of the screen as it appears after the Enter key has been pressed. Labels have been added to the figure to identify a, b, σ_a and σ_b. The best values for intercept and slope and their respective standard errors may be written:

$$\alpha = (3.0 \pm 1.1) \qquad \beta = (-4.99 \pm 0.22)$$

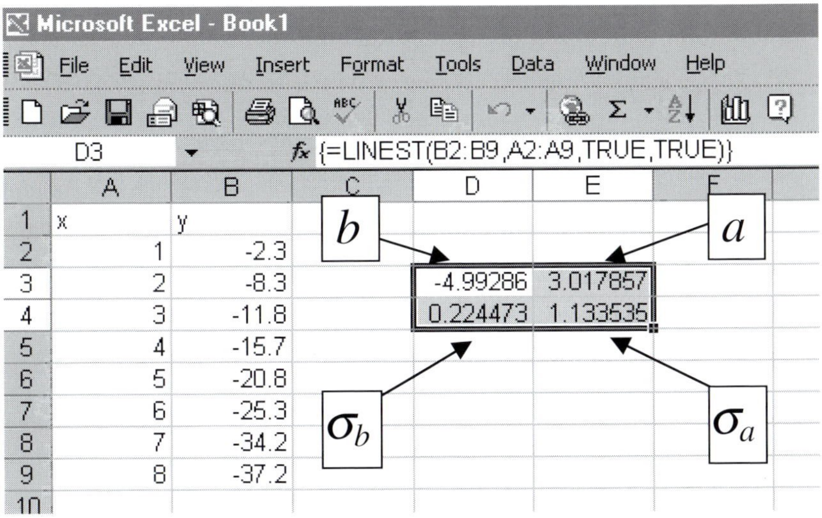

Figure 6.7. Screen shot of Excel® showing use of LINEST() function.

Exercise D

Consider the x–y data in table 6.8.

(i) Use the LINEST() function to determine the intercept and slope of the best line through the x–y data, and the standard error in intercept and slope.

(ii) Plot the data on an x–y graph and show the line of best fit.

Table 6.8. x–y data for exercise D.

x	0.1	0.2	0.3	0.4	0.5	0.6	0.7	0.8	0.9
y	3.7	7.7	8.5	12.1	13.5	15.5	15.3	15.0	18.7

6.4 Using the line of best fit

There are several reasons why a straight line might be fitted to data. These include:

(i) A visual inspection of data presented as an x–y graph appears to indicate that there is a linear relationship between the dependent and independent variables. A line through the points helps to confirm this and may reveal anomalies such as points deviating systematically from the line.

(ii) A model or theory predicts a linear relationship between the independent and the dependent variables. The best line through the points can provide strong evidence to support or refute the theory.[12]

(iii) The best line may be required for interpolation or extrapolation purposes. That is a value of y may be found for a particular value of x, x_0, where x_0 lies within the range of x values used in the determination of the line of best fit (interpolation) or outside the range (extrapolation). Occasionally, the equation of the best line is used to determine an x value for a given y value.

We consider the use of the line of best fit next, but will return to the important matter of comparing models and data in later sections of this chapter and again in chapter 7.

---

[12] In section 6.7 we will see how the scatter of residuals can provide convincing evidence for the appropriateness, or otherwise, of fitting a straight line to data.

6.4.1 Comparing a 'physical' equation to $y = a + bx$

In some situations, a and b derived from fitting a straight line to experimental data can be directly related to a physical constant or parameter. In fact, it may be that something is known of the 'theoretical' relationship between variables prior to performing the experiment and that the main purpose of the experiment is to establish the value of a particular physical constant appearing in an equation derived from theory.

Take, as an example, a thermal expansion experiment in which the length of a rod is measured at different temperatures. Studies on the expansion of solids indicate that the relationship between the length of the rod, l (in metres), and temperature, θ (in degrees Celsius), may be written as

$$l = l_0(1 + \alpha\theta) \tag{6.13}$$

where l_0 is the length of the rod at $0\,°C$, and α is the temperature coefficient of expansion. The right hand side of equation (6.13) can be expanded to give

$$l = l_0 + l_0\alpha\theta \tag{6.14}$$

Equation (6.14) is of the form $y = a + bx$, where $l = y$ and $\theta = x$. Comparing equation (6.14) to the equation of a straight line, we see that

$$a = l_0 \quad b = l_0\,\alpha$$

It follows that

$$\alpha = \frac{b}{a} \tag{6.15}$$

i.e. the ratio b/a gives the temperature coefficient of expansion of the material being studied. It would be usual to compare the value of α obtained through analysis of the length–temperature data with values reported by other experimenters who have studied the same, or similar materials.

Exercise E

The pressure, P, at the bottom of a water tank is related to the depth of water, h, in the tank by the equation

$$P = \rho gh + P_A \tag{6.16}$$

where P_A is the atmospheric pressure, g is the acceleration due to gravity and ρ is the density of the water.

In an experiment, P is measured as the depth of water, h, in the tank increases. Assuming equation (6.16) to be valid:

(i)   What would you choose to plot on each axis of an x–y graph in order to obtain a straight line?

(ii) How are the intercept and slope of that line related to ρ, g and P_A in equation (6.16)?

6.4.1.1   UNCERTAINTIES IN PARAMETERS WHICH ARE FUNCTIONS OF a AND b

The thermal expansion example discussed in the previous section brings up an important question: How may we establish the uncertainty in the temperature coefficient of expansion, α, given the uncertainties in a and b? α is a function of a and b, so at first sight it seems reasonable to apply the usual relationship for propagation of uncertainties (as derived in appendix 2) and write

$$\sigma_\alpha^2 = \left(\frac{\partial \alpha}{\partial a}\right)^2 \sigma_a^2 + \left(\frac{\partial \alpha}{\partial b}\right)^2 \sigma_b^2 \tag{6.17}$$

However, equation (6.17) is only valid if there is no correlation between the errors in a and b. It turns out that the errors in a and b *are* correlated[13] and so we must consider this matter in more detail.

The line of best fit (for an unweighted fit) always passes through the point $(\bar{x}, \bar{y})$, where[14]

$$\bar{x} = \frac{\sum x_i}{n} \qquad \bar{y} = \frac{\sum y_i}{n}$$

Using these relationships, we can write the intercept, a, as

$$a = \bar{y} - b\bar{x} \tag{6.18}$$

Any equation which is a function of a and b (such as equation (6.15)) can now be written as a function of $\bar{y}$ and b, by replacing a by $\bar{y} - b\bar{x}$. The advantage in this is that the errors in $\bar{y}$ and b are *not* correlated. This allows us to apply equation (5.26) to find the uncertainty in a parameter whose calculation involves both a and b.

---

[13] An estimated slope, b, that is slightly larger than the true slope will consistently coincide with an estimated intercept, a, that is slightly smaller than the true intercept, and vice versa. See Weisberg (1985) for a discussion of correlation between a and b.

[14] See appendix 3.

As an example, consider the thermal expansion problem described in section 6.4.1. The temperature coefficient of expansion, α, is given by equation (6.15). Replacing a in equation (6.15) by $\bar{y} - b\bar{x}$ gives

$$\alpha = \frac{b}{\bar{y} - b\bar{x}} \tag{6.19}$$

Expressing the uncertainty in α as the standard error, σ_α, we write

$$\sigma_\alpha^2 = \left(\frac{\partial \alpha}{\partial \bar{y}}\right)^2 \sigma_{\bar{y}}^2 + \left(\frac{\partial \alpha}{\partial b}\right)^2 \sigma_b^2 \tag{6.20}$$

After determining the partial derivatives in equation (6.20) and substituting for the variances in $\bar{y}$ and b we obtain[15]

$$\sigma_\alpha = \frac{\sigma}{a^2}\left[\frac{b^2}{n} + \frac{n\bar{y}^2}{n\sum x_i^2 - \left(\sum x_i\right)^2}\right]^{\frac{1}{2}} \tag{6.21}$$

where n is the number of data points, σ is given by equation (6.10), a is given by equation (6.6) and b is given by equation (6.7).

Exercise F

In an experiment to study thermal expansion, the length of an alumina rod is measured at various temperatures. Assume that the relationship between length and temperature is given by equation (6.13). Using the data in table 6.9:

(i) calculate the intercept and slope of the best straight line through the length–temperature data and the standard errors in the intercept and slope;

(ii) determine the temperature coefficient of expansion, α, for the alumina using equation (6.15);

(iii) calculate the standard error in α.

Table 6.9. *Length–temperature data for an alumina rod.*

T (°C)	100	200	300	400	500	600	700	800	900	1000
l (m)	1.2019	1.2018	1.2042	1.2053	1.2061	1.2064	1.2080	1.2078	1.2102	1.2122

6.4.2 Estimating *y* for a given *x*

Once the intercept and slope of the best line through the points have been determined, it is an easy matter to find the predicted value of y, $\hat{y}_0$, at an arbitrary x value, x_0, using the relationship

---

[15] $\sigma_{\bar{y}} = \sigma/\sqrt{n}$ and σ_b is given by equation (6.9).

$$\hat{y}_0 = a + bx_0 \tag{6.22}$$

$\hat{y}_0$ is the best estimate of the population mean of the y quantity at $x=x_0$. The population mean of y at $x=x_0$ is sometimes written $\mu_{y|x_0}$. Just as the uncertainties in the measured y values contribute to the uncertainties in a and b, so the uncertainties in a and b contribute to the uncertainty in $\hat{y}_0$. As in section 6.4.1.1, we avoid the problem of correlation of errors in a and b by replacing a by $\bar{y} - b\bar{x}$, so that equation (6.22) becomes

$$\hat{y}_0 = \bar{y} + b(x_0 - \bar{x}) \tag{6.23}$$

As the errors in $\bar{y}$ and b are independent, the standard error in $\hat{y}_0$, written as $\sigma_{\hat{y}_0}$, is given by

$$\sigma_{\hat{y}_0} = \sigma \left(\frac{1}{n} + \frac{n(x_0 - \bar{x})^2}{n\sum x_i^2 - \left(\sum x_i\right)^2} \right)^{\frac{1}{2}} \tag{6.24}$$

where σ is given by equation (6.10).

We conclude that $\sigma_{\hat{y}_0}$ and hence any confidence interval for $\mu_{y|x_0}$ depends upon the value of x at which the estimate of $\mu_{y|x_0}$ (i.e. $\hat{y}_0$) is calculated. Note that the closer x_0 is to $\bar{x}$, the smaller is the second term inside the brackets of equation (6.24).

In general, the $X\%$ confidence interval for $\mu_{y|x_0}$ may be written as

$$\hat{y}_0 \pm t_{X\%,\nu}\sigma_{\hat{y}_0} \tag{6.25}$$

where $t_{X\%,\nu}$ is the critical t value corresponding to the $X\%$ confidence level evaluated with ν degrees of freedom.[16]

Example 5

Consider the data in table 6.10.

(i) Assuming a linear relationship between the x–y data determine:

   (a) intercept and slope, a and b, of the best line through the points;
   (b) $\hat{y}_0$ for $x_0 = 12$ and $x_0 = 22.5$.
   (c) the standard error in $\hat{y}_0$ when $x_0 = 12$ and $x_0 = 22.5$.

(ii) Plot the data on an x–y graph showing the line of best fit and the 95% confidence limits for $\mu_{y|x_0}$ for values of x_0 between 0 and 45.

---

[16] $t_{X\%,\nu}$ may be found using table 2 in appendix 1 or the TINV() function in Excel®, as explained in section 3.9.1.

Table 6.10. *x–y data for example 5.*

x	5	10	15	20	25	30	35	40
y	28.1	18.6	−0.5	−7.7	−14.8	−27.7	−48.5	−62.9

ANSWER

(i) (a) a and b are found using equations (6.6) and (6.7). We find $a=42.43$ and $b=-2.527$, so that the equation for $\hat{y}_0$ can be written

$$\hat{y}_0 = 42.43 - 2.527x_0$$

(b) When $x_0 = \sigma_{\hat{y}_0} 12$, $\hat{y}_0 = 12.11$. When $x_0 = 22.5$, $\hat{y}_0 = -14.43$.

(c) Using the data in table 6.10:

$$\bar{x} = 22.5 \quad \sum x_i^2 = 5100 \quad \sum x_i = 180 \quad \sigma = 4.513$$

(σ is calculated using equation (6.10)). Substituting values into equation (6.24) gives for $x_0 = 12$, $\sigma_{\hat{y}_0} = 2.2$. When $x_0 = 22.5$, $\sigma_{\hat{y}_0} = 1.6$.

(ii) When the number of degrees of freedom equals 6, the critical t value for the 95% confidence interval, given by table 2 in appendix 1, is $t_{95\%,6} = 2.447$. Equation (6.25) is used to find lines which represent the 95% confidence intervals for $\mu_{y|x_0}$ for values of x_0 between 0 and 45. These are indicated on the graph in figure 6.8.

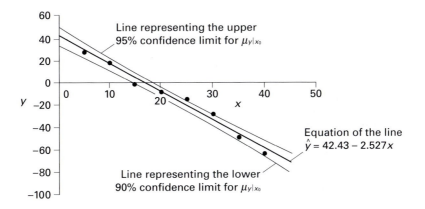

Figure 6.8. Line of best fit and 95% confidence limits for data in table 6.10.

Exercise G

For the *x–y* data in example 5, calculate the 99% confidence interval for $\mu_{y|x_0}$ when $x_0 = 15$.

6.4.2.1 UNCERTAINTY IN PREDICTION OF y AT A PARTICULAR VALUE OF x

Let us assume that we have fitted the best straight line to a set of x–y data. If we make a measurement of y at $x = x_0$, between what limits would we expect the measured value of y to lie? This is different from considering confidence limits associated with the estimate of the population mean at $x = x_0$ because two factors must be taken into consideration:

   (i) the uncertainty in the line of best fit to the data;
  (ii) the uncertainty in the measurement of y made at $x = x_0$.

Using addition of independent uncertainties as discussed in section 5.7.3, we combine the uncertainty in the measurement of y with the uncertainty in the line of best fit to give

$$\sigma_{P\hat{y}_0} = \sigma \left[1 + \frac{1}{n} + \frac{n(x_0 - \bar{x})^2}{n \sum x_i^2 - \left(\sum x_i \right)^2} \right]^{\frac{1}{2}} \tag{6.26}$$

where $\sigma_{P\hat{y}_0}$ represents that standard error in the predicted value of y at $x = x_0$. The $X\%$ prediction interval for y at $x = x_0$ is written

$$\hat{y}_0 \pm t_{X\%,\,\nu}\sigma_{P\hat{y}_0} \tag{6.27}$$

$\hat{y}_0$ is the best estimate of the predicted value. $\hat{y}_0$ is the same as the best estimate of the population mean at $x = x_0$ and is given by equation (6.22). $t_{X\%,\,\nu}$ is the critical t value corresponding to the $X\%$ confidence level evaluated with ν degrees of freedom.

Equation (6.26) is very similar to equation (6.24). However, the inclusion of the unity term within the brackets of equation (6.26) leads to a prediction interval for a y value at $x = x_0$ which is much larger than the confidence interval of the population mean $\mu_{y|x_0}$ at $x = x_0$.

Exercise H

Using the information supplied in example 5, calculate the 95% prediction interval for y if a measurement of y is to be made at $x_0 = 12$.

6.4.3 Estimating x for a given y

Using the best straight line through points, we can estimate a value of x for any particular value of y. The equation of the best straight line through points is rearranged to give

$$\hat{x}_0 = \frac{\bar{y}_0 - a}{b} \tag{6.28}$$

where $\hat{x}_0$ is the value of x when $y=\bar{y}_0$ and $\bar{y}_0$ is the mean of repeated measurements of the dependent variable. The question arises: as there is uncertainty in a, b and $\bar{y}_0$, what will be the uncertainty in $\hat{x}_0$? As discussed in section 6.4.1.1, the uncertainties in a and b are correlated. Replacing a in equation (6.28) by $\bar{y} - b\bar{x}$, we have

$$\hat{x}_0 = \bar{x} + \frac{\bar{y}_0 - \bar{y}}{b} \tag{6.29}$$

Assuming the uncertainties in $\bar{y}_0$, $\bar{y}$ and b to be uncorrelated (and that there is no uncertainty in $\bar{x}$) we write

$$\sigma_{\hat{x}_0}^2 = \left(\frac{\partial \hat{x}_0}{\partial \bar{y}_0}\right)^2 \sigma_{\bar{y}_0}^2 + \left(\frac{\partial \hat{x}_0}{\partial \bar{y}}\right)^2 \sigma_{\bar{y}}^2 + \left(\frac{\partial \hat{x}_0}{\partial b}\right)^2 \sigma_b^2 \tag{6.30}$$

where $\sigma_{\bar{y}_0} = \sigma/\sqrt{m}$ (m is the number of repeat measurements made of y and σ is given by equation 6.10)), $\sigma_{\bar{y}} = \sigma/\sqrt{n}$ and σ_b is given by equation (6.9).

Calculating the partial derivatives in equation (6.30) and substituting the expressions for the variances in $\bar{y}_0$, $\bar{y}$ and b gives

$$\sigma_{\hat{x}_0} = \frac{\sigma}{b}\left\{\frac{1}{m} + \frac{1}{n} + \frac{n(\bar{y}_0 - \bar{y})^2}{b^2\left[n\sum x_i^2 - \left(\sum x_i\right)^2\right]}\right\}^{\frac{1}{2}} \tag{6.31}$$

The $X\%$ confidence interval for x at $y=\bar{y}_0$ can be written

$$\hat{x}_0 \pm t_{X\%, \nu}\sigma_{\hat{x}_0} \tag{6.32}$$

where ν is the number of degrees of freedom.

Example 6

A spectrophotometer is used to measure the concentration of arsenic in solution. Table 6.11 shows calibration data of the variation of absorbance[17] with arsenic concentration. Assuming that the absorbance is linearly related to the arsenic concentration:

(i) Plot a calibration graph of absorbance versus concentration.
(ii) Find the intercept and the slope of the best straight line through the data.

Three values of absorbance are obtained from repeat measurements on a sample of unknown arsenic concentration. The mean of these values is found to be 0.3520.

(iii) Calculate the concentration of arsenic corresponding to this absorbance and the standard error in the concentration.

[17] Absorbance is proportional to the amount of light absorbed by the solution as the light passes from source to detector within the spectrophotometer.

Table 6.11. *Absorbance as a function of arsenic concentration.*

Concentration (ppm)	Absorbance
2.151	0.0660
9.561	0.2108
16.878	0.3917
23.476	0.5441
30.337	0.6795

ANSWER

(i) Figure 6.9 shows a plot of the variation of absorbance versus concentration data contained in table 6.11.

(ii) a and b are determined using equations (6.6) and (6.7):

$$a = 0.01258, \ b = 0.02220 \text{ ppm}^{-1}$$

(iii) Using the relationship, $\hat{x}_0 = (\bar{y}_0 - a)/b$

$$\hat{x}_0 = \left(\frac{0.3520 - 0.01258}{0.02220} \right) \text{ppm} = 15.29 \text{ ppm}$$

We use equation (6.31) to obtain the standard error in $\hat{x}_0$. Using the information in the question and the data in table 6.11 we find

$$n=5 \quad m=3 \quad \bar{y}_0 = 0.3520 \quad \bar{y} = 0.37842 \quad \sum x_i^2 = 1852.363$$
$$\sum x_i = 82.403 \quad \sigma = 0.01153$$

(σ is calculated using equation (6.10)). Substituting these values into equation (6.31) gives

$$\sigma_{\hat{x}_0} = 0.38 \text{ ppm}$$

It is worth remarking that the third term in the brackets of equation (6.31) becomes large for y values far from the mean of the y values obtained during the calibration procedure. The third term is zero when the y value of the sample under test is equal to the mean of the y values obtained during calibration.

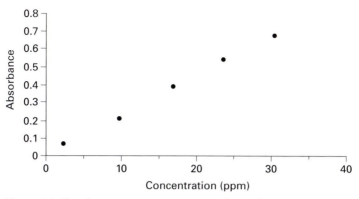

Figure 6.9. Absorbance versus concentration for a solution containing arsenic.

Exercise I

The data shown in table 6.12 were obtained in an experiment to determine the amount of nitrite in solution using high performance liquid chromatography (HPLC).

(i) Regarding the peak area as the dependent variable, determine the equation of the best straight line through the data.

(ii) Four repeat measurements are made on a solution with unknown nitrite concentration. The mean peak area is found to be 57156. Use the equation of the best line to find the concentration corresponding to this peak area and the standard error in the concentration.

Table 6.12. *Variation of peak area with nitrite concentration.*

Concentration (ppm)	2.046	3.068	5.114	7.160	10.23
Peak area (arbitrary units)	37752	47658	75847	105499	154750

6.5 Fitting a straight line to data when random errors are confined to the x quantity

One of the assumptions we rely upon consistently in least squares analysis is that errors in x–y data are confined to the y quantity. The validity of this assumption must be considered on a case by case basis, but it is reasonable to argue that all measurements have some error so that there will be some error in the x values. It is possible to derive equations for slope, intercept and the uncertainties in these quantities in situations in which there are uncertainties in both the x and the y values.[18] If the errors in the x values are constant and errors in the y values are negligible, we can use the results already discussed in this chapter to find the best line through the data. We write the equation of the line through the data as

$$x = a* + b*y \qquad\qquad (6.33)$$

where x is regarded as the dependent variable and y as the independent variable, $a*$ is the intercept (i.e. the value of x when $y=0$) and $b*$ is the slope of the line. To find $a*$ and $b*$ we must minimise the sum of squares of the residuals of the observed values of x from the predicted values based on a line drawn through the points. In essence we recreate the argument begun in section 6.2.2, but with y replacing x and x replacing y. The equation for the best line through the x–y data in this case is given when the intercept, $a*$, is

$$a* = \frac{\sum y_i^2 \sum x_i - \sum y_i \sum x_i y_i}{n \sum y_i^2 - \left(\sum y_i\right)^2} \qquad\qquad (6.34)$$

and the slope, $b*$, is

$$b* = \frac{n \sum x_i y_i - \sum x_i \sum y_i}{n \sum y_i^2 - \left(\sum y_i\right)^2} \qquad\qquad (6.35)$$

Compare these equations with equations (6.6) and (6.7) for a and b when the sum of squares of residuals in the y values is minimised.

Equation (6.33) can be rewritten as

$$y = \frac{-a*}{b*} + \frac{x}{b*} \qquad\qquad (6.36)$$

It is tempting to compare equation (6.36) with $y = a + bx$ and reach the conclusion that

$$a = \frac{-a*}{b*} \qquad\qquad (6.37)$$

and

$$b = \frac{1}{b*} \qquad\qquad (6.38)$$

However, a and b given by equations (6.37) and (6.38) are equal to a and b given by equations (6.6) and (6.7) only if both least squares fits (i.e. that which minimises the sum of the squares of the x residuals and that which minimises the sum of the squares of the y residuals) produce the same

---

[18] This is beyond the scope of this text. For a good review of least squares methods when both x and y variables are affected by error, see Macdonald and Thompson (1992).

Table 6.13. *x–y data.*

x	y
2.52	2
3.45	4
3.46	6
4.25	8
4.71	10
5.47	12
6.61	14

straight line through the points. The only situation in which this happens is when there are *no* errors in the x and y values, i.e. all the data lie exactly on a straight line!

As example, consider the $x–y$ data in table 6.13. We can perform a least squares analysis assuming that:

(i) The errors are in the x values only. Using equations (6.34) and (6.35) we find $a^* = 1.844$ and $b^* = 0.3136$. Using equations (6.37) and (6.38) we obtain $a = -5.882$ and $b = 3.189$.

(ii) The errors are in the y values only. Using equations (6.6) and (6.7) we find $a = -5.348$ and $b = 3.067$.

As anticipated, the parameter estimates, a and b, depend on whether minimisation occurs in the sum of the squares of the x residuals or the sum of the squares of the y residuals.

Exercise J

With the current through a silicon diode held constant, the voltage across the diode, V, is measured as a function of temperature, θ. Data from the experiment are shown in table 6.14. Theory suggests that the relationship between V and θ is

$$V = k_0 + k_1 \theta \qquad\qquad (6.39)$$

As V is the dependent variable and θ the independent variable, it would be usual to plot V on the y axis and θ on the x axis. However, the experimenter has evidence that the measured values of diode voltage have less error than those of temperature.

(i) Use least squares to obtain best estimates for k_0 and k_1, where only values of V are assumed to have error.

(ii) Use least squares to obtain best estimates for k_0 and k_1, where only values of θ are assumed to have error.

Table 6.14. *Voltage across a diode, V, as a function of temperature, θ.*

θ (°C)	V(V)
2.0	0.6859
10.0	0.6619
19.0	0.6379
26.4	0.6139
40.9	0.5899
48.8	0.5659
59.7	0.5419
65.0	0.5179
80.0	0.4939
91.0	0.4699
101.3	0.4459

6.6 Linear correlation coefficient, *r*

When the influence of random errors on values is slight, it is usually easy to establish 'by eye' whether x and y quantities are related. However, when data are scattered it is often quite difficult to be sure of the extent to which y is dependent on x. It is sometimes useful to define a parameter directly related to the extent of the correlation between x and y. That parameter is called the linear correlation coefficient, ρ. The estimate of this parameter, obtained from data, is represented by the symbol r.

When all points lie on the same straight line, i.e. there is perfect correlation between x and y, then equations (6.37) and (6.38) correctly relate a and b (which are determined assuming errors are confined to the y quantity) with a^* and b^* (which are determined assuming errors are confined to the x quantity). Focussing on equation (6.38) we can say that, for perfect correlation,

$$bb^* = 1$$

When there is no correlation between x and y, there is no tendency for y to increase with increasing x nor x to increase with increasing y. Put another way, with no correlation we would expect both b and b^* to be close to zero, so that $bb^* \approx 0$. This suggests that it may be useful to use the product bb^* as a measure of the correlation between x and y. We define the linear correlation coefficient, r, as

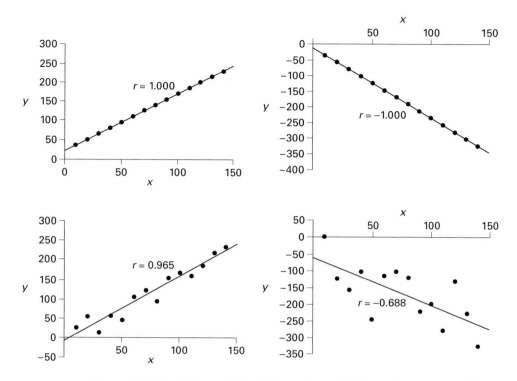

Figure 6.10. Correlation coefficients for x–y data exhibiting various amounts of scatter.

$$r = \sqrt{bb^*} \qquad\qquad (6.40)$$

Substituting for b and b^* using equations (6.7) and (6.35) respectively, we get

$$r = \frac{n\sum x_i y_i - \sum x_i \sum y_i}{\left[n\sum x_i^2 - \left(\sum x_i\right)^2\right]^{\frac{1}{2}}\left[n\sum y_i^2 - \left(\sum y_i\right)^2\right]^{\frac{1}{2}}} \qquad (6.41)$$

For perfect correlation, r is either $+1$ or -1. We note that r has the same sign as that of the slope (b or b^*). Figure 6.10 shows graphs of x–y data along with the value of r for each. As $|r|$ decreases from 1 to 0, the correlation between x and y becomes less and less convincing. Notions of 'goodness' relating to values of r can be misleading. A value of $|r|$ close to unity does indicate good correlation; however, it is possible that x and y are not *linearly* related but still give a value for $|r|$ in excess of 0.99. This is illustrated by the next example.

Example 7

Thermoelectric coolers (TECs) are devices widely used to cool electronic components, such as laser diodes. A TEC consists of a hot and a cold surface with the temperature difference between the surfaces maintained by an electric current. In an experiment, the temperature difference between the hot and the cold surface, ΔT, is measured as a function of the electrical current, I, passing through the TEC. The data gathered are shown in table 6.15.

(i) Calculate the value of the correlation coefficient, r.

(ii) Plot a graph of ΔT versus I and show the line of best fit through the points.

ANSWER

(i) We begin by drawing up table 6.16 which contains all the values needed to calculate r using equation (6.41). Summing the values in each column gives $\Sigma x_i = 4.2$, $\Sigma y_i = 113.2$, $\Sigma x_i y_i = 93.02$, $\Sigma x_i^2 = 3.64$, $\Sigma y_i^2 = 2402.22$. Substituting the summations into equation (6.41) gives

$$r = \frac{175.7}{2.8 \times 63.2558} = 0.992$$

A fit of the equation $y = a + bx$ to the x–y data given in table 6.16 gives intercept $a = 2.725\,°C$ and slope $b = 22.41\,°C/A$.

(ii) Figure 6.11 shows the data points and the line of best fit.

Table 6.15. *Temperature difference, ΔT, of a TEC as a function of current, I.*

I (A)	ΔT (°C)
0.0	0.8
0.2	7.9
0.4	12.5
0.6	17.1
0.8	21.7
1.0	25.1
1.2	28.1

Table 6.16. *Values needed to calculate r using equation (6.41).*

$x_i\,(=I)$	$y_i\,(=\Delta T)$	x_iy_i	x_i^2	y_i^2
0.0	0.8	0.0	0.0	0.64
0.2	7.9	1.58	0.04	62.41
0.4	12.5	5.0	0.16	156.25
0.6	17.1	10.26	0.36	292.41
0.8	21.7	17.36	0.64	470.89
1.0	25.1	25.1	1.0	630.01
1.2	28.1	33.72	1.44	789.61

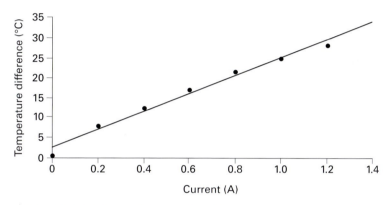

Figure 6.11. Temperature difference for a TEC versus current.

A value of $r = 0.992$ indicates a high degree of correlation between x and y. That x and y are correlated is revealed by the graph, but we might be overlooking something more important. Is the relationship between temperature difference and current *really* linear? We assumed that to be the case when calculating the line of best fit, though a close inspection of the graph above seems to indicate that a curve through the points is more appropriate than a straight line. This is a point worth emphasising: a correlation coefficient with a magnitude close to unity is no guarantee that the x–y data are linearly related. We need to inspect an x–y graph of the raw data, the line of best fit and preferably a plot of the residuals (dealt with in section 6.7) in order to be satisfied that the assumption of linearity between x and y is reasonable.

Exercise K

The temperature reached by the cold surface of a TEC cooler depends on the size of the heat sink to which it is attached. Table 6.17 shows the temperature, T, of the cold surface of the TEC for heat sinks of various volumes, V.

(i)   Calculate the linear correlation coefficient for these data.
(ii) Find the equation of the line of best fit through the data assuming that a linear relationship is appropriate. Take T to be the dependent variable and V to be the independent variable.
(iii) Plot a graph of temperature versus volume and indicate the line of best fit.
(iv) Do you think that the assumption of linearity between T and V is valid?

Table 6.17. *Variation of temperature of the cold surface of TEC with size of heat sink.*

Volume, V (cm³)	Temperature, T (°C).
5	37.0
10	25.5
15	17.1
25	11.5
50	6.4

6.6.1 Calculating *r* using Excel®

The CORREL() function in Excel® calculates the correlation coefficient, r, and returns the value into a cell. The syntax of the function is

CORREL(*y* values, *x* values)

Consider the calculation of r for the data shown in sheet 6.2. To calculate r:

Sheet 6.2. *x–y data.*

	A	B
1	x	y
2	42	458
3	56	420
4	56	390
5	78	380
6	69	379
7	92	360
8	102	351
9	120	300
10		

(i) enter the data shown in sheet 6.2 into an Excel® spreadsheet;

(ii) type =**CORREL(B2:B9,A2:A9)** into cell B10;

(iii) press the Enter key;

(iv) the value –0.9503 is returned in cell B10.

Exercise L

Use the CORREL() function to determine the correlation coefficient of the x–y data shown in table 6.18.

Table 6.18. *x–y data for exercise L.*

x	2	4	6	8	10	12	14	16	18	20	22	24
y	16.5	49.1	65.2	71.6	101.5	90.1	101.4	113.7	127.7	156.5	203.6	188.4

6.6.2   Is the value of r significant ?

We have seen that we need to be cautious when using r to infer the extent to which there is a linear relationship between x–y data, as values of $|r| > 0.8$ may be obtained when the underlying relationship between x and y is not linear. There is another problem: values of $|r| > 0.8$ may be obtained when x and y are totally uncorrelated, especially when the number of x–y values is small. To illustrate this, consider the values in table 6.19.

The first column of table 6.19 contains five values of x from 0.2 to 1. The remainder of the columns contain numbers between 0 and 1 that have been randomly generated so that there is *no underlying correlation between x and y*. The bottom row shows the correlation coefficients calculated when each column of y is correlated in turn with the column containing x. The column of values headed $y8$, when correlated with the column

Table 6.19. *Correlation coefficient for ten sets of randomly generated y values correlated with the x column values.*

x	$y1$	$y2$	$y3$	$y4$	$y5$	$y6$	$y7$	$y8$	$y9$	$y10$
0.2	0.020	0.953	0.508	0.324	0.233	0.872	0.446	0.673	0.912	0.602
0.4	0.965	0.995	0.231	0.501	0.265	0.186	0.790	0.911	0.491	0.186
0.6	0.294	0.159	0.636	0.186	0.227	0.944	0.291	0.153	0.780	0.832
0.8	0.561	0.096	0.905	0.548	0.187	0.002	0.331	0.051	0.862	0.255
1.0	0.680	0.936	0.783	0.034	0.860	0.363	0.745	0.083	0.239	0.363
r	0.400	-0.323	0.743	-0.392	0.655	-0.455	0.094	-0.822	-0.541	-0.242

Table 6.20. *Probabilities of obtaining calculated r values when the x–y data are uncorrelated.*

	|r| (calculated from x–y data)					
n	0.5	0.6	0.7	0.8	0.9	0.95	1.0
3	0.667	0.590	0.506	0.410	0.287	0.202	0
4	0.500	0.400	0.300	0.200	0.100	0.050	0
5	0.391	0.285	0.188	0.104	0.037	0.013	0
6	0.313	0.208	0.122	0.056	0.014	0.004	0
7	0.253	0.154	0.080	0.031	0.006	0.001	0
8	0.207	0.116	0.053	0.017	0.002	<0.001	0
9	0.170	0.088	0.036	0.010	0.001	<0.001	0
10	0.141	0.067	0.024	0.005	<0.001	<0.001	0

headed x, gives a value of $|r| > 0.8$, which might be considered in some circumstances as 'good' but in this case has just happened 'by chance'. The basic message is that, if there are only a few points (say ≤ 5), it is possible that consecutive numbers will be in ascending or descending order. If this is the case then the magnitude of the correlation coefficient can easily exceed 0.8. How, then, do you know if the correlation coefficient is significant? The question can be put like this:

> On the assumption that a set of x–y data are uncorrelated, what is the probability of obtaining the value of r at least as large as that calculated?

If that probability is small (say <0.05) then it is highly unlikely that the x–y data are uncorrelated, i.e. the data *are* likely to be correlated. Table 6.20 shows the probability of obtaining a particular value for $|r|$ when the x–y data are uncorrelated. The probability of obtaining a particular value for $|r|$ decreases when the x–y values are uncorrelated as the number of values, n, increases.[19]

A correlation coefficient is considered significant if the probability of it occurring when the data are uncorrelated[20] is less than 0.05. Referring to table 6.20 we see that, for example, with only four data points ($n=4$), a value of r of 0.9 is not significant as the probability that this could happen with uncorrelated data is 0.10. When the number of data values is 10 or more, then values of r greater than about 0.6 become significant.

[19] See Bevington and Robinson (1992) for a discussion of the calculation of the probabilities in table 6.20.
[20] This value of probability is often used in tests to establish statistical significance, as discussed in chapter 8.

Exercise M

Consider the *x–y* data in table 6.21.

(i) Calculate the value of the correlation coefficient, r.
(ii) Is the value of r significant?

Table 6.21. *x–y data.*

x	y
2.67	1.54
1.56	1.60
0.89	1.56
0.55	1.34
−0.25	1.33

6.7 Residuals

Another indicator of 'goodness of fit', complementary to the correlation coefficient and often more revealing, is the distribution of residuals, Δy. The residual, $\Delta y = y - \hat{y}$, is plotted on the vertical axis against *x*. If we assume that a particular equation is 'correct' in that it accurately describes the data that we have collected, and only random errors exist to obscure the true value of *y* at any value of *x*, then the residuals would be expected to be scattered randomly about the $\Delta y = 0$ axis. Additionally, if the residuals are plotted in the form of a histogram, they would appear to be normally distributed if the errors are normally distributed.

If we can discern a pattern in the residuals then one or more factors must be at play to cause the pattern to occur. It could be:

(i) We have chosen the wrong equation to fit to the data.
(ii) There is an outlier in the data which has a large influence on the line of best fit.
(iii) We should be using a weighted fit, i.e., the uncertainties in the measured *y* values are not constant and we need to take this into account when applying least squares to obtain parameter estimates (section 6.10 considers weighted fitting of a straight line to data).

Typical plots of residuals[21] versus *x*, which illustrate patterns that can emerge, are shown in figures 6.12 to 6.15. No discernible pattern is

---

[21] Another way to display residuals is to plot Δy_i versus $\hat{y}_i$.

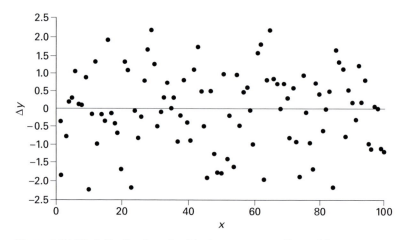

Figure 6.12. Ideal distribution of residuals – no pattern discernible.

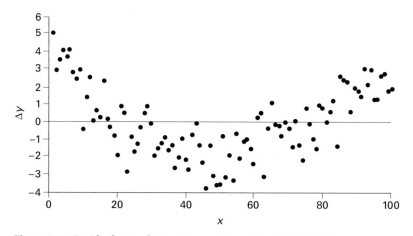

Figure 6.13. Residuals revealing an incorrect equation fitted to data.

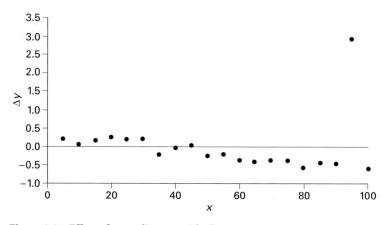

Figure 6.14. Effect of an outlier on residuals.

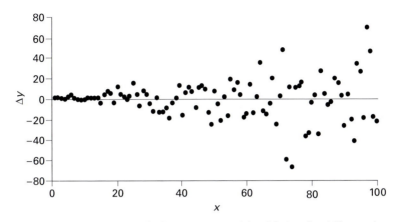

Figure 6.15. Pattern in residuals suggesting weighted fitting should be used.

observable in the residuals in figure 6.12. This offers strong evidence to support the notion that the fit of the equation to the data is good. In figure 6.13 the residuals reveal a systematic variation as x increases. This should lead us to question the appropriateness of the equation we have fitted to the data. Figure 6.14 shows an extreme example of an outlier existing in x–y data. If there are many data, the intercept and slope, a and b, are little affected by such an outlier. However, the situation is quite different if there are far fewer data points (say ten or so). In such a situation an outlier can have a dramatic effect on a and b (and the uncertainties in a and b). In figure 6.15 the residuals increase steadily with increasing x. This occurs when the scatter in the data (i.e. due to random errors in the y quantity) increases with increasing x. In this situation an unweighted fit of the equation to the data is not appropriate and a *weighted* fit using weighted least squares should be used instead. Weighted least squares is dealt with in section 6.10.

Though residuals can be very revealing with regard to identifying outliers, inappropriate equations and incorrect weighting, they can be misleading if there are only a few data (say, <10) as a pattern can emerge within the residuals 'by chance'.

Exercise N

The period, T, of oscillation of a body on the end of a spring is measured as the mass, m, of the body increases. Data obtained are shown in table 6.22.

(i) Plot a graph of period versus mass of body.
(ii) Assuming that the period is linearly related to the mass, use least squares to find the equation of the best line through the points.

(iii)   Calculate the residuals and plot a graph of residuals versus mass.

(iv)   Is there any 'pattern' discernible in the residuals? If so, suggest a possible cause of the pattern.

Table 6.22. *Variation of period, T, of a body of mass, m, on a spring.*

m(kg)	T (s)
0.2	0.39
0.4	0.62
0.6	0.72
0.8	0.87
1.0	0.92
1.2	1.07
1.4	1.13
1.6	1.16
1.8	1.23
2.0	1.32

6.7.1   Standardised residuals

If each residual, Δy_i, is divided by the standard deviation in each y value, σ_i, we refer to the quantity

$$\Delta y_{is} = \frac{\Delta y_i}{\sigma_i} \tag{6.42}$$

as the *standardised* residual.[22] In situations in which the standard deviation in each y value is the same, we replace σ_i by σ, so that

$$\Delta y_{is} = \frac{\Delta y_i}{\sigma} \tag{6.43}$$

where σ is given by equation (6.10).

Scaling residuals in this manner is very useful as the standardised residuals should be normally distributed with a standard deviation of 1 if the errors causing scatter in the data are normally distributed. If we were to plot Δy_s against x, we should find that not only should the standardised residuals be scattered randomly about the $\Delta y_s = 0$ axis, but that (based on

---

[22] See Devore (1991) for a fuller discussion of standardised residuals.

properties of the normal distribution[23]) about 70% of the standardised residuals should lie between $\Delta y_s = \pm 1$ and about 95% should lie between $\Delta y_s = \pm 2$. If this is not the case, it suggests that the scatter of y values does not follow a normal distribution.

Example 8

Consider the x–y data in table 6.23.

(i) Assuming a linear relationship between x and y, determine the equation of the best line through the data in table 6.23.
(ii) Calculate the standard deviation, σ, of the data about the fitted line.
(iii) Determine the standardised residuals and plot a graph of standardised residuals versus x.

ANSWER

(i) Applying equations (6.6) and (6.7) to the data in table 6.23, we find $a = 7.849$ and $b = 2.830$, so that the equation of the best line can be written

$$\hat{y} = 7.849 + 2.830x \tag{6.44}$$

(ii) Applying equation (6.10) gives $\sigma = 5.376$.
(iii) Table 6.24 includes the predicted y values found using equation (6.44), the residuals and the standardised residuals. Figure 6.16 shows a plot of the standardised residual, Δy_s, versus x. As anticipated, the standardised residuals lie between $\Delta y_s = \pm 2$.

Table 6.23. *x–y data for example 8.*

x	2	4	6	8	10	12	14	16	18	20	22
y	8.2	23.0	26.3	31.4	39.5	36.7	56.7	46.8	53.4	63.2	74.7

Table 6.24. *Predicted values, residuals and standardised residuals.*

x	2	4	6	8	10	12	14	16	18	20	22
y	8.2	23.0	26.3	31.4	39.5	36.7	56.7	46.8	53.4	63.2	74.7
$\hat{y}$	13.509	19.169	24.829	30.489	36.149	41.809	47.469	53.129	58.789	64.449	70.109
Δy	−5.309	3.831	1.471	0.911	3.351	−5.109	9.231	−6.329	−5.389	−1.249	4.591
Δy_s	−0.988	0.713	0.274	0.169	0.623	−0.950	1.717	−1.177	−1.002	−0.232	0.854

---

[23] See section 3.5.

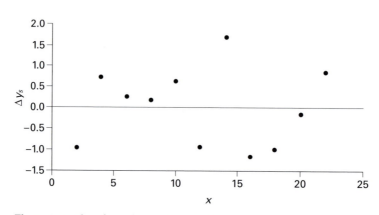

Figure 6.16. Plot of standardised residuals.

6.8   Data rejection

A topic of some importance and controversy is that of data rejection. If we gather x–y data and most of the points lie along a straight line, but one point lies well away from the line, should we reject that point? This is a difficult question to which to give a simple yes or no answer. It may be that a 'slip-up' occurred during the recording of the data and therefore we are justified in ignoring the point. However, it could be that there is a true deviation from straight line behaviour and that the 'outlier' is revealing something really important – perhaps we have observed a new effect! To reject that point may be to eliminate the most important value you have gathered. If possible, repeating the experiment is always preferable to rejecting data with little justification other than it 'doesn't fit'.

The decision to reject data largely depends on what relationship we believe underlies the data that has been obtained. For example, figure 6.17 shows data obtained in an experiment to study the specific heat of a metal at low temperature. Though the underlying trend between x and y may or may not be linear, it *does* appear that there is a spurious value for the specific heat at about 3 K. Performing a statistically based test, such as that described next, would indicate that the point should be removed. However, the 'jump' in specific heat at about 3 K is a real effect and to eliminate the point would mean that a very important piece of evidence giving insight into the behaviour of the material would have been discarded. Nevertheless, there may be situations in which large deviations can be attributed to spurious effects. In these cases it is possible to use the trend exhibited by the majority of the data as a basis for rejecting data that appear to be outliers.

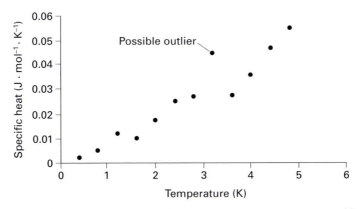

Figure 6.17. Variation of specific heat of tin with temperature. A possible outlier is indicated.

In section 5.8.2 we introduced Chauvenet's criterion as a way of deciding whether a value within a sample is so far from the mean of the sample that it should be considered for rejection. The basic assumption when applying the criterion is that data are normally distributed. If the assumption is valid, the probability can be determined by obtaining a value at least as far from the mean as a 'suspect' value. Multiplying that probability by the number of data gives the number of values *expected* to lie at least as far from the mean as a 'suspect' value. If the expected number is less than or equal to 0.5, the suspect point should be rejected and the sample mean and standard deviation recalculated.

We can apply Chauvenet's criterion to x–y data by considering how far an observed y value is from the line of best fit. The 'distance' between the observed value of y and the predicted value of y based on the line of best fit is expressed in terms of the number of standard deviations between the observed and the predicted value. Specifically, we write

$$z_{OUT} = \frac{y_{OUT} - \hat{y}}{\sigma} \tag{6.45}$$

where y_{OUT} is the outlier (i.e. that furthest from the line of best fit), $\hat{y}$ is the predicted value, found using $\hat{y} = a + bx$. The standard deviation, σ, is calculated using equation (6.10).

Once z_{OUT} has been calculated, we determine the expected number of values, N, at least as far away from $\hat{y}$ as y_{OUT}. To do this:

(i) Determine the probability, P, of obtaining a value at least as far from $\hat{y}$ as y_{OUT}, i.e. $P = P(z < -|z_{OUT}|) + P(z > |z_{OUT}|)$.

(ii) Calculate the expected number of values, N, at least as far from $\hat{y}$ as y_{OUT} using $N = nP$, where n is the number of data.

If N is less than 0.5, consider rejecting the point. If a point *is* rejected, then a and b should be recalculated (as well as other related quantities, such as σ_a and σ_b).

Example 9

Consider x–y values in table 6.25.

(i) Plot an x–y graph and use unweighted least squares to fit a line of the form $y=a+bx$ to the data.

(ii) Identify any suspect point(s).

(iii) Calculate the standard deviation of the y values.

(iv) Apply Chauvenet's criterion to the suspect point – should it be rejected?

ANSWER

(i) A plot of data is shown in figure 6.18 with line of best fit attached ($a=-1.610$ and $b=2.145$).

(ii) A suspect point would appear to be $x=6$, $y=8.5$ as this point is furthest from the line of best fit.

(iii) Using the data in table 6.25 and equation (6.10), $\sigma=1.895$.

(iv) Using equation (6.45), we have

$$z_{OUT}=\frac{8.5-(-1.610+2.145\times 6)}{1.895}=-1.46$$

Now $P=P(z<-1.46)+P(z>1.46)=2\times 0.072\,15=0.144$. The expected number of points at least this far from the line is expected to be $5\times 0.144=0.72$. As this value is greater than 0.5, do not reject the point.

Table 6.25. *x–y data with a 'suspect' value.*

x	y
2.0	3.5
4.0	7.2
6.0	8.5
8.0	17.1
10.0	20.0

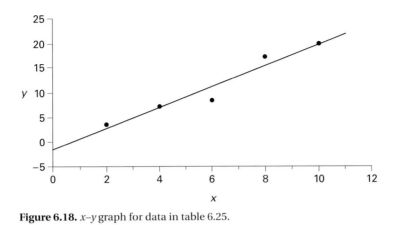

Figure 6.18. x–y graph for data in table 6.25.

Exercise O

Consider the x–y data in table 6.26. When a straight line is fitted to these data, it is found that $a = 4.460$ and $b = 0.6111$. Assuming that the data point $x = 10$, $y = 13$ is 'suspect', apply Chauvenet's criterion to decide whether this point should be rejected.

Table 6.26. x–y data for exercise O.

x	y
5	7
6	8
8	9
10	13
12	11
14	12
15	14

6.9 Transforming data for least squares analysis

Using equations (6.6) and (6.7) we can establish best estimates for the intercept, a, and slope, b, of a line through x–y data. These equations should only be applied when we are confident that there is a linear relationship between the quantities plotted on the x and y axes. What do we do if x–y data are clearly non-linearly related, such as those in figure 6.19? It may be possible to apply a mathematical operation to the x or y data (or to

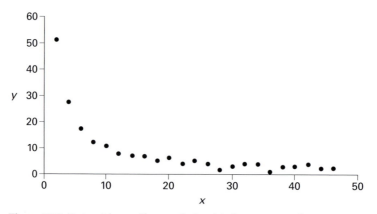

Figure 6.19. Data with non-linear relationship between x and y.

both x and y) so that the *transformed* data appear linearly related. The next stage is to fit the equation $y = a + bx$ to the transformed data. How do we choose which mathematical operation to apply? If we have little or no idea what the relationship between the dependent and independent variable is likely to be, we may be forced into a 'trial and error' approach to transforming data. As an example, the rapid decrease in y with x indicated for $x < 10$ in figure 6.19 suggests that there might be an exponential relationship between x and y, such as

$$y = Ae^{Bx} \tag{6.46}$$

where A and B are constants. Assuming this to be the case, taking natural logs of each side of the equation gives

$$\ln y = \ln A + Bx \tag{6.47}$$

Comparing equation (6.47) with the equation of a straight line predicts that plotting $\ln y$ versus x will produce a straight line with intercept $\ln A$ and slope B. Figure 6.20 shows the effect of applying the transformation suggested by equation (6.47) to the data in figure 6.19. The transformation has not been successful in producing a linear x–y graph, indicating that equation (6.46) is not appropriate to these data and that other transformation options should be considered (for example $\ln y$ versus $\ln x$). Happily, in many situations in the physical sciences the work of others (either experimental or theoretical) provides clues as to how the data should be treated in order to produce a linear relationship between data plotted as an x–y graph. Without such clues we must use 'intelligent guesswork'.

As an example, consider a ball falling freely under gravity. The distance, s, through which a ball falls in a time t is measured and the values

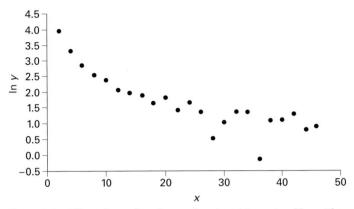

Figure 6.20. Effect of transforming y values by taking natural logarithms.

Table 6.27. *Distance– time
data for a freely falling
ball.*

t (s)	s (m)
1	7
2	22
3	52
4	84
5	128
6	200

obtained are shown in table 6.27. The data are shown in graphical form in figure 6.21. The relationship between s and t is not linear. Can the data be transformed so that a linear graph is produced? The starting point is to look for an equation which might describe the motion of the ball. When the acceleration, g, is constant, we can write the relationship between s and t as[24]

$$s = ut + \tfrac{1}{2}gt^2 \tag{6.48}$$

where u is the initial velocity of the body. To linearise equation (6.48), divide throughout by t to give

$$\frac{s}{t} = u + \tfrac{1}{2}gt \tag{6.49}$$

[24] See Young and Freedman (1996).

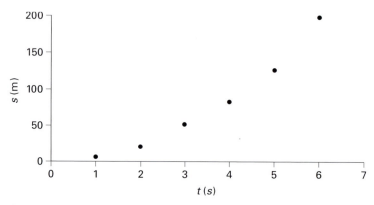

Figure 6.21. Distance–time data for a falling ball.

Table 6.28.
*Transformation of data
given in table 6.27.*

t (s)	s/t (m/s)
1	7
2	11
3	17.3
4	21
5	25.6
6	33.3

Compare the quantities in equation (6.49) with $y = a + bx$:

$$\frac{s}{t} = u + \tfrac{1}{2}g \; t$$

$$y = a + b \; x$$

If the relationship between the displacement and time can be represented by equation (6.48), then plotting s/t versus t should transform the experimental data so that they lie on (or close to) a straight line which has intercept u and slope $\tfrac{1}{2}g$. Table 6.28 contains the transformed data and figure 6.22 the corresponding graph.

It certainly does appear that transforming the data has produced a graph in which the quantities plotted on the x and y axes are linearly related. Fitting a line to the transformed data using least squares gives

$$a = 1.298 \text{ m/s} \qquad b = 5.118 \text{ m/s}^2$$

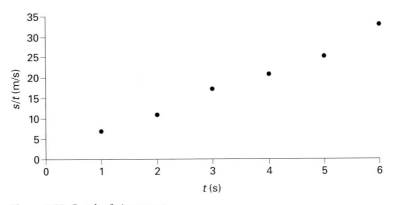

Figure 6.22. Graph of *s/t* versus *t*.

When transforming data, the dependent variable should remain on the left hand side of the equation (so that the assumption of uncertainties restricted to the quantity plotted on the *y* axis is valid). Usually the independent variable only appears on the right hand side of the equation. However, there are situations, such as in the linearisation of equation (6.48), where this condition must be relaxed.

Exercise P

1. Transform the equations shown in table 6.29 into the form $y = a + bx$, and indicate how the constants in each equation are related to a and b.

2. The capacitance of a semiconductor diode decreases as the reverse bias voltage applied to the junction increases. An important diode parameter, namely the contact potential, can be found if the capacitance of the junction is measured as a function of reverse bias voltage. Table 6.30 shows experimental capacitance/voltage data for a particular diode. Assume that the relationship between C_j and V can be written

$$\frac{1}{C_j^2} = k(V + \phi) \tag{6.50}$$

where ϕ is the contact potential and k is a constant. Use least squares to find k and ϕ.

Table 6.29. *Equations to be transformed into the form, $y = a + bx$.*

	Equation	Dependent variable	Independent variable	Constant(s)	Hint
(i)	$R = Ae^{-BT}$	R	T	A,B	Take the natural logs of both sides of the equation.
(ii)	$v = u + gt$	v	t	u,g	
(iii)	$H = C(T - T_o)$	H	T	C, T_o	Multiply out the brackets.
(iv)	$T_w = T_c - kR^2$	T_w	R	T_c, k	
(v)	$T = 2\pi\sqrt{\dfrac{m}{k}}$	T	m	k	
(vi)	$E = I(R + r)$	I	R	E, r	Move I to the LHS of equation, and E to the RHS.
(vii)	$\dfrac{1}{u} + \dfrac{1}{v} = \dfrac{1}{f}$	u	v	f	Move $1/u$ to the RHS of the equation.
(viii)	$N = kC^{1/n}$	N	C	k, n	Take logs of both sides of equation.
(ix)	$n^2 = 1 + \dfrac{A\lambda^2}{\lambda^2 - B}$	n	λ	A,B	Subtract 1 from both sides of equation then take the reciprocals of both sides of the equation.
(x)	$t = \dfrac{A}{D^2}(1 + BD)$	t	D	A,B	Multiply both sides of equation by D^2.

6.9.1 Consequences of data transformation

Our least squares analysis to this point has assumed that the uncertainty in the y values is constant. Is this assumption still valid if the data are transformed? Very often the answer is no and to see this let us consider a situation in which data transformation requires that the natural logarithms of the y quantity be calculated.

Table 6.30. *Junction capacitance, C_j, of a diode as a function of bias voltage, V.*

V (V)	C_j (pF)
6.0	248
8.1	217
10.1	196
14.1	169
18.5	149
24.6	130
31.7	115
38.1	105
45.6	96.1
50.1	92.1

When light is absorbed by a medium, the intensity of light, I, after it has passed a distance x through the medium is given by

$$I = I_0 e^{-kx} \tag{6.51}$$

where I_0 is the intensity of light incident on the medium (i.e. at $x=0$) and k is the absorption coefficient. Measurements of I are made at different values of x. We linearise equation (6.51) by taking natural logarithms of both sides of the equation giving

$$\ln I = \ln I_0 - kx \tag{6.52}$$

Assuming that equation (6.51) is valid for the experimental data, plotting ln I versus x should produce a plot in which the transformed data lie close to a straight line. A straight line fitted to the transformed data would have intercept $\ln I_0$ and slope $-k$. As $\ln I$ is taken as the 'y quantity' when fitting a line to data using least squares, we must determine the uncertainty in $\ln I$. We write

$$y = \ln I \tag{6.53}$$

If the uncertainty in I, u_I, is small, then the uncertainty in y, u_y, is given by[25]

$$u_y \approx \left| \frac{\partial y}{\partial I} \right| u_I \tag{6.54}$$

Now $\partial y / \partial I = 1/I$, so that

---

[25] Section 5.7.2 deals with propagation of uncertainties.

$$u_y \approx \left| \frac{u_I}{I} \right| \tag{6.55}$$

Equation (6.55) indicates that if u_I is constant, the uncertainty in $\ln I$ decreases as I increases. The consequence of this is that the assumption of constant uncertainty in the y values used in least squares analysis is no longer valid and unweighted least squares must be abandoned in favour of an approach that takes into account changes in the uncertainties in the y values. There are many situations in which data transformation leads to a similar outcome, i.e. that the uncertainty in y values is not constant and so requires a straight line to be fitted using *weighted* least squares. This is dealt with in the next section.

Exercise Q

For the following equations, determine y and the uncertainty in y, u_y, given that $V = 56$ and $u_V = 2$ in each case. Express u_y to two significant figures.

(i) $y = V^{\frac{1}{2}}$, (ii) $y = V^2$, (iii) $y = \dfrac{1}{V}$, (iv) $y = \dfrac{1}{V^2}$, (v) $y = \log_{10} V$.

6.10 Weighted least squares

When the uncertainties in y values are constant, unweighted least squares analysis is appropriate, as discussed in section 6.2.2. However, there are many situations in which the uncertainty in the y values does not remain constant. These include when:

(i) measurements are repeated at a particular value of x, thus reducing the uncertainty in the corresponding value of y;

(ii) data are transformed, for example by 'squaring' or 'taking logs';

(iii) the size of the uncertainty in y is some function of y (such as in counting experiments).

How do you know if you should use a weighted fit? A good starting point is to perform unweighted least squares to find the line of best fit through the data (data should be transformed if necessary, as discussed in section 6.9). A plot of the residuals should reveal whether a weighted fit is required. Figure 6.23 shows a plot of residuals in which the residuals decrease with increasing x (figure 6.23(a)) and increase with increasing x (figure 6.23(b)). Such patterns in residuals are 'tell tale' signs that weighted least squares fitting should be used.

In order to find the best line through the points when weighted fitting

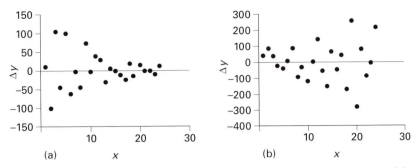

Figure 6.23. Residuals indicating weighted fit is required: (a) indicates the size of the residual that decreases with increasing x; (b) indicates the size of residual that increases with increasing x.

is required, we must include explicitly the uncertainty in each y value in the equations for a and b. This is due to the fact that for weighted fitting we must minimise the weighted sum of squares of residuals, χ^2, where[26]

$$\chi^2 = \sum \left(\frac{y_i - \hat{y}_i}{\sigma_i} \right)^2 \tag{6.56}$$

where σ_i is the standard deviation in y_i. The intercept, a, and slope, b, are given by

$$a = \frac{\sum \frac{x_i^2}{\sigma_i^2} \sum \frac{y_i}{\sigma_i^2} - \sum \frac{x_i}{\sigma_i^2} \sum \frac{x_i y_i}{\sigma_i^2}}{\Delta} \tag{6.57}$$

$$b = \frac{\sum \frac{1}{\sigma_i^2} \sum \frac{x_i y_i}{\sigma_i^2} - \sum \frac{x_i}{\sigma_i^2} \sum \frac{y_i}{\sigma_i^2}}{\Delta} \tag{6.58}$$

where

$$\Delta = \sum \frac{1}{\sigma_i^2} \sum \frac{x_i^2}{\sigma_i^2} - \left(\sum \frac{x_i}{\sigma_i^2} \right)^2 \tag{6.59}$$

Equations (6.57) and (6.58) give more weight to the points that have smaller uncertainty thereby ensuring that the fitted line will pass closer to these points than those with large uncertainty.

With weighted fitting, the best line no longer passes through $(\bar{x}, \bar{y})$ but through the *weighted* centre of gravity of the points, $(\bar{x}_w, \bar{y}_w)$. The coordinates of the weighted centre of gravity are given by[27]

---

[26] Refer to appendix 3 for more details.

[27] For a derivation of equation (6.60) see section A3.1.

$$\bar{x}_w = \frac{\sum \dfrac{x_i}{\sigma_i^2}}{\sum \dfrac{1}{\sigma_i^2}}$$

(6.60)

and

$$\bar{y}_w = \frac{\sum \dfrac{y_i}{\sigma_i^2}}{\sum \dfrac{1}{\sigma_i^2}}$$

(6.61)

One difficulty with finding a and b is that in many situations we do not know σ_i. When calculating a and b, a knowledge of the absolute values of σ_i is not crucial. As long as we know the relative magnitudes of σ_i for all data, the values of a and b are unaffected.

Example 10

x–y data along with associated uncertainties are shown in table 6.31. Using weighted least squares, find the intercept and slope of the best line through the points.

ANSWER

Table 6.32 contains all the quantities necessary to calculate a and b. Summing the appropriate columns gives

$\sum 1/\sigma_i^2 = 0.178403$ $\sum x_i/\sigma_i^2 = 15.00986$ $\sum y_i/\sigma_i^2 = 15.46528$
$\sum x_i y_i /\sigma_i^2 = 1242.986$ $\sum x_i^2 /\sigma_i^2 = 1379.559$

Substituting the summations into equations (6.57) to (6.59) gives

$a = 128.6$ $b = -0.4985$

Table 6.31. *x–y data.*

x	y
18	125 ± 10
42	108 ± 8
67	91 ± 6
89	84 ± 4
108	76 ± 4

Table 6.32. *Quantities for weighted least squares calculation.*

x_i	y_i	σ_i	$1/\sigma_i^2$	x_i/σ_i^2	y_i/σ_i^2	x_iy_i/σ_i^2	x_i^2/σ_i^2
18	125	10	0.01	0.18	1.25	22.5	3.24
42	108	8	0.015625	0.65625	1.6875	70.875	27.5625
67	91	6	0.027778	1.861111	2.527778	169.3611	124.6944
89	84	4	0.0625	5.5625	5.25	467.25	495.0625
108	76	4	0.0625	6.75	4.75	513	729

Exercise R

Repeat example 10 with every value of σ_i multiplied by 5 (so, for example, when $y_i = 125$, $\sigma_i = 50$). Show that the values of a and b remain unchanged. (Suggestion: use a spreadsheet!)

6.10.1 Weighted uncertainty in a and b

When a weighted fit is required, equations (6.62) and (6.63) can be used to calculate the uncertainties (in this case taken to be the standard errors) in a and b[28]

$$\sigma_a = \left(\frac{\sum \frac{x_i^2}{\sigma_i^2}}{\Delta}\right)^{\frac{1}{2}} \tag{6.62}$$

$$\sigma_b = \left(\frac{\sum \frac{1}{\sigma_i^2}}{\Delta}\right)^{\frac{1}{2}} \tag{6.63}$$

where Δ is given by equation (6.59). Equations (6.62) and (6.63) are applicable as long as actual values of σ_i are known, as relative magnitudes will not do in this case. Although this might seem unduly restrictive, there is one case in which σ_i may be estimated fairly accurately and that is in counting experiments (such as those involving radioactivity or X rays), where the Poisson distribution is valid. If the number of counts recorded is C_i, then the standard deviation in C_i, σ_i, is given by

$$\sigma_i = \sqrt{C_i} \tag{6.64}$$

---

[28] See appendix 4 for more details.

Exercise S

In a diffusion experiment, a radiotracer diffuses into a solid when the solid is heated to a high temperature for a fixed period of time. The solid is sectioned and the number of gamma counts is recorded by a particle counter over a period of 1 minute for each section. Table 6.33 shows the number of counts, C, as a function of depth, d, cut in the material. Assume that the equation that relates C to d is

$$C = A\exp(-\lambda d^2)$$

where A and λ are constants.

(i) Transform this equation into the form $y = a + bx$.
(ii) Assuming the uncertainty in C to be $\sqrt{C}$, perform a weighted fit to find values of A, λ, σ_A and σ_λ.

Table 6.33. *Variation of counts with depth in a solid.*

C (counts)	1.86×10^4	1.18×10^4	4.33×10^3	1.00×10^3	1.36×10^2
d (mm)	10	30	50	70	90

6.10.2 Weighted standard deviation, σ_w

When only the relative magnitudes of σ_i are known, it is still possible to determine σ_a and σ_b. In this case it is necessary to calculate the weighted standard deviation of y values (i.e. the weighted equivalent to equation (6.10)). Writing the weighted standard deviation as σ_w, we have[29]

$$\sigma_w = \frac{\left(\frac{n}{n-2}\right)^{\frac{1}{2}}}{\sum\frac{1}{\sigma_i^2}}\left[\sum\frac{1}{\sigma_i^2}\sum\frac{y_i^2}{\sigma_i^2}-\left(\sum\frac{y_i}{\sigma_i^2}\right)^2-\frac{\left(\sum\frac{1}{\sigma_i^2}\sum\frac{x_iy_i}{\sigma_i^2}-\sum\frac{x_i}{\sigma_i^2}\sum\frac{y_i}{\sigma_i^2}\right)^2}{\Delta}\right]^{\frac{1}{2}}$$

(6.65)

where Δ is given by equation (6.59). σ_a and σ_b are written

$$\sigma_a = \sigma_w\left(\frac{\sum\frac{1}{\sigma_i^2}\sum\frac{x_i^2}{\sigma_i^2}}{n\Delta}\right)^{\frac{1}{2}}$$

(6.66)

[29] See Dietrich (1991) for a derivation of equation (6.65).

and

$$\sigma_b = \frac{\sigma_w \sum \dfrac{1}{\sigma_i^2}}{(n\Delta)^{\frac{1}{2}}}$$
(6.67)

For completeness, we include the expression for the weighted linear correlation coefficient, r_w, which should be used whenever weighted least squares fitting occurs. r_w is given as

$$r_w = \frac{\sum \dfrac{1}{\sigma_i^2} \sum \dfrac{x_i y_i}{\sigma_i^2} - \sum \dfrac{x_i}{\sigma_i^2} \sum \dfrac{y_i}{\sigma_i^2}}{\left[\sum \dfrac{1}{\sigma_i^2} \sum \dfrac{x_i^2}{\sigma_i^2} - \left(\sum \dfrac{x_i}{\sigma_i^2}\right)^2\right]^{\frac{1}{2}} \left[\sum \dfrac{1}{\sigma_i^2} \sum \dfrac{y_i^2}{\sigma_i^2} - \left(\sum \dfrac{y_i}{\sigma_i^2}\right)^2\right]^{\frac{1}{2}}}$$
(6.68)

Table 6.34. *Current–voltage data for a tunnel diode.*

V (V)	0.01	0.02	0.03	0.04	0.05	0.06	0.07	0.08
I (A)	0.082	0.130	0.136	0.210	0.181	0.180	0.190	0.172

V (V)	0.09	0.10	0.11	0.12	0.13	0.14	0.15
I (A)	0.136	0.150	0.108	0.119	0.110	0.070	0.080

Example 11

The data shown in table 6.34 were obtained from a study of the relationship between current and voltage for a tunnel diode.[30] For the range of voltage in table 6.34, the relationship between current, I, and voltage, V, for a tunnel diode can be written

$$I = CV \exp\left(\frac{-V}{B}\right)$$
(6.69)

(i)   Transform equation (6.69) into the form $y = a + bx$.
(ii) Determine the summations, $\Sigma 1/\sigma_i^2$, $\Sigma x_i/\sigma_i^2$ etc. required in the calculation of the weighted standard deviation, σ_w, given by equation (6.65). Assume that the uncertainties in the values of current are constant.
(iii) Calculate the weighted standard deviation.

[30] A tunnel diode is a semiconductor device with unusual electrical characteristics. It is used in high frequency oscillator circuits.

Answer

(i) Dividing both sides of equation (6.69) by V gives

$$\frac{I}{V} = C\exp\left(\frac{-V}{B}\right) \tag{6.70}$$

Taking the natural logarithms of both sides of equation (6.70),

$$\ln\frac{I}{V} = \ln C - \frac{V}{B} \tag{6.71}$$

Comparing equation (6.71) with $y = a + bx$, we find that $y = \ln(I/V)$, $x = V$, $a = \ln C$ and $b = -1/B$. Plotting $\ln(I/V)$ versus V would be expected to give a straight line with intercept equal to $\ln C$ and slope equal to $-1/B$.

(ii) In order to determine the standard deviation in the ith y value, σ_i, we must consider the left hand side of equation (6.71).

$$y = \ln\frac{I}{V} \tag{6.72}$$

The standard deviation in the ith value of y, σ_i, is given by

$$\sigma_i \approx \left|\frac{\partial y}{\partial I}\right|\sigma_I \tag{6.73}$$

Now $\partial y/\partial I = 1/I$, so that

$$\sigma_i \approx \frac{\sigma_I}{I} \tag{6.74}$$

Table 6.35 shows the raw data, the transformed data and the sums of the columns necessary to calculate the weighted standard deviation, σ_w. For convenience we take the standard deviation in I, σ_I, to be equal to 1, so that using equation (6.74), $\sigma_i = 1/I$.

(iii) The weighted standard deviation, calculated using equation (6.65) and the sums of numbers appearing in the bottom row of table 6.35, is $\sigma_w = 0.1064$.

Exercise T

Determine the intercept and slope of the transformed data in example 11. Calculate also the standard errors in intercept and slope.

Table 6.35. *Weighted fitting of data in table 6.34.*

$V(V)=x$	$I(A)$	$\ln(I/V)=y$	$\sigma_i=\dfrac{1}{I}$	$\dfrac{1}{\sigma_i^2}$	$\dfrac{x_i}{\sigma_i^2}$	$\dfrac{y_i}{\sigma_i^2}$	$\dfrac{x_iy_i}{\sigma_i^2}$	$\dfrac{x_i^2}{\sigma_i^2}$	$\dfrac{y_i^2}{\sigma_i^2}$
0.01	0.082	2.104134	12.19512	0.006724	6.72E−05	0.014148	0.000141	6.72E−07	0.02977
0.02	0.13	1.871802	7.692308	0.0169	0.000338	0.031633	0.000633	6.76E−06	0.059212
0.03	0.136	1.511458	7.352941	0.018496	0.000555	0.027956	0.000839	1.66E−05	0.042254
0.04	0.21	1.658228	4.761905	0.0441	0.001764	0.073128	0.002925	7.06E−05	0.121263
0.05	0.181	1.286474	5.524862	0.032761	0.001638	0.042146	0.002107	8.19E−05	0.05422
0.06	0.18	1.098612	5.555556	0.0324	0.001944	0.035595	0.002136	0.000117	0.039105
0.07	0.19	0.998529	5.263158	0.0361	0.002527	0.036047	0.002523	0.000177	0.035994
0.08	0.172	0.765468	5.813953	0.029584	0.002367	0.022646	0.001812	0.000189	0.017334
0.09	0.136	0.412845	7.352941	0.018496	0.001665	0.007636	0.000687	0.00015	0.003152
0.1	0.15	0.405465	6.666667	0.0225	0.00225	0.009123	0.000912	0.000225	0.003699
0.11	0.108	−0.01835	9.259259	0.011664	0.001283	−0.00021	−2.4E−05	0.000141	3.93E−06
0.12	0.119	−0.00837	8.403361	0.014161	0.001699	−0.00012	−1.4E−05	0.000204	9.92E−07
0.13	0.11	−0.16705	9.090909	0.0121	0.001573	−0.00202	−0.00026	0.000204	0.000338
0.14	0.07	−0.69315	14.28571	0.0049	0.000686	−0.0034	−0.00048	9.6E−05	0.002354
0.15	0.08	−0.62861	12.5	0.0064	0.00096	−0.00402	−0.0006	0.000144	0.002529
			Sums	0.307286	0.021316	0.290285	0.013336	0.001824	0.411229

6.10.3   Weighted least squares and Excel®

If the uncertainties in y values are small, then it matters little whether weighted or unweighted least squares is used to fit a straight line to data, as both will yield very similar values for intercept and slope. In fact, due to its comparative computational simplicity, it is wise to apply unweighted least squares first. By examining the residuals we can determine whether fitting an equation using weighted least squares is justified. An added advantage of performing unweighted least squares first is that we have values for the 'unweighted' intercept and slope against which weighted estimates of intercept and slope can be compared. If there is a large difference between unweighted and weighted estimates this might indicate a mistake in the weighted analysis.

If weighted fitting is required, it is quite daunting to attempt such an analysis equipped with only a pocket calculator. The evaluation of the many summations required for the determination of a and b (see equations (6.57) and (6.58)) is enough to deter all but the most tenacious individual. Most 'built in' least squares facilities available with calculators and spreadsheets offer only unweighted least squares. This is also true of the LINEST() feature in Excel® which does not provide for the weighting of the fit. Nevertheless, a table, such as table 6.35, can be quite quickly drawn up in Excel®. The table should include the 'raw x–y data', the standard deviation in each y value, σ_i, and the other quantities, such as $x_i y_i / \sigma_i^2$. Using the AutoSum button, $\boxed{\Sigma}$, on the standard toolbar permits the sums to be calculated with the minimum of effort. The calculations required for the examples in this chapter involving weighted least squares were carried out in this manner.

6.11   Review

An important goal in science is to progress from a qualitative understanding of phenomena to a quantitative description of the relationship between physical variables. Least squares is a powerful and widely used technique in the physical sciences (and in many other disciplines) for establishing a quantitative relationship between two or more variables. In this chapter we have focussed upon fitting an equation representing a straight line to data, as linearly related data frequently emerge from experiments.

Due to the ease with which modern analysis tools such as spreadsheets can fit an equation to data, it is easy to overlook the question: 'Should we really fit a straight line to data?' To assist in answering this question we introduced the correlation coefficient and residual plots as quantitative and qualitative indicators of 'goodness of fit' and have indicated situations in which each can be misleading. We have also considered situations in which data transformation is required before least squares is applied. Often, after completing the data transformation we must forsake unweighted least squares in favour of weighted least squares.

The technique of fitting an equation to data using least squares can be extended to situations in which there are more than two parameters to be estimated, for example $y = \alpha + \beta x^{-1} + \gamma x$, or where there are more than two independent variables. This will be considered in the next chapter along with situations in which linear least squares cannot be applied to the analysis of data.

Problems

1. The acceleration due to gravity, g, is measured at various heights, h, above sea level. Table 6.36 contains the value of g corresponding to the various heights. Taking the height as the independent variable (x) and the acceleration due to gravity as the dependent variable (y):

   (i) Plot a graph of g versus h.
  (ii) Assuming g to be linearly related to h, calculate the intercept and slope of the line of best fit.
 (iii) Calculate the sum of squares of the residuals and the standard deviation of the y values.
 (iv) Calculate the standard errors in a and b.
  (v) Determine the correlation coefficient.
 (vi) The data pair $h = 50$ km, $g = 9.73$ m/s^2 was incorrectly recorded and should be $h = 50$ km, $g = 9.65$ m/s^2. Repeat parts (i) to (v) with the corrected value.

2. When a rigid tungsten sphere presses on a flat glass surface, the surface of the glass deforms. The relationship between the mean contact pressure, p_m, and the indentation strain, s, is

$$p_m = ks \tag{6.75}$$

Table 6.37 shows s–p_m data for indentations made into glass. k is a constant dependent upon the mechanical stiffness of the material. Taking p_m to be the dependent variable and s to be the independent variable, use least squares to determine k and the standard error in k.

Table 6.36. *Variation of acceleration due to gravity with height.*

h (km)	g (m/s^2)
10	9.76
20	9.74
30	9.70
40	9.69
50	9.73
60	9.62
70	9.59
80	9.55
90	9.54
100	9.51

Table 6.37. *Variation of contact pressure with indentation strain.*

s	p_m (MPa)
0.0280	1.0334
0.0354	1.2295
0.0383	1.1851
0.0415	1.3574
0.0521	1.6714
0.0568	1.7977
0.0570	1.9303
0.0754	2.3957
0.0764	2.4258
0.0975	3.0208
0.1028	2.8824
0.1149	3.5145
0.1210	3.5072
0.1430	3.8888
0.1703	5.0148

Table 6.38. *Speed/current values in a water flow experiment.*

u (m/s)	0.1	0.2	0.3	0.4	0.5	0.6	0.7	0.8	0.9	1.0	1.1
$I\,(\mu A)\pm 10\ \mu A$	270	300	300	330	300	330	350	330	330	360	340

u (m/s)	1.2	1.3	1.4	1.5	1.6	1.7	1.8	1.9	2.0	2.1	2.2
$I\,(\mu A)\pm 10\ \mu A$	340	360	350	370	340	370	360	360	380	380	390

u (m/s)	2.3	2.4	2.5	2.6	2.7	2.8	2.9	3.0
$I\,(\mu A)\pm 10\ \mu A$	390	400	410	410	400	400	410	410

3. A thermal convection flowmeter measures the local speed of a fluid by measuring the heat loss from a heated element in the path of the flowing fluid. Assume the relationship between the speed of the fluid, u, and the electrical current, I, supplied to the element is

$$u = K_1(I^2 - K_2)^2 \tag{6.76}$$

where K_1 and K_2 are constants. Values for u and I obtained in a water flow experiment are shown in table 6.38. Assume I to be the *dependent* variable, and u the *independent* variable.

(i) Show that equation (6.76) can be rearranged into the form

$$I^2 = K_2 + \frac{1}{K_1^{1/2}}\,u^{\frac{1}{2}} \tag{6.77}$$

(ii) Compare equation (6.77) with $y = a + bx$, and use a spreadsheet to perform a *weighted* least squares fit to data to find K_1 and K_2.

(iii) Plot the standardised residuals. Can anything be concluded from the pattern of the residuals? (For example, is the weighting you have used appropriate?)

4. Acetic acid is adsorbed from solution by activated charcoal. The amount of acetic acid adsorbed, Y, is given by

$$Y = kC^{\frac{1}{n}} \tag{6.78}$$

where C is the concentration of the acetic acid. k and n are constants. Y–C data are given in table 6.39.

Table 6.39. *Variation of adsorbed acetic acid with concentration.*

C (mol/L)	0.017	0.032	0.060	0.125	0.265	0.879
Y (mol)	0.46	0.61	0.79	1.10	1.53	2.45

    (i) Transform equation (6.78) into the form $y = a + bx$.

   (ii) Use unweighted least squares to find best estimates for k and n and the standard errors in k and n.

  (iii) Plot the standardised residuals. Do you think fitting by unweighted least squares is appropriate?

  (iv) Determine the 95% confidence interval for Y when $C = 0.085$ mol/L.

5. The variation of intensity, I, of light passing through a lithium fluoride crystal is measured as a function of the angular position of a polariser placed in the path of the light emerging from the crystal. Table 6.40 gives the intensity for various polariser angles, θ. Assuming that the relationship between I and θ is

$$I = (I_{max} - I_{min})\cos(2\theta) + I_{min} \tag{6.79}$$

where I_{max} and I_{min} are constants:

    (i) Perform an *unweighted* fit to find the intercept and slope of a graph of I versus $\cos(2\theta)$.

   (ii) Use the values for the slope and intercept to find I_{max} and I_{min}.

  (iii) Determine the standard error in I_{max}.

Table 6.40. *Variation of intensity with polariser angle.*

θ (degrees)	0	20	40	60	80	100	120	140
Intensity, I (arbitrary units)	1.86	1.63	1.13	0.52	0.16	0.00	0.57	1.11

6. The Langmuir adsorption isotherm can be written

$$\frac{P}{X} = \frac{1}{A} - \frac{B}{A}P \tag{6.80}$$

where X is the mass of gas adsorbed per unit area of surface, P is the pressure of the gas, and A and B are constants. Table 6.41 shows data gathered in an adsorption experiment.

    (i) Write equation (6.80) in the form $y = a + bx$. How do the constants in equation (6.80) relate to a and b?

Table 6.41. *Gas adsorbed as a function of pressure.*

P (N/m^2)	X (kg/m^2)
0.27	13.9×10^{-5}
0.39	17.8×10^{-5}
0.62	22.5×10^{-5}
0.93	27.5×10^{-5}
1.72	32.9×10^{-5}
3.43	38.6×10^{-5}

    (ii) Using least squares, find values for a and b and standard errors in a and b.

    (iii) Estimate A and B and their respective standard errors.

7. The relationship between the mean free path, λ, of electrons moving through a gas and the pressure of the gas, P, can be written

$$\frac{1}{\lambda} = \left(\frac{\pi d^2}{4kT} \right) P \tag{6.81}$$

where d is the diameter of the gas molecules, k is Boltzmann's constant ($=1.38 \times 10^{-23}$ J/K) and T is the temperature of the gas in kelvins. Table 6.42 shows data gathered in an experiment in which λ was measured as a function of P.

Table 6.42. *Variation of free path, λ, with pressure, P.*

P(Pa)	λ (mm)
1.5	35
5.8	20
5.8	30
6.5	25
8.0	16
13.1	9
14.5	10
18.9	7
24.7	6
27.6	6.5

(i) Fit the equation $y = a + bx$ to equation (6.81) to find b and the standard error in b.

(ii) If the temperature of the gas is 298 K, use equation (6.81) to estimate the diameter of the gas molecules and the standard error in the diameter.

8. The intercept on the x axis x_{INT} of the best straight line through data is found by setting $\hat{y} = 0$ in equation (6.3). x_{INT} is given by

$$x_{INT} = -\frac{a}{b}$$

Given that standard errors in $\bar{y}$ and b are,

$$\sigma_{\bar{y}} = \frac{\sigma}{\sqrt{n}} \qquad \sigma_b = \frac{\sigma n^{\frac{1}{2}}}{\left[n \sum x_i^2 - \left(\sum x_i \right)^2 \right]^{\frac{1}{2}}}$$

show that the uncertainty in the intercept, $\sigma_{x_{INT}}$, can be expressed as

$$\sigma_{x_{INT}} = \frac{\sigma}{b} \left\{ \frac{1}{n} + \frac{n \bar{y}^2}{b^2 \left[n \sum x_i^2 - \left(\sum x_i \right)^2 \right]} \right\}^{\frac{1}{2}} \tag{6.82}$$

9. In section 6.4.2, we found that

$$\hat{y}_0 = \bar{y} + b(x_0 - \bar{x})$$

Show that if the uncertainties in $\bar{y}$ and b are independent, the standard error in $\hat{y}_0$, $\sigma_{\hat{y}_0}$, can be written

$$\sigma_{\hat{y}_0} = \sigma \left[\frac{1}{n} + \frac{n(x_0 - \bar{x})^2}{n \sum x_i^2 - \left(\sum x_i \right)^2} \right]^{\frac{1}{2}} \tag{6.83}$$

where expressions for standard errors in $\bar{y}$ and b are given in problem 8.

10. The equation of a straight line passing through the origin can be written

$$y = bx \tag{6.84}$$

Use the method outlined in appendix 3 to show that the slope, b, of the best line to pass through the points and the origin is given by

$$b = \frac{\sum x_i y_i}{\sum x_i^2}$$

Use the method outlined in appendix 4 to show that the standard error in b, σ_b, is given by

$$\sigma_b = \frac{\sigma}{\left(\sum x_i^2 \right)^{\frac{1}{2}}} \tag{6.85}$$

where σ is given by equation (6.10).

A word of caution: In situations involving 'real' data it is advisable not to treat the origin as a special point by forcing a line through it, even if 'theoretical' considerations predict that the line *should* pass through the origin. In most situations, random and systematic errors conspire to move the line so that it does not pass through the origin.

11. Data obtained in an experiment designed to study whether the amount of calcium oxide in rocks drawn from a geothermal field is correlated to the amount of magnesium oxide in those rocks are shown in table 6.43.

   (i) Plot a graph of concentration of calcium oxide versus concentration of magnesium oxide.

  (ii) Calculate the linear correlation coefficient, r.

 (iii) Is the value of r obtained in (ii) significant?

Table 6.43. *Concentrations of calcium oxide and magnesium oxide in rock specimens.*

CaO (wt%)	MgO (wt%)
2.43	0.72
4.90	2.42
10.43	4.75
15.64	3.99
16.62	5.39
21.12	8.65

Chapter 7

Least squares II

7.1 Introduction

In chapter 6 we considered fitting a straight line to x–y data in situations in which the equation representing the line is written

$$y = a + bx$$

where a is the intercept of the line and b is its slope. Other situations that we need to consider, as they occur regularly in science, require fitting equations to x–y data where the equations:

- possess more than two parameters that must be estimated, for example $y = a + bx + cx^2$ or $y = a + b/x + cx$;
- contain more than one independent variable, for example $y = a + bx + cz$, where x and z are the independent variables;
- cannot be written in a form suitable for analysis by linear least squares, for example $y = a + be^{cx}$.

As in chapter 6, we apply the technique of least squares to obtain best estimates of the parameters in equations to be fitted to data. As the number of parameters increases, so does the number of the calculations required to estimate those parameters. Matrices are introduced as a means of solving for parameter estimates efficiently. Matrix manipulation is tedious to carry out 'by hand', but the built in matrix functions in Excel® make it well suited to assist in extending the least squares technique. In addition, Excel®'s dedicated least squares function, LINEST(), is capable of fitting equations to data where those equations contain more than two parameters.

We also consider in this chapter the important issue of choosing the

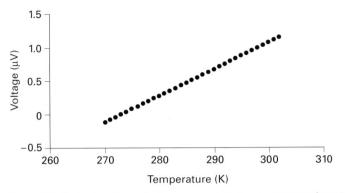

Figure 7.1. Output voltage of a thermocouple between 270 K and 303 K.

best equation to fit to data, and what steps to take if two (or more) equations fitted to the same data must be compared.

7.2 Extending linear least squares

In chapter 6 we introduced α and β as (population) parameters which, in the absence of systematic errors, are regarded as the 'true' intercept and slope respectively of a straight line drawn through x–y data. Estimates of these parameters, obtained using least squares, were written as a and b. As we extend the least squares technique, we must deal with an increased number of parameters. Consistent with our earlier symbolism, we write estimates of parameters obtained by least squares as a, b, c, d and so on.

Fitting a straight line to data has a central position in the analysis of experimental data in the physical sciences. This is due to the fact that many physical quantities (at least over a limited range) are related to each other in a linear fashion. For example, consider the relationship between the output voltage of a thermocouple and the temperature of one junction of a thermocouple[1] as shown in figure 7.1. There is little departure from linearity exhibited by the data in figure 7.1. However, the linearity of output voltage with temperature is no longer maintained when a much larger range of temperature is considered, as shown in figure 7.2. If 'raw' x–y data recorded during an experiment are not linearly related, then they can often be 'linearised' by some simple transformation (such as taking the logarithms of the dependent variable, as described in section 6.9). In other situations, such as for the thermocouple data shown in figure 7.2, no amount

[1] The reference junction of the thermocouple was held at 273 K.

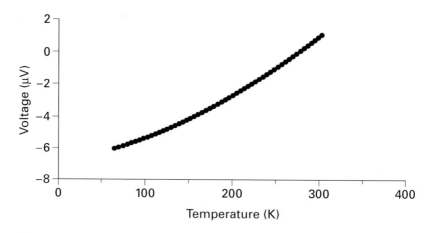

Figure 7.2. Output voltage of a thermocouple between 70 K and 303 K.

of legitimate mathematical manipulation will linearise these data, and we must consider fitting an equation other than $y = a + bx$ to the data.

The technique of linear least squares is not confined to finding the best straight line through data, but can be extended to any function linear in the parameters appearing in the equation. As examples, a, b and c in the following equations can be determined using linear least squares:

$$y = a + bx + cx^2$$
$$y = a + bx + cx\ln x$$
$$y = a + \frac{b}{x} + cx$$

In fitting functions with more than two parameters to data, we continue to assume that

- random errors affecting y values are normally distributed;
- there are no random errors in the independent (x) values;
- systematic errors in both x and y values are negligible.

The approach to finding an equation which best fits the data, where the equation incorporates more than two parameters follows that described in appendix 3. Though the complexity of the algebraic manipulations increases as the number of parameters increases, this can be overcome by using matrices to assist in parameter estimation.

7.3 Formulating equations to solve for parameter estimates

To determine best estimates of the parameters in an equation using the technique of least squares, we begin with the weighted sum of squares of residuals, χ^2, given by[2]

$$\chi^2 = \sum \left(\frac{y_i - \hat{y}_i}{\sigma_i} \right)^2 \tag{7.1}$$

where y_i is the ith value of y (obtained through measurement), $\hat{y}_i$ is the corresponding predicted value of y found using an equation relating y to x and σ_i is the standard deviation in the ith value of y.

As an example of fitting an equation with more than two parameters to data, consider the equation

$$y = a + bx + cx^2 \tag{7.2}$$

To determine a, b and c, replace $\hat{y}_i$ in equation (7.1) by $a + bx_i + cx_i^2$, so that

$$\chi^2 = \sum \left[\frac{y_i - (a + bx_i + cx_i^2)}{\sigma_i} \right]^2 \tag{7.3}$$

If the uncertainty in y_i is the same for all x_i then σ_i is replaced by σ, so that

$$\chi^2 = \frac{1}{\sigma^2} \sum (y_i - a - bx_i - cx_i^2)^2 \tag{7.4}$$

To find a, b and c that minimise χ^2, differentiate χ^2 with respect to a, b and c in turn then set the resulting equations equal to zero.

Differentiating equation (7.4) gives

$$\frac{\partial \chi^2}{\partial a} = \frac{-2}{\sigma^2} \sum (y_i - a - bx_i - cx_i^2) = 0 \tag{7.5}$$

$$\frac{\partial \chi^2}{\partial b} = \frac{-2}{\sigma^2} \sum x_i(y_i - a - bx_i - cx_i^2) = 0 \tag{7.6}$$

$$\frac{\partial \chi^2}{\partial c} = \frac{-2}{\sigma^2} \sum x_i^2(y_i - a - bx_i - cx_i^2) = 0 \tag{7.7}$$

Rearranging equations (7.5) to (7.7) gives

$$na + b\sum x_i + c\sum x_i^2 = \sum y_i \tag{7.8}$$

$$a\sum x_i + b\sum x_i^2 + c\sum x_i^3 = \sum x_i y_i \tag{7.9}$$

[2] Appendix 3 explains the origin of χ^2.

$$a\sum x_i^2 + b\sum x_i^3 + c\sum x_i^4 = \sum x_i^2 y_i \qquad (7.10)$$

We can rearrange equations (7.8) to (7.10) and substitute from one equation into another to solve for a, b and c. However, this approach is labour intensive and time consuming and the likelihood of a numerical or algebraic mistake is quite high. If the number of parameters to be estimated increases to four or more, then solving for these estimates by 'elimination and substitution' becomes even more formidable. An effective way to proceed is to write the equations in matrix form, as solving linear equations using matrices is quite efficient, especially if software is available that can manipulate matrices.

Writing equations (7.8) to (7.10) in matrix form gives

$$\begin{bmatrix} n & \sum x_i & \sum x_i^2 \\ \sum x_i & \sum x_i^2 & \sum x_i^3 \\ \sum x_i^2 & \sum x_i^3 & \sum x_i^4 \end{bmatrix} \begin{bmatrix} a \\ b \\ c \end{bmatrix} = \begin{bmatrix} \sum y_i \\ \sum x_i y_i \\ \sum x_i^2 y_i \end{bmatrix} \qquad (7.11)$$

Equation (7.11) can be written concisely as[3]

$$\mathbf{AB} = \mathbf{P} \qquad (7.12)$$

where

$$\mathbf{A} = \begin{bmatrix} n & \sum x_i & \sum x_i^2 \\ \sum x_i & \sum x_i^2 & \sum x_i^3 \\ \sum x_i^2 & \sum x_i^3 & \sum x_i^4 \end{bmatrix} \qquad \mathbf{B} = \begin{bmatrix} a \\ b \\ c \end{bmatrix} \qquad \mathbf{P} = \begin{bmatrix} \sum y_i \\ \sum x_i y_i \\ \sum x_i^2 y_i \end{bmatrix}$$

To determine elements, a, b and c of $\mathbf{B}$ (which are the parameter estimates appearing in equation (7.2)), equation (7.12) is manipulated to give[4]

$$\mathbf{B} = \mathbf{A}^{-1}\mathbf{P} \qquad (7.13)$$

where $\mathbf{A}^{-1}$ is the inverse matrix of the matrix $\mathbf{A}$. Matrix inversion and matrix multiplication are tedious to perform 'by hand', especially if matrices are large. The built in matrix functions in Excel® are well suited to estimating parameters in linear least squares problems.

Exercise A

The relationship between the voltage across a semiconductor diode, V, and the temperature of the diode, T, is given by

$$V = \alpha + \beta T + \gamma T \ln T \qquad (7.14)$$

---

[3] By convention, matrices are indicated in bold type.
[4] See appendix 5 for an introduction to matrices.

where α, β and γ are constants. An experiment is performed in which the voltage across the diode is measured as the temperature of the diode changes. Express, in matrix form, the equations that must be solved to obtain best estimates of α, β and γ (assume an unweighted fit is appropriate).

7.4 Matrices and Excel®

Among the many built in functions provided by Excel® are those related to matrix inversion and matrix multiplication. These functions can be used to solve for the parameter estimates (and the standard errors in the estimates) appearing in an equation fitted to data.

7.4.1 The MINVERSE() function

An important step in parameter estimation using matrices is that of matrix inversion. The Excel® function MINVERSE() may be used to invert a matrix such as that appearing in sheet 7.1. The syntax of the function is

=MINVERSE(array)

where array contains the elements of the matrix to be inverted.

Sheet 7.1. *Cells containing a square matrix.*

	A	B	C
1	2.6	7.3	3.4
2	9.5	4.5	5.5
3	6.7	2.3	7.8
4			

Example 1
Use the MINVERSE() function to invert the matrix shown in sheet 7.1.

ANSWER
Inverting a 3×3 matrix creates another 3×3 matrix. As the MINVERSE() function returns an array with nine elements, we highlight nine cells (cells E1:G3) into which Excel® can return those elements. Sheet 7.2 shows the highlighted cells along with the function =MINVERSE(A1:C3) typed into cell E1. To complete the inversion of the matrix shown in columns A to C in sheet 7.2, it is necessary to hold down the CTRL and Shift keys then press the Enter key. Sheet 7.3 shows the elements of the inverted matrix in columns E to G.

Sheet 7.2. *Example of matrix inversion using the MINVERSE() function in Excel®.*

	A	B	C	D	E	F	G
1	2.6	7.3	3.4		=MINVERSE(A1:C3)		
2	9.5	4.5	5.5				
3	6.7	2.3	7.8				
4							

Sheet 7.3. *Outcome of matrix inversion.*

	A	B	C	D	E	F	G
1	2.6	7.3	3.4		-0.09285	0.203164	-0.10278
2	9.5	4.5	5.5		0.154069	0.01034	-0.07445
3	6.7	2.3	7.8		0.034329	-0.17756	0.238445
4							

Exercise B

Use Excel® to invert the following matrices

$$(i) \begin{bmatrix} 2.5 & 9.9 & 6.7 \\ 7.8 & 3.0 & 6.9 \\ 8.8 & 4.5 & 3.2 \end{bmatrix}, (ii) \begin{bmatrix} 62 & 41 & 94 & 38 \\ 42 & 65 & 41 & 87 \\ 102 & 45 & 76 & 32 \\ 91 & 83 & 14 & 21 \end{bmatrix}, (iii) \begin{bmatrix} 2.3 & 1.8 & 4.4 & 9.4 & 6.9 \\ 6.7 & 9.9 & 3.4 & 3.3 & 7.0 \\ 8.4 & 3.4 & 8.2 & 3.9 & 2.9 \\ 9.4 & 6.6 & 9.0 & 6.6 & 5.6 \\ 10.4 & 1.5 & 1.7 & 5.7 & 4.6 \end{bmatrix}.$$

7.4.2   The MMULT() function

Matrix multiplication is required when performing least squares using matrices. The MMULT() function in Excel® is an array function which returns the product of two matrices.

Suppose $\mathbf{P} = \mathbf{AB}$, where

$$\mathbf{A} = \begin{bmatrix} 2.6 & 7.3 & 3.4 \\ 9.5 & 4.5 & 5.5 \\ 6.7 & 2.3 & 7.8 \end{bmatrix} \qquad \mathbf{B} = \begin{bmatrix} 34.4 \\ 43.7 \\ 12.3 \end{bmatrix}$$

The elements of the $\mathbf{P}$ matrix can be found using the MMULT() function. The syntax of the function is,

=MMULT(array1, array2)

where array1 and array2 contain the elements of the matrices to be multiplied together.

Example 2

Use the MMULT() function to determine the product, **P**, of the matrices **A** and **B** shown in sheet 7.4.

ANSWER

Multiplying the 3×3 matrix by the 3×1 matrix in sheet 7.4 produces another matrix of dimension 3×1. As the MMULT() function returns an array containing the elements of the matrix **P**, we highlight cells (shown in column G of sheet 7.5) into which those elements can be returned.

To determine **P**, type =**MMULT(A2:C4,E2:E4)** into cell G2. Holding down the CTRL and Shift keys, then pressing the Enter key returns the elements of the **P** matrix into cells G2 to G4 as shown in sheet 7.6.

Sheet 7.4. *Worksheet containing two matrices,* **A** *and* **B**.

	A	B	C	D	E	F	G
1		A			B		
2	2.6	7.3	3.4		34.4		
3	9.5	4.5	5.5		43.7		
4	6.7	2.3	7.8		12.4		

Sheet 7.5. *Matrix multiplication using MMULT().*

	A	B	C	D	E	F	G	H
1		A			B		P	
2	2.6	7.3	3.4		34.4		=MMULT(A2:C4,E2:E4)	
3	9.5	4.5	5.5		43.7		██████████	
4	6.7	2.3	7.8		12.4		██████████	

Sheet 7.6. *Outcome of multiplying* **A** *and* **B** *appearing in sheet 7.4.*

	G
1	P
2	450.61
3	591.65
4	427.71

Exercise C

Use Excel® to carry out the following matrix multiplications:

(i) $\begin{bmatrix} 56.8 & 123.5 & 67.8 \\ 87.9 & 12.5 & 54.3 \\ 23.6 & 98.5 & 56.7 \end{bmatrix} \begin{bmatrix} 23.1 \\ 34.6 \\ 56.8 \end{bmatrix}$,

(ii) $\begin{bmatrix} 12 & 45 & 67 & 56 \\ 34 & 54 & 65 & 43 \\ 12 & 54 & 49 & 31 \\ 84 & 97 & 23 & 99 \end{bmatrix} \begin{bmatrix} 32 \\ 19 \\ 54 \\ 12 \end{bmatrix}$.

Table 7.1. *x–y data.*

x	y
4	30.4
6	51.2
8	101.6
10	184.4
12	262.6
14	369.6
16	479.4
18	601.5
20	764.9

7.4.3 Fitting the polynomial $y = a + bx + cx^2$ to data

We now combine the matrix approach to solving for estimated parameters discussed in section 7.3 with the built in matrix functions in Excel® to determine a, b and c in the equation $y = a + bx + cx^2$. Consider the data in table 7.1. Assuming it is appropriate to fit the equation $y = a + bx + cx^2$ to the data in table 7.1, we can find a, b and c by constructing the matrices

$$A = \begin{bmatrix} n & \sum x_i & \sum x_i^2 \\ \sum x_i & \sum x_i^2 & \sum x_i^3 \\ \sum x_i^2 & \sum x_i^3 & \sum x_i^4 \end{bmatrix} \tag{7.15}$$

$$B = \begin{bmatrix} a \\ b \\ c \end{bmatrix} \tag{7.16}$$

and

$$P = \begin{bmatrix} \sum y_i \\ \sum x_i y_i \\ \sum x_i^2 y_i \end{bmatrix} \tag{7.17}$$

Using the data in table 7.1 (and with the assistance of Excel®), we find $\sum x_i = 108$, $\sum y_i = 2845.6$, $\sum x_i^2 = 1536$, $\sum x_i^3 = 24192$, $\sum x_i^4 = 405312$, $\sum x_i y_i = 45206.6$, $\sum x_i^2 y_i = 761100.4$, $n = 9$.

Matrices A and P become

$$A = \begin{bmatrix} 9 & 108 & 1536 \\ 108 & 1536 & 24192 \\ 1536 & 24192 & 405312 \end{bmatrix} \quad P = \begin{bmatrix} 2845.6 \\ 46206.6 \\ 761100.4 \end{bmatrix}$$

As $\mathbf{B} = \mathbf{A}^{-1}\mathbf{P}$, we find (using Excel® for matrix inversion and multiplication)[5]

$$\mathbf{B} = \begin{bmatrix} a \\ b \\ c \end{bmatrix} = \begin{bmatrix} 7.780 \times 10^{-1} & -1.380 \times 10^{-1} & 5.285 \times 10^{-3} \\ -1.380 \times 10^{-1} & 3.532 \times 10^{-2} & -1.586 \times 10^{-5} \\ 5.285 \times 10^{-3} & -1.586 \times 10^{-5} & 7.708 \times 10^{-5} \end{bmatrix} \begin{bmatrix} 2845.6 \\ 46206.6 \\ 761100.4 \end{bmatrix}$$

$$= \begin{bmatrix} 1.9157 \\ -2.7458 \\ 2.0344 \end{bmatrix}$$

so that the relationship between x and y can be written, $y = 1.916 - 2.746x + 2.034x^2$.

Exercise D

The variation of the electrical resistance, R, with temperature, T, of a wire made from high purity platinum is shown in table 7.2. Assuming the relationship between R and T can be written

$$R = A + BT + CT^2 \tag{7.18}$$

use least squares to determine best estimates for A, B and C.

Table 7.2. *Variation of resistance with temperature of a platinum wire.*

T(K)	R (Ω)
70	17.1
100	30.0
150	50.8
200	71.0
300	110.5
400	148.6
500	185.0
600	221.5
700	256.2
800	289.8
900	322.2
1000	353.4

[5] For convenience, the elements of $\mathbf{A}^{-1}$ are shown to four figures, but full precision (to 15 figures) is used 'internally' when calculations are carried out with Excel®.

7.5 Multiple least squares[6]

To this point we have considered situations in which the dependent variable, y, is a function of only one independent variable, x. However, there are situations in which the effect of two or more independent variables must be accounted for. For example, when vacuum depositing a thin film, the thickness, T, of the film depends upon:

1. deposition time, t,
2. the gas pressure in the vacuum chamber, P,
3. the distance, d, between deposition source and surface to be coated with the thin film.

To find the relationship between T and the independent variables, we could consider one independent variable at a time and determine the relationship between that variable and T. Another approach is to allow all the variables t, P and d to vary and to use multiple least squares to establish the relationship between T and all the independent variables.

An example of an equation possessing two independent variables is

$$y = a + bx + cz \tag{7.19}$$

As usual, y is the dependent variable. x and z are the independent variables. Can we determine a, b and c by fitting equation (7.19) to experimental data using linear least squares? The answer is yes and the problem is no more complicated than fitting a polynomial to data.

We begin with χ^2, given by

$$\chi^2 = \sum \left(\frac{y_i - \hat{y}_i}{\sigma_i} \right)^2 \tag{7.20}$$

To fit the equation $y = a + bx + cz$ to the data, replace $\hat{y}_i$ in equation (7.20) by $a + bx_i + cz_i$ to give

$$\chi^2 = \sum \left[\frac{y_i - (a + bx_i + cz_i)}{\sigma_i} \right]^2 \tag{7.21}$$

If the uncertainty in the y values is constant, σ_i is replaced by σ, so that

$$\chi^2 = \frac{1}{\sigma^2} \sum (y_i - a - bx_i - cz_i)^2 \tag{7.22}$$

To minimise χ^2, differentiate χ^2 with respect to a, b and c in turn and set the resulting equations equal to zero. Following the approach described in section 7.3, the matrix equation to be solved for a, b and c is

---

[6] Often referred to as 'Multiple Regression'.

$$\begin{bmatrix} n & \sum x_i & \sum z_i \\ \sum x_i & \sum x_i^2 & \sum x_i z_i \\ \sum z_i & \sum x_i z_i & \sum z_i^2 \end{bmatrix} \begin{bmatrix} a \\ b \\ c \end{bmatrix} = \begin{bmatrix} \sum y_i \\ \sum x_i y_i \\ \sum x_i y_i \end{bmatrix} \qquad (7.23)$$

$$\underset{\mathbf{A}}{\big\downarrow} \qquad \underset{\mathbf{B}}{\big\downarrow} \quad \underset{\mathbf{P}}{\big\downarrow}$$

To solve for a, b, and c, we determine $\mathbf{B} = \mathbf{A}^{-1}\mathbf{P}$.

Table 7.3. *Variation of resonance frequency with string tension and string length.*

f (Hz)	T (N)	L (m)
28	0.49	1.72
48	0.98	1.43
66	1.47	1.24
91	1.96	1.06
117	2.45	0.93
150	2.94	0.82
198	3.43	0.67

Example 3

In an experiment to study waves on a stretched string, the frequency, f, at which the string resonates is measured as a function of the tension, T, of the string and the length, L, of the string. The data gathered in the experiment are shown in table 7.3. Assuming that the frequency, f, can be written in the form

$$\ln f = \ln K + B \ln T + C \ln L \qquad (7.24)$$

where K is a constant, use multiple least squares to determine best estimates of K, B and C.

ANSWER

Equation (7.24) can be written

$$y = a + bx + cz \qquad (7.25)$$

where

$$y = \ln f \qquad (7.26)$$

$$x = \ln T \qquad (7.27)$$

and

$$z = \ln L \qquad (7.28)$$

Also,

$$a = \ln K \qquad b = B \qquad c = C$$

Using the data in table 7.3, and the transformations given by equations (7.26) to (7.28), the matrices $\mathbf{A}$ and $\mathbf{P}$ in equation (7.23) become

$$\mathbf{A} = \begin{bmatrix} 7 & 3.532 & 0.5019 \\ 3.532 & 4.596 & -1.045 \\ 0.5019 & -1.045 & 0.6767 \end{bmatrix} \qquad \mathbf{P} = \begin{bmatrix} 30.97 \\ 18.38 \\ 0.8981 \end{bmatrix}$$

The inverse of matrix $\mathbf{A}$ is

$$\mathbf{A}^{-1} = \begin{bmatrix} 2.433 & -3.512 & -7.226 \\ -3.512 & 5.406 & 10.95 \\ -7.226 & 10.95 & 23.74 \end{bmatrix}$$

Determining $\mathbf{B} = \mathbf{A}^{-1}\mathbf{P}$ gives

$$\mathbf{B} = \begin{bmatrix} 4.281 \\ 0.4476 \\ -1.157 \end{bmatrix}$$

so that

$$a = 4.281 = \ln K, \text{ so } K = 72.31,$$
$$b = 0.4476 = B, \text{ and}$$
$$c = -1.157 = C.$$

Finally, we write

$$\ln f = 4.281 + 0.4476 \ln T - 1.157 \ln L$$

Exercise E
Consider the data shown in table 7.4. Assuming that y is a function of x and z such that $y = a + bx + cz$, use linear least squares to solve for a, b and c.

Table 7.4. *y, x and z data for exercise E.*

y	x	z
20.7	1	1
23.3	2	2
28.2	3	4
35.7	4	5
48.2	5	11
56.0	6	14

7.6   Standard errors in parameter estimates

We can extend the calculation of standard errors in intercept and slope given in appendix 4 to include situations in which there are more than two parameters. In section 7.3 we found $\mathbf{B} = \mathbf{A}^{-1}\mathbf{P}$, where the elements of $\mathbf{B}$ are the parameter estimates. The standard error in each parameter estimate is obtained by forming the product of the diagonal elements of $\mathbf{A}^{-1}$ and the standard deviation, σ, in each value, y_i. The diagonal elements of $\mathbf{A}^{-1}$ are written as A_{11}^{-1}, A_{22}^{-1}, A_{33}^{-1}. The standard errors of the elements a, b and c of $\mathbf{B}$, written as σ_a, σ_b, σ_c respectively, are given by[7]

$$\sigma_a = \sigma(A_{11}^{-1})^{\frac{1}{2}} \tag{7.29}$$

$$\sigma_b = \sigma(A_{22}^{-1})^{\frac{1}{2}} \tag{7.30}$$

$$\sigma_c = \sigma(A_{33}^{-1})^{\frac{1}{2}} \tag{7.31}$$

σ is determined using

$$\sigma = \left[\frac{1}{n-M} \sum (y_i - \hat{y}_i)^2 \right]^{\frac{1}{2}} \tag{7.32}$$

where n is the number of data and M is the number of parameters in the equation.

---

[7] See Bevington and Robinson (1992) for a derivation of equations (7.29) to (7.31).

Table 7.5. *x–y data for example 4.*

x	y
2	−7.27
4	−17.44
6	−25.99
8	−23.02
10	−16.23
12	−15.29
14	1.85
16	21.26
18	46.95
20	71.87
22	105.97

Example 4

Consider the data in table 7.5. Assuming it is appropriate to fit the function $y = a + bx + cx^2$ to these data determine, using linear least squares,

(i) a, b and c;

(ii) σ;

(iii) standard errors, σ_a, σ_b and σ_c.

ANSWER

(i) Writing equations to solve for a, b and c in matrix form gives (see section 7.3)

$$\underbrace{\begin{bmatrix} n & \sum x_i & \sum x_i^2 \\ \sum x_i & \sum x_i^2 & \sum x_i^3 \\ \sum x_i^2 & \sum x_i^3 & \sum x_i^4 \end{bmatrix}}_{A} \underbrace{\begin{bmatrix} a \\ b \\ c \end{bmatrix}}_{B} = \underbrace{\begin{bmatrix} \sum y_i \\ \sum x_i y_i \\ \sum x_i^2 y_i \end{bmatrix}}_{P} \tag{7.33}$$

To find the elements of **B**, write $\mathbf{B} = \mathbf{A}^{-1}\mathbf{P}$. Using the MINVERSE() and MMULT() functions in Excel® gives

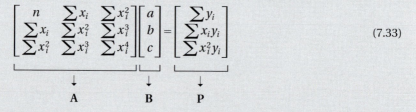

$$\underbrace{\begin{bmatrix} 1.206 & -0.2091 & 0.007576 \\ -0.2091 & 0.04423 & -0.001748 \\ 0.007576 & -0.001748 & 7.284 \times 10^{-5} \end{bmatrix}}_{A^{-1}} \underbrace{\begin{bmatrix} 142.6 \\ 4210 \\ 94510 \end{bmatrix}}_{P} = \underbrace{\begin{bmatrix} 7.847 \\ -8.862 \\ 0.6058 \end{bmatrix}}_{B} \tag{7.34}$$

so that $a = 7.847$, $b = -8.862$ and $c = 0.6058$.

(ii) σ is determined using equation (7.32) with $n = 11$, $M = 3$ and $\hat{y}_i = a + bx_i + cx_i^2$ so that

$$\sigma = \left\{ \frac{1}{8} \sum [y_i - (a + bx_i + cx_i^2)]^2 \right\}^{\frac{1}{2}} = 2.423.$$

(iii) σ_a, σ_b and σ_c are found using equations (7.29) to (7.31)

$$\sigma_a = \sigma(A_{11}^{-1})^{\frac{1}{2}} = 2.423 \times (1.2061)^{\frac{1}{2}} = 2.7$$

$$\sigma_b = \sigma(A_{22}^{-1})^{\frac{1}{2}} = 2.423 \times (0.04423)^{\frac{1}{2}} = 0.51$$

$$\sigma_c = \sigma(A_{33}^{-1})^{\frac{1}{2}} = 2.423 \times (7.284 \times 10^{-5})^{\frac{1}{2}} = 0.021$$

Exercise F

Consider the thermocouple data in table 7.6. Assuming the relationship between V and T can be written as

$$V = A + BT + CT^2 + DT^3 \tag{7.35}$$

use linear least squares to determine best estimates for A, B, C, D and the standard errors in these estimates.

Table 7.6. *Output voltage of a thermocouple as a function of temperature.*

T (K)	V (μV)
75	−5.936
100	−5.414
125	−4.865
150	−4.221
175	−3.492
200	−2.687
225	−1.817
250	−0.892
275	0.079
300	1.081

7.6.1 Confidence intervals for parameters

If a, b and c are best estimates of the parameters α, β and γ respectively obtained using least squares, then the $X\%$ confidence interval for each parameter can be written

$$\alpha = a \pm t_{X\%,\nu}\,\sigma_a$$

$$\beta = b \pm t_{X\%,\nu}\,\sigma_b$$

$$\gamma = c \pm t_{X\%,\nu}\,\sigma_c$$

where σ_a, σ_b and σ_c are the standard errors in a, b and c respectively. $t_{X\%,\nu}$ is the critical t value for ν degrees of freedom at the $X\%$ confidence level. ν is found using

$$\nu = n - M \tag{7.36}$$

where n is the number of data and M is the number of parameters in the equation.

Example 5

In fitting the equation $y = \alpha + \beta x + \gamma x^2 + \delta x^3$ to 15 data points, it is found that estimates of the parameters α, β, γ and δ (written as a, b, c and d respectively) and their standard errors[8] are

$a = 14.65$ $\sigma_a = 1.727$

$b = 0.4651$ $\sigma_b = 0.02534$

$c = 0.1354$ $\sigma_c = 0.02263$

$d = 0.04502$ $\sigma_d = 0.01018$

Use this information to determine the 95% confidence interval for each parameter.

ANSWER

For the parameter α we write

$$\alpha = a \pm t_{X\%,\nu}\,\sigma_a$$

The number of degrees of freedom, ν $(= n - M)$ is $15 - 4 = 11$. Referring to table 2 in appendix 1, we have for a confidence level of 95%,

$$t_{95\%,\,11} = 2.201$$

---

[8] Standard errors are given to four significant figures to avoid rounding errors in subsequent calculations.

It follows that

$$\alpha = 14.65 \pm 2.201 \times 1.727 = 14.65 \pm 3.801$$

(which we normally round to 14.7 ± 3.8). Similarly for β, γ and δ

$$\beta = 0.4651 \pm 2.201 \times 0.02534 = 0.465 \pm 0.056$$

$$\gamma = 0.1354 \pm 2.201 \times 0.02263 = 0.135 \pm 0.050$$

$$\delta = 0.04502 \pm 2.201 \times 0.01018 = 0.045 \pm 0.022$$

Exercise G

A force, F, is required to displace the string of an archer's bow by an amount d. Assuming that the relationship between F and d can be written

$$F = \alpha + \beta d + \gamma d^2 \tag{7.37}$$

where α, β and γ are parameters to be estimated using linear least squares, estimates of the parameters were determined by fitting equation (7.37) to F versus d data (not shown here) which consisted of 12 data points. Estimates of α, β and γ (written as a, b and c respectively) and their standard errors are

$$a = 0.01047 \qquad \sigma_a = 0.005993 \text{ (unit N)}$$

$$b = 3.931 \qquad \sigma_b = 1.834 \text{ (unit N/m)}$$

$$c = 426.8 \qquad \sigma_c = 13.03 \text{ (unit N/m}^2\text{)}$$

Use this information to determine the 90% confidence interval for α, β and γ.

7.7 Weighting the fit

If a weighted fit is required,[9] we return to the weighted sum of squares, χ^2, given by

$$\chi^2 = \sum \left[\frac{y_i - \hat{y}_i}{\sigma_i} \right]^2 \tag{7.38}$$

To allow for uncertainty that varies from one y value to another, we retain σ_i^2 in the subsequent analysis. Equation (7.38) is differentiated with respect to each parameter in turn and the resulting equations are set equal to zero.

[9] Weighted fitting is required if the uncertainty in each y value is not constant – see section 6.10.

Next we solve the equations for estimates of the parameters which mini-mise χ^2. As an example, if the function to be fitted to the data is the poly-nomial $y = a + bx + cx^2$, then a, b and c may be determined using matrices by writing $\mathbf{B} = \mathbf{A}^{-1}\mathbf{P}$, where

$$\mathbf{B} = \begin{bmatrix} a \\ b \\ c \end{bmatrix} \tag{7.39}$$

$\mathbf{A}^{-1}$ is the inverse of $\mathbf{A}$, where $\mathbf{A}$ is written

$$\mathbf{A} = \begin{bmatrix} \sum \frac{1}{\sigma_i^2} & \sum \frac{x_i}{\sigma_i^2} & \sum \frac{x_i^2}{\sigma_i^2} \\ \sum \frac{x_i}{\sigma_i^2} & \sum \frac{x_i^2}{\sigma_i^2} & \sum \frac{x_i^3}{\sigma_i^2} \\ \sum \frac{x_i^2}{\sigma_i^2} & \sum \frac{x_i^3}{\sigma_i^2} & \sum \frac{x_i^4}{\sigma_i^2} \end{bmatrix} \tag{7.40}$$

and $\mathbf{P}$ is given by

$$\mathbf{P} = \begin{bmatrix} \sum \frac{y_i}{\sigma_i^2} \\ \sum \frac{x_i y_i}{\sigma_i^2} \\ \sum \frac{x_i^2 y_i}{\sigma_i^2} \end{bmatrix} \tag{7.41}$$

The diagonal elements of the $\mathbf{A}^{-1}$ matrix can be used to determine the stan-dard errors in a, b and c. We have[10]

$$\sigma_a = (A_{11}^{-1})^{\frac{1}{2}} \tag{7.42}$$

$$\sigma_b = (A_{22}^{-1})^{\frac{1}{2}} \tag{7.43}$$

$$\sigma_c = (A_{33}^{-1})^{\frac{1}{2}} \tag{7.44}$$

Exercise H

Consider the data in table 7.7. Assuming that it is appropriate to fit the equation $y = a + bx + cx^2$ to these data and that the standard deviation, σ_i, in each y value is given by $\sigma_i = 0.1y_i$:

(i) use weighted least squares to determine a, b and c;
(ii) determine the standard errors in a, b and c.

---

[10] See Neter, Kutner, Nachtsheim and Wasserman (1996).

Table 7.7. *x–y data for exercise H.*

x	y
−10.5	143
−9.5	104
−8.3	79
−5.3	34
−2.1	12
1.1	19
2.3	29

7.8 Coefficients of multiple correlation and multiple determination

In section 6.6 we introduced the linear correlation coefficient, r, which quantifies how well x and y are correlated. r close to 1 or –1 indicates excellent correlation, while r close to 0 indicates poor correlation. We can generalise the correlation coefficient to take into account equations fitted to data which incorporate more than one independent variable or where there are more than two parameters in the equation. The coefficient of multiple correlation, R, is written

$$R = \left[1 - \frac{\sum (y_i - \hat{y}_i)^2}{\sum (\bar{y} - \hat{y}_i)^2} \right]^{\frac{1}{2}} \tag{7.45}$$

where y_i is the ith observed value of y, $\hat{y}_i$ is the predicted y value found using the equation representing the best line through the points and $\bar{y}$ is the mean of the observed y values.[11] We argue that equation (7.45) is plausible by considering the summation $\sum (y_i - \hat{y}_i)^2$ which is the sum of squares of the residuals, SSR. If the line passes near to all the points then SSR is close to 0 (and is equal to 0 if all the points lie on the line) and so R tends to 1, as required by perfectly correlated data.

The square of the coefficient of multiple correlation is termed the *coefficient of multiple determination*, R^2, and is written

$$R^2 = 1 - \frac{\sum (y_i - \hat{y}_i)^2}{\sum (\bar{y} - \hat{y}_i)^2} \tag{7.46}$$

R^2 gives the fraction of the square of the deviations of the observed y values about their mean which can be explained by the equation fitted to the data. So for example, if $R^2 = 0.92$ this indicates that 92% of the scatter of the deviations can be explained by the equation fitted to the data.

[11] See Walpole, Myers and Myers (1998) for a discussion of equation (7.45).

Exercise I

Consider the data in table 7.8. Fitting the equation $y = a + bx + cz$ to these data using least squares gives

$$\hat{y}_i = 2.253 + 1.978x_i - 0.3476z_i$$

where $\hat{y}_i$ is the predicted value of y for a given x_i. Use this information to determine the coefficient of multiple determination, R^2, for the data in table 7.8.

Table 7.8. *Data for exercise I.*

y	x	z
0.84	0.24	5.5
1.47	0.35	4.2
2.23	0.55	3.2
2.82	0.78	2.8
3.60	0.98	1.5
4.25	1.1	0.7

7.9 The LINEST() function for multiple least squares

The LINEST() function introduced in section 6.3 can be used to determine:

- estimates of parameters;
- standard errors in the estimates;
- the coefficient of multiple determination, R^2;
- the standard deviation in the y values, σ, as given by equation (7.32).

When fitting a straight line to data using LINEST(), a single column of numbers representing the x values is entered into the function whose syntax is[12]

LINEST(y values, x values, constant, statistics)

By contrast, when there are two independent variables (such as x and z), an array with two columns of numbers must be entered into LINEST(). That array consists of one column of x values and an adjacent column of z values in the case where $y = a + bx + cz$ is fitted to data. Where the same independent variable appears in two terms of an equation, such as in

---

[12] More details of the syntax of this function appear in section 6.3.

$y = a + bx + cx^2$, one column in Excel® contains x values and an adjacent column contains x^2.

The use of the LINEST() function to fit an equation containing more than two parameters is best illustrated by example. Consider the data in sheet 7.7. Assuming it is appropriate to fit the equation $y = a + bx + cz$ to these data, we proceed as follows:

- Highlight a range of cells (cells A9 to C11 in sheet 7.7) into which Excel® can return the estimates a, b and c, the standard errors in these estimates, the coefficient of multiple determination and the standard deviation in the y values.
- Type =**LINEST(A2:A7,B2:C7,TRUE,TRUE)** into cell A9.
- Hold down the Ctrl and Shift keys, then press the Enter key.

Sheet 7.7. *Using the* LINEST() *function.*

	A	B	C
1	y	x	z
2	28.8	1.2	12.8
3	42.29	3.4	9.8
4	50.69	4.5	6.7
5	66.22	7.4	4.5
6	73.12	8.4	2.2
7	81.99	9.9	1
8			
9	=LINEST(A2:A7, B2:C7,TRUE,TRUE)		
10			
11			

The numbers returned by Excel® are shown in sheet 7.8.

Sheet 7.8. *Numbers returned by the* LINEST() *function.*

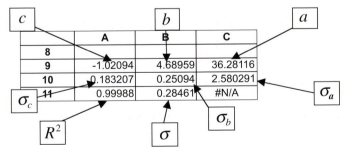

	A	B	C
8			
9	-1.02094	4.68959	36.28116
10	0.183207	0.25094	2.580291
11	0.99988	0.28461	#N/A

A limitation of the LINEST() function is that it does not allow for weighted fitting. If weighted fitting of linear functions containing several parameters is required, then the matrix method described in section 7.7 is recommended.

Table 7.9. *Data for exercise J.*

x	y
0.02	−0.0261
0.04	−0.0202
0.06	−0.0169
0.08	−0.0143
0.10	−0.0124
0.12	−0.0105
0.14	−0.0094
0.16	−0.0080
0.18	−0.0067

Exercise J

Consider the data in table 7.9. Use LINEST() to fit the equation $y = a + bx + c\ln x$ to these data and so determine:

(i) a, b, and c;
(ii) the standard errors in a, b, and c;
(iii) the coefficient of multiple determination, R^2;
(iv) the standard deviation, σ, in the y values.

7.10   Choosing equations to fit to data

In some situations, selecting an equation to fit to data is not difficult. If consideration of the physical principles underlying the x–y data predicts that the relationship between y and x should be linear, and if a visual inspection of the data lends support to that prediction, then fitting the equation $y = a + bx$ is appropriate.

If there is little or no past experience to act as a guide as to which equation to fit to data we may need to rely on an examination of the data presented in graphical form to suggest one or more equations which can be trialled. However, it is worth emphasising that an equation relating the dependent to independent variables which can be supported by an application of the physical principles involved is to be preferred to one that simply gives the smallest sum of squares of residuals.

Another situation in which the choice of equation to fit to data is 'clear cut' is when the purpose of data gathering is to calibrate an instrument. Previous experience, manufacturers' guidelines or other published information are determining factors in the choice of the equation. When calibration is the aim of the least squares analysis, the physical interpretation of the parameter estimates that emerge from the fitting process is often not important, as the main aim is to use the fitted function for the purpose of interpolation (and occasionally extrapolation).

Whatever equation is fitted to data, we should be suspicious if standard errors of parameter estimates are of comparable size to (or larger than) the estimates themselves. For example, if some parameter, β, appearing in an equation is estimated by least squares to be $\beta = 3.6 \pm 5.5$, then it is likely that the term containing β can be eliminated with little adverse effect on the quality of the fit of the equation to data. In the next chapter we will see that it is possible to make a judgement, based upon statistical considerations as to whether a particular parameter estimate is 'significant' and we will leave discussion of this until we have considered 'tests of significance'.

7.10.1 Comparing equations fitted to data

Occasionally a situation arises in which, based on a consideration of the physical processes believed to underlie the relationship being studied, two or more equations may be fit to data. For example, the electrical resistance, r, of a material at a temperature, T, may be described by

$$r = A + BT \tag{7.47}$$

or

$$r = \alpha + \beta T + \gamma T^2 \tag{7.48}$$

where A, B, α, β and γ are constants. If we use χ^2 as given by equation (7.1) as a means of determining which of the two equations better fits the data, we will always favour equation (7.48) over equation (7.47). This is due to the fact that the addition of the term in T^2 in equation (7.48) gives extra flexibility to the line so that predicted values found using the equation lie closer to the data values than if that term were omitted. However, the weighted sum of squares obtained when fitting equation (7.48) to data may only be marginally less than when equation (7.47) is fitted to the same data.

Another possibility as an indicator of the equation that better fits the data is the coefficient of multiple determination, R^2, such that the equation yielding the larger value of R^2 is regarded as the better equation.

Regrettably, as with the weighted sum of squares, R^2 favours the equation with the greater number of parameters so that, for example, the equation $y = a + bx + cx^2 + dx^3$ would always be favoured over the equation $y = a + bx + cx^2$. To take into account the number of parameters such that any marginal increase in R^2 is offset by the number of parameters used, the adjusted coefficient of multiple determination, R^2_{ADJ}, is sometimes used, where[13]

$$R^2_{ADJ} = \frac{(n-1)R^2 - (M-1)}{n-M} \tag{7.49}$$

R^2 is given by equation (7.46), n is the number of data and M is the number of parameters. Once R^2_{ADJ} is calculated for each equation fitted to data, the equation is preferred that has the larger value of R^2_{ADJ}.

Another way of comparing two (or more) equations fitted to data where the equations have different numbers of parameters is to use the Akaike information criterion[14] (AIC). This criterion takes into account the sum of squares of residuals SSR, but also includes a term proportional to the number of parameters used. AIC may be written

$$AIC = n \ln SSR + 2M \tag{7.50}$$

where n is the number of data and M is the number of parameters in the equation.

AIC depends on the sum of squares which appears in the first term of the right hand side of equation (7.50). The second term on the right hand side can be considered as a 'penalty' term. If the addition of another parameter in an equation reduces SSR then the first term on the right hand side of equation (7.50) becomes smaller. However, the second term on the right hand side increases by 2 for every extra parameter used. It follows that a modest decrease in SSR which occurs when an extra term is introduced into an equation may be more than offset by the increase in AIC by using another parameter. We conclude that if two or more equations are fitted to data, then the equation producing the *smallest* value for AIC should be preferred.

If weighted least squares is required, we replace SSR in equation (7.50) by the weighted sum of squares of residuals, χ^2, where χ^2 is given by equation (7.1).

---

[13] See Neter, Kutner, Nachtsheim and Wasserman (1996) for a discussion of equation (7.49).

[14] See Akaike (1974) for a discussion on model identification.

Example 6

Table 7.10 shows the variation of the resistance of an alloy with temperature. Using (unweighted) linear least squares, fit both equation (7.47) and equation (7.48) to these data and determine for each equation:

   (i) estimates for the parameters;
   (ii) the standard error in each estimate;
   (iii) the standard deviation, σ, in each y value;
   (iv) the multiple correlation coefficient, R^2;
   (v) the adjusted correlation coefficient, R^2_{ADJ};
   (vi) the sum of squares of the residuals, SSR;
   (vii) the Akaike information criterion, AIC.

ANSWER

The extent of computation required in this problem is a strong encouragement to use the LINEST() function in Excel®. As usual, we write estimates of parameters as a, b and c etc. In the case of equation (7.47) we determine a and b where the equation is of the form $y = a + bx$, and for equation (7.48) we determine a, b and c where $y = a + bx + cx^2$. Most numbers shown in table 7.11 are rounded to four significant figures, though intermediate calculations utilise the full precision of Excel®.

Returning to equations (7.47) and (7.48), we can now write[15]

$$A = (12.41 \pm 0.49)\ \Omega \qquad B = (4.30 \pm 0.19) \times 10^{-2}\ \Omega/K$$

$$\alpha = (14.0 \pm 2.2)\ \Omega \qquad \beta = (3.0 \pm 1.8) \times 10^{-2}\ \Omega/K \qquad \gamma = (2.7 \pm 3.6) \times 10^{-5}\ \Omega/K^2$$

R^2_{ADJ} for equation (7.47) fitted to data is greater than R^2_{ADJ} for equation (7.48), indicating that equation (7.47) is the better fit. The AIC for equation (7.47) fitted to data is lower than for equation (7.48), further supporting equation (7.47) as the more appropriate equation. Finally, an inspection of the standard errors of the parameter estimates suggests that, for the equation with three parameters, the standard error in c is so large in comparison to c that γ in equation (7.48) is 'redundant'.

Table 7.10. *Data for example 6.*

r (Ω)	19.5	18.4	20.2	20.1	20.9	20.8	21.2	21.8	21.9	23.6	23.2
T (K)	150	160	170	180	190	200	210	220	230	240	250

r (Ω)	23.9	23.2	24.1	24.2	26.3	25.5	26.1	26.3	27.1	28.0
T (K)	260	270	280	290	300	310	320	330	340	350

[15] We write the standard errors in the parameters to two significant figures in line with the convention adopted in chapter 1.

Table 7.11. *Parameter estimates found by fitting equations (7.47) and (7.48) to data in table 7.10.*[16]

	Fitting $y=a+bx$ to data	Fitting $y=a+bx+cx^2$ to data
Parameter estimates	$a=12.41, b=4.299\times10^{-2}$	$a=13.97, b=2.970\times10^{-2}, c=2.657\times10^{-5}$
Standard errors	$\sigma_a=0.49, \sigma_b=1.9\times10^{-3}$	$\sigma_a=2.2, \sigma_b=1.8\times10^{-2}, \sigma_c=3.6\times10^{-5}$
σ	0.5304	0.5368
R^2	0.9638	0.9649
R^2_{ADJ}	0.9619	0.9610
SSR	5.344	5.186
AIC	39.20	40.57

Exercise K

Using the data in table 7.10 verify that the parameter estimates and the standard errors in the estimates in table 7.11 are correct.

7.11 Non-linear least squares

The equations that we have fitted to data so far have been linear in the parameters and we have determined best estimates of those parameters using linear least squares. However, there are many equations that occur in the physical sciences that are not linear in the parameters, for example

$$I=I_0\exp(-\mu x)+B \tag{7.51}$$

(where I_0, μ and B are parameters, I is the dependent variable and x is the independent variable) and

$$I=\frac{AV}{B-V}+C \tag{7.52}$$

(where A, B and C are parameters, I is the dependent variable and V is the independent variable).

We cannot apply the technique of linear least squares to either equation (7.51) or equation (7.52) as they do not permit the construction of a set of simultaneous equations which are linear in the unknown parameters. In

---

[16] In the interests of conciseness, units of measurement have been omitted from the table.

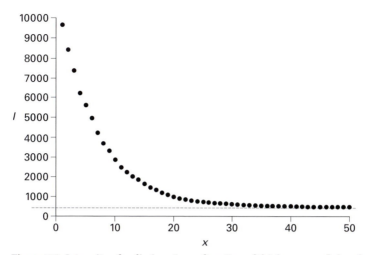

Figure 7.3. Intensity of radiation, I, as a function of thickness, x, of absorber.

some cases it is possible to estimate one or more parameters by inspecting a graph of the data. While this is not as satisfactory as applying least squares, it can produce quite reasonable values for an unknown parameter. Consider, for example, equation (7.51) which relates to the intensity of radiation, I, passing through a material of thickness, x, with a background radiation, B. Data gathered in an absorption experiment are shown in figure 7.3. When x is large, equation (7.51) predicts that I will tend to B. Figure 7.3 indicates that, for this set of data, $B \approx 500$. Using this value for B we can write equation (7.51) as

$$I - 500 = I_0 \exp(-\mu x) \tag{7.53}$$

It is now possible to linearise equation (7.53) by taking natural logarithms of each side, and then perform least squares fitting to data to find I_0 and μ. It should be emphasised that this method is fairly crude as there some guess work in finding B, especially if the data are noisy. Nevertheless, this is a fair way of obtaining rough estimates of the parameters involved. How may we determine parameter estimates more 'rigorously' if we are unable to apply linear least squares?

We return to the equation which represents the weighted sum of squares, χ^2, given by

$$\chi^2 = \sum \left(\frac{y_i - \hat{y}_i}{\sigma_i} \right)^2 \tag{7.54}$$

In contrast to the equations we have so far fitted to data, differentiating χ^2 with respect to each of the parameters in turn, then setting the derivatives to 0, does not produce a set of linear equations that can be solved for 'best estimates'. The technique that is adopted to find parameter estimates is to begin with equation (7.54) and replace $\hat{y}_i$ by the equation to be fitted. For example we might have

$$\hat{y}_i = a\exp(-bx_i) + c \tag{7.55}$$

so that

$$\chi^2 = \sum \left[\frac{y_i - (a\exp(-bx_i) + c)}{\sigma_i} \right]^2 \tag{7.56}$$

The next stage is to 'guess' values (usually referred to as starting values) for a, b and c and then to calculate χ^2. Assuming that the guessed values are not those that will minimise χ^2, we need to begin a search for the best values of a, b and c by modifying the starting values in some systematic manner and at each stage determine whether χ^2 has reduced. If χ^2 *has* reduced then the modified starting values are closer to the best parameter estimates. The parameters are further adjusted until no more reduction in χ^2 is obtained. At this point we have the best estimates of the parameters we seek. Due to the iterative nature of fitting demanded by non-linear least squares, such fitting is done using a computer based mathematics or statistics package.[17] We will not discuss the method of non-linear least squares in detail here but just point to two important issues to be aware of when using any package to fit an equation to data using the technique of non-linear least squares.

(i) If the starting values are quite different from the best estimates, then the non-linear least squares technique may converge very slowly to the best estimates. Worse still, convergence may never occur, possibly due to the fact that, when starting values are varied, the value obtained for χ^2 is beyond the range of numbers that the computer program can cope with and the program returns an error message.

(ii) It is possible, especially with noisy data, for a minimum in χ^2 to be obtained such that any further small changes in the parameter estimates causes χ^2 to increase. However what has been found by the computer is, in fact, a 'local minimum'. This means that there is another combination of parameter estimates that produces an even smaller value for χ^2 (often referred to as a 'global' minimum). Avoiding being trapped in a local minimum can be quite difficult and often relies on the user of the program 'knowing' that the parameter estimates obtained by the package are nonsense.

[17] Excel® does not provide built in facilities for fitting equations to data using non-linear least squares.

7.12 Review

Fitting equations to experimental data is a common occupation of experimenters in the physical sciences. Whether the aim of the comparison is to establish the values of parameters so that they can be compared with those predicted by theory, or to use the parameter estimates for calibration purposes, finding 'best' estimates of the parameters and the standard errors in the estimates is very important.

In this chapter we considered how the technique of least squares can be extended beyond fitting the equation $y = a + bx$ to data. Increasing the number of parameters that must be determined in turn increases the complexity of the calculations that must be carried out. Applying matrix methods considerably eases the process of determining best estimates of parameters. The facility to produce a better fit by adding more and more terms must be applied cautiously as adding more terms reduces the sum of squares of residuals, but the reduction may only be marginal. We introduced the adjusted coefficient of multiple determination and the Akaike information criterion as a means of establishing whether the extra terms introduced do improve the fit. The idea of significance in the context of data analysis is pursued in the next chapter when we consider tests of significance.

Problems

1. The equation

$$y = a + bx + ce^x \qquad (7.57)$$

is to be fit to x–y data.

   (i) Substitute equation (7.57) into equation (7.1). By differentiating χ^2 with respect to each parameter in turn, obtain three equations that may be solved for a, b and c. Assume that an unweighted fit using least squares is appropriate.
   (ii) Write the equations obtained in part (i) in matrix form.

Consider the data in table 7.12.

   (iii) Assuming that equation (7.57) is appropriate to the data in table 7.12, use matrices to solve for a, b and c.

2. The movement of a solute through a chromatography column can be described by the van Deemter equation,

$$H = A + \frac{B}{v} + Cv \qquad (7.58)$$

Table 7.12. *x–y data.*

x	y
1	8.37
2	3.45
3	1.70
4	0.92
5	0.53
6	0.62
7	−0.06
8	0.63
9	0.16
10	−0.27

Table 7.13. *Variation of plate height, H, with flow rate of solute v.*

v (mL/minute)	H (mm)
3.5	9.52
7.5	5.46
15.7	3.88
20.5	3.48
25.8	3.34
36.7	3.31
41.3	3.13
46.7	3.78
62.7	3.55
78.4	4.24
96.7	4.08
115.7	4.75
125.5	4.89

where H is the plate height, v is the rate at which the mobile phase of the solute flows through the column. A, B and C are constants. For a particular column, H varies with v as given in table 7.13.

(i)   Write the equations in matrix form that must be solved for best estimates of A, B and C, assuming unweighted fitting using least squares is appropriate. (Hint: follow the steps given in section 7.3.)

Table 7.14. *Variation of displacement of an object with time.*

t (s)	s (m)
0.0	135.2
0.5	159.8
1.0	163.1
1.5	183.6
2.0	181.2
2.5	181.6
3.0	189.5
3.5	175.2
4.0	162.2
4.5	136.2
5.0	113.8
5.5	83.0
6.0	49.6

    (ii) Solve for best estimates of A, B and C.

    (iii) Determine the standard errors for the estimates obtained in (ii).

3. An object is thrown off a building and its displacement above the ground at various times after it is released is shown in table 7.14. The predicted relationship between s and t is

$$s = s_0 + ut + gt^2 \tag{7.59}$$

where s_0 is the displacement of the object at $t=0$, u is its initial vertical velocity and g is the acceleration of the body falling freely under the action of gravity. Use least squares to determine best estimates for s_0, u and g and the standard errors in the estimates.

4. To illustrate the way in which a real gas deviates from perfect gas behaviour, PV/RT is often plotted against $1/V$, where P is the pressure, V the volume and T the temperature (in kelvin) of the gas. R is the gas constant. Values for PV/RT and V are shown in table 7.15 for argon gas at 150 K. Assuming the relationship between PV/RT and $1/V$ can be written

$$\frac{PV}{RT} = A + \frac{B}{V} + \frac{C}{V^2} + \frac{D}{V^3} \tag{7.60}$$

use least squares to obtain best estimates for A, B, C and D and standard errors in the estimates. (Suggestion: let $y = PV/RT$, and $x = 1/V$.)

Table 7.15. *Variation of PV/RT with V for argon gas.*

V (cm^3)	PV/RT
35	1.21
40	0.89
45	0.72
50	0.62
60	0.48
70	0.45
80	0.51
90	0.50
100	0.53
120	0.57
150	0.61
200	0.69
300	0.76
500	0.84
700	0.89

5. Consider the x–y data in table 7.16. Fit equations $y = a + bx$ and $y = a + bx + cx^2$ to these data and use the adjusted multiple correlation coefficient and the Akaike information criterion to establish which equation better fits the data.

6. Table 7.17 shows data from an experiment to determine how the molar heat capacity (at constant pressure), C_p, of oxygen varies with temperature, T. Assuming that the relationship between C_p and T can be written

$$C_p = A + BT + \frac{C}{T^2} \tag{7.61}$$

determine:

   (i)   best estimates of A, B and C;
   (ii) the standard errors in the estimates;
  (iii) the 95% confidence intervals for A, B and C.

7. As part of a study into the behaviour of electrical contacts made to a ceramic conductor, the data in table 7.18 were obtained for the temperature variation of the electrical resistance of the contacts. It is suggested that there are two possible models that can be used to describe the variation of the contact resistance with temperature.

Table 7.16. *x–y data.*

x	y
5	361.5
10	182.8
15	768.6
20	822.5
25	1168.2
30	1368.6
35	1723.3
40	1688.7
45	1800.9
50	2124.5
55	2437.9
60	2641.2

Table 7.17. *Measured molar heat capacity of oxygen in the temperature range 300 K to 1000 K.*

T (K)	C_p (J$\cdot$mol$^{-1}\cdot$K^{-1})
300	29.43
350	30.04
400	30.55
450	30.86
500	31.52
550	31.71
600	32.10
650	32.45
700	32.45
750	32.80
800	33.11
850	33.38
900	33.49
950	33.85
1000	34.00

Table 7.18. *Resistance versus temperature for electrical contacts on a ceramic conductor.*

T (K)	R(Ω)	T (K)	R(Ω)
50	4.41	190	0.69
60	3.14	200	0.85
70	2.33	210	0.94
80	2.08	220	0.78
90	1.79	230	0.74
100	1.45	240	0.77
110	1.36	250	0.68
120	1.20	260	0.66
130	0.86	270	0.84
140	1.12	280	0.77
150	1.05	290	0.75
160	1.05	300	0.86
170	0.74		
180	0.88		

Model 1: The first model assumes that contacts show semiconducting behaviour, where the relationship between R and T can be written

$$R = A \exp\left(\frac{B}{T}\right) \tag{7.62}$$

where A and B are constants.

Model 2: Another equation proposed to describe the data is

$$R = \alpha + \beta T + \gamma T^2 \tag{7.63}$$

where α, β and γ are constants.

Using plots of residuals, the adjusted coefficient of multiple determination, the Akaike information criterion and any other indicators of 'goodness of fit', determine whether equation (7.62) or equation (7.63) better fits the data.

Assistance: To allow a straight line based on equation (7.62) to be fitted to data, the data need to be transformed. Transform back to the original units after fitting a line to data before calculating SSR, R^2_{ADJ} and AIC. This will allow for direct comparison of SSR, R^2_{ADJ} and AIC between equations (7.62) and (7.63).

Chapter 8

Tests of significance

8.1 Introduction

What can reasonably be inferred from data gathered in an experiment? This simple question lies at the heart of experimentation, as an experiment can be judged by how much insight can be drawn from data. Experiments may have a broad or narrow focus, and may be designed to:

- challenge relationships that have an established theoretical basis;
- confirm discoveries that result from 'chance' observations;
- check for drift in an instrument;
- compare analysis of materials carried out in two or more laboratories.

Such general goals give way to specific questions that we hope can be answered by careful analysis of data gathered in well designed experiments. Questions that might be asked include:

- Is there a linear relationship between quantities measured in an experiment?
- Could the apparent correlation between variables have occurred 'by chance'?
- Does a new manufacturing process produce lenses with focal lengths that are less variable than the old manufacturing process?
- Is there agreement between two methods used to determine the concentration of iron in a specimen?
- Has the gain of an instrument changed since it was calibrated?

It is usually not possible to answer these questions with a definite 'yes' or definite 'no'. Though we hope evidence gathered during an experiment will give strong clues as to which reply to favour, we must be satisfied with answers expressed in terms of probability.

Consider a situation in which a manufacturer supplies an instrument containing an amplifier with a gain specified as 1000. Would it be reasonable to conclude that the instrument is faulty or needs recalibrating if the gain is determined by experiment to be 995? It is possible that random errors inherent in the measurement process, as revealed by the scatter of the data, would be sufficient to explain the discrepancy between the 'expected' value of gain of 1000 and the 'experimental' value of 995. What we would really like to know is whether, after taking into account the scatter in the data, the difference between the value we have reason to expect will occur and that actually obtained through experiment or observation is 'significant'.

The purpose of this chapter is to introduce the notion of significance and how significance may be quantified so that it can be useful in the analysis of experimental data. This requires we make assumptions about which probability distribution is applicable to the experimental data. While the t distribution is applicable to many 'tests of significance', we will introduce other distributions including the F distribution which describes the probability of obtaining a particular ratio of *variances* of two samples. With the aid of this distribution we will be able to test whether there is a significant difference between the variability in data as exhibited by two samples.

8.2   Confidence levels and significance testing

In chapter 3 we introduced confidence levels and confidence intervals. As these are closely related to tests of significance, it is helpful to consider again the main points.

To determine the best estimate of the population mean, μ, of a quantity, we make repeat measurements of the quantity.[1] The values obtained through measurement constitute a sample upon which subsequent analysis is based. If the number of values, n, that make up the sample is large (say in excess of 30) so that the standard deviation, s, (given by equation (1.16)) is a good estimate of the population standard deviation, σ, then the $X\%$ confidence interval for μ is given by

---

[1] When repeat measurements are made of a quantity, such as the diameter of a wire, the population mean is taken to be the true value of the quantity, so long as systematic errors are negligible (see section 5.9 for more details).

Table 8.1. *z value corresponding to X% confidence level.*

X%	50%	68%	90%	95%	99%
$z_{X\%}$	0.67	0.99	1.64	1.96	2.58

Table 8.2. *Forty repeat values of focal length of a lens.*

Focal length (cm)							
15.1	14.6	15.0	15.9	15.9	16.2	14.1	15.2
15.8	14.7	14.9	14.4	14.3	14.8	14.9	14.2
15.0	15.1	15.3	15.1	15.1	15.1	16.0	15.2
15.3	15.0	16.3	15.7	16.5	14.9	16.1	14.4
15.6	15.7	16.3	15.3	15.0	15.6	15.1	15.7

$$\bar{x} - z_{X\%}\sigma_{\bar{x}} \leq \mu \leq \bar{x} + z_{X\%}\sigma_{\bar{x}} \tag{8.1}$$

This can also be expressed as

$$\mu = \bar{x} \pm z_{X\%}\sigma_{\bar{x}} \tag{8.2}$$

where $\bar{x}$ is the sample mean and $\sigma_{\bar{x}} \approx s/\sqrt{n}$ is the standard error of the mean. $z_{X\%}$ is the z value corresponding to the X% confidence level as given in table 8.1. The probability that the true value lies in the interval given by equation (8.2) is X%/100 %.

As an example of determining a confidence interval, consider the 40 values of focal length of a convex lens in table 8.2 obtained by repeat measurements on one lens. The mean, $\bar{x}$, of the values in table 8.2 and standard deviation, s, are

$$\bar{x} = 15.26 \text{ cm} \qquad s = 0.6080 \text{ cm}$$

To determine the 95% confidence interval for the true value of the focal length, we use table 8.1 which gives $z_{95\%} = 1.96$. Using equation (8.1) gives

$$15.26 - 1.96 \times \frac{0.6080}{\sqrt{40}} \leq \mu \leq 15.26 + 1.96 \times \frac{0.6080}{\sqrt{40}}$$

i.e. the 95% confidence interval for μ is 15.07 cm to 15.45 cm.

If the lens in the experiment has been supplied by the manufacturer with the assurance that the focal length of the lens is 15.00 cm, should we

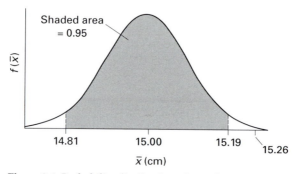

Figure 8.1. Probability distribution of sample means.

be concerned that the mean of the measured values differs by 0.26 cm from the 'assured' value? The answer is yes and to justify this, consider figure 8.1 which shows the probability distribution of sample means assuming the population mean $= 15.00$ cm and the standard error of the mean is $0.6080/\sqrt{40} = 0.09613$ cm. There is a probability of 0.95 that the mean of a sample consisting of 40 values lies between $(15.00 - 1.96 \times 0.09613)$ cm and $(15.00 + 1.96 \times 0.09613)$ cm, i.e. between 14.81 cm and 15.19 cm. Put another way, the probability that the mean of 40 values would lie *outside* the interval 14.81 to 15.19 cm is $1 - 0.95 = 0.05$. It appears unlikely (i.e. the probability is less than 0.05) that a sample consisting of 40 values with a mean of 15.26 cm has been drawn from a population which has a mean of 15.00 cm. In short, there is a *significant difference* between the anticipated value of focal length of the lens (15.00 cm) and the mean of the values in table 8.2 (15.26 cm).

Though there is a significant difference between anticipated focal length and the sample mean, we must decide whether the difference is important. For example, if the lens in question is expensive, or perhaps was purchased to replace another lens of focal length 15.00 cm, we may show the data to the manufacturer and request a replacement. On the other hand, if the lens is to be used to demonstrate the principles of image formation, a difference of 0.26 cm between the focal length as specified by the manufacturer and the true focal length may not be regarded as important. This is a point easy to overlook in hypothesis testing: a difference between two numbers may be 'statistically significant', but too small to be important in a practical sense.

Another question we might ask regarding the focal length data in table 8.2 is:

If the population mean is 15.00 cm, what is the probability that we would obtain by chance, a mean based on 40 repeat measurements that differs as much as 0.26 cm from the population mean?

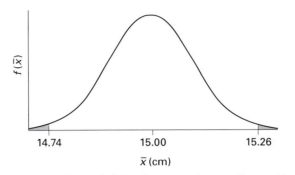

Figure 8.2. The probability that a sample mean lies outside the interval 14.74 cm to 15.26 cm is equal to the sum of the shaded areas.

In order to answer this question we redraw figure 8.1, and shade the areas corresponding to the probability that a sample mean differs from 15.00 cm by at least 0.26 cm, as shown in figure 8.2. The sum of the shaded areas is the probability that a sample mean would differ from a population with a mean of 15.00 cm by 0.26 cm or more when the standard error of the mean is 0.09613 cm. The shaded areas can be determined with the aid of tables, or a computer package such as Excel®. Notice that due to the symmetry of the normal distribution, the shaded area in each tail of the distribution shown in figure 8.2 is the same, and so long as we calculate the area in one tail, doubling that area gives the total area required.

One method of finding the area in the tails is to find the z value corresponding to the cumulative probability $P(-\infty \leq \bar{x} \leq 15.26 \text{ cm})$, where z is given by

$$z = \frac{\bar{x} - \mu}{\sigma_{\bar{x}}} = \frac{15.26 - 15.00}{0.09613} = 2.70$$

Table 1 in appendix 1 gives the cumulative probability between $z = -\infty$ and $z = 2.70$ as 0.99653. The area in one tail is therefore $1 - 0.99653 = 0.0035$. The sum of the areas in both tails is $2 \times 0.0035 = 0.0070$. A probability of 0.0070 that the sample comes from a population with a mean of 15.00 cm is sufficiently small that it would be reasonable to conclude that the population from which the sample was drawn does *not* have a mean of 15.00 cm.

The process we have just carried out in which a hypothesised population mean is compared with a sample mean is an example of 'hypothesis testing'.

Table 8.3. *Population parameters and their corresponding sample statistics.*

Population parameter	Sample statistic
Mean, μ	Mean, $\bar{x}$
Standard deviation, σ	Standard deviation, s
Correlation coefficient, ρ	Correlation coefficient, r
Intercept, α (of a line through x–y data)	Intercept, a
Slope, β (of a line through x–y data)	Slope, b

8.3 Hypothesis testing

Formally, hypothesis testing involves comparing a hypothesised population parameter with the corresponding number (or statistic) determined from sample data. The testing process takes into account both the variability in the data as well as the number of values in the sample. Table 8.3 shows several population parameters and their corresponding sample statistics.[2] For example, we can compare a hypothesised population mean,[3] μ_0, with a sample mean, $\bar{x}$, and address the question:

> What is the probability that a sample with mean $\bar{x}$ has come from a population with mean μ_0?

Consider an example: An experimenter adds weights to one end of a wire suspended from a fixed point, causing the wire to extend. The experimenter wishes to be satisfied that the weights have been drawn from a population with mean mass 50 g (as specified by the manufacturer). None of the weights will be 'exactly' 50 g, but it is reasonable to expect that, if many weights are tested, the mean will be close to 50 g. The experimenter 'hypothesises' that the population from which the sample is drawn has a mean of 50 g. This is a conservative hypothesis in the sense that, owing to the manufacturer's assurances, the experimenter expects that the population mean *is* 50 g. Such a conservative hypothesis is normally referred to as the 'null hypothesis' and is represented symbolically by H_0. Formally, we write:

> H_0: $\mu = 50$ g. This is read as 'the null hypothesis is that the population mean is 50 g'.

[2] As the number of values, n, in the sample tends to infinity then a sample statistic tends to its corresponding population parameter. So, for example, as $n \to \infty$, $\bar{x} \to \mu$.

[3] It is usual to denote hypothesised population parameters by attaching the subscript '0' to the symbol used for that parameter.

It is usual to state an alternative hypothesis, H_a, which is favoured if the probability of the null hypothesis being true is small. In this example we would write:

> H_a: $\mu \neq 50$ g. This is read as 'the alternative hypothesis is that the population mean is *not* equal to 50 g'.

The next stage is to decide what probability distribution is appropriate for the sample means. The central limit theorem[4] predicts that (so long as the sample size is large enough) the distribution of sample means is normal with a population mean μ, and a standard error $\sigma_{\bar{x}}$, where $\sigma_{\bar{x}} = \sigma / \sqrt{n}$. The number of values in the sample is n, and σ is the population standard deviation. In the vast majority of situations σ is not known so we use the approximation[5] $\sigma_{\bar{x}} \approx s / \sqrt{n}$, where s is the best estimate of σ.

Next we decide under what circumstances we will reject the null hypothesis and so favour the alternative hypothesis. How small must be the probability, p, (often referred to as the p value) of the null hypothesis being true before we reject the null hypothesis? In principle, any value of p could be chosen, but conventionally a null hypothesis is rejected when the likelihood of it being true is less than 0.05. The probability of 0.05 is to an extent arbitrary, but it does represent a 'bench mark' recognised and widely adopted by the scientific community when hypothesis testing is carried out. If the probability of the null hypothesis being true is less than 0.05, the difference between the hypothesised value for the population mean and the sample mean is regarded as *significant*. If the probability of the null hypothesis being true is less than 0.01, then the difference between the hypothesised population mean and the sample mean is regarded as *highly significant*.

The result of a hypothesis test may be that the null hypothesis is rejected at the level of significance,[6] $\alpha = 0.05$. The 'α significance level' refers to the sum of the areas in the tails of the probability distribution shown in figure 8.3.[7] α can be related directly to the confidence level, $X\%$, as shown in figure 8.3. As the total area under the curve in figure 8.3 equals 1, we write

$$\frac{X\%}{100\%} + \alpha = 1 \tag{8.3}$$

---

[4] See section 3.8.

[5] The approximation is good for n in excess of 30.

[6] α is sometimes expressed as a percentage, so that a significance level of 0.05 may be stated as $\alpha = 5\%$.

[7] This is true for a 'two tailed' test. Section 8.3.3 discusses one and two tailed tests.

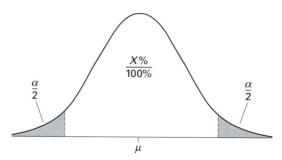

Figure 8.3. Relationship between level of significance, α, and confidence level, $X\%$.

8.3.1   Distribution of the test statistic, z

If we consider a population with a hypothesised mean, μ_0, and a standard deviation, σ, then samples consisting of n values should have means that are distributed normally about μ_0 such that the standard deviation of the means is $\sigma/\sqrt{n}$. A convenient way to express this is to use the continuous random variable, z, given by

$$z = \frac{\bar{x} - \mu_0}{\sigma/\sqrt{n}} \tag{8.4}$$

The variable, z, is normally distributed with a mean of zero and a standard deviation of 1.

A value for z can be calculated using data gathered in an experiment. If data have been drawn from a population with mean μ_0, then large values of $|z|$ are unlikely. We can determine a 'critical' value of z, z_{crit}, for each level of significance, α. If $|z| > z_{crit}$, the probability of the null hypothesis being true (i.e. that the sample has been drawn from a population with a mean μ_0) is *less than* the probability corresponding to the significance level α, and the null hypothesis is rejected. This is shown pictorially in figure 8.4. By contrast, if $|z| < z_{crit}$, then the test has not revealed any significant difference (at the chosen α) between the hypothesised population mean, μ_0, and the sample mean and the null hypothesis cannot be rejected. Values of z_{crit} for particular significance levels may be determined using table 1 in appendix 1.

As always, it is not possible to determine the population standard deviation σ and we must 'make do' with the estimate of the population standard deviation, s, found using equation (1.16).

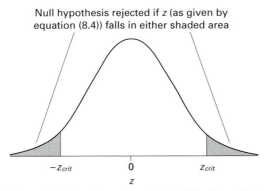

Null hypothesis rejected if z (as given by equation (8.4)) falls in either shaded area

$-z_{crit}$ 0 z_{crit}

z

Figure 8.4. Rejection regions (shaded) for the null hypothesis.

Example 1

The masses of a sample of 40 weights measured with an electronic balance are shown in table 8.4. The experimenter assumes that the masses have been drawn from a population with a mean of 50.00 g. Is this assumption reasonable?

ANSWER

We perform a test to determine whether there is a significant difference between the hypothesised population mean and the mean of the values in table 8.4. Table 8.5 describes the steps in the test.

Table 8.4. *Masses of 40 weights.*

Weights (g)									
50.06	50.02	50.18	50.05	50.05	50.12	50.05	50.07	50.20	49.98
50.13	50.05	49.99	49.99	49.94	50.20	50.03	49.97	50.09	49.77
50.10	49.93	49.99	50.01	50.12	50.16	50.12	50.22	50.13	50.09
50.22	50.09	50.07	50.02	50.05	50.14	50.10	50.18	50.08	50.02

Exercise A

Bandgap reference diodes are used extensively in electronic measuring equipment as they provide highly stable voltages against which other voltages can be compared.[8] The manufacturer's specification indicates that the nominal voltage across the diodes should be 1.260 V. A sample of 36 diodes is tested and the voltage across each diode is shown in table 8.6. Using these data, test whether the diodes have been drawn from a population with a mean of 1.260 V. Reject the null hypothesis at $\alpha = 0.05$.

[8] As examples, bandgap reference diodes are found in voltmeters and on interfacing cards in microcomputers.

Table 8.5. *Step by step description of the hypothesis test for example 1.*

Step	Details/comment
Decide the purpose of the test.	To compare a sample mean with a hypothesised population mean.
State the null hypothesis.	H_0: $\mu = 50.00$ g.
State the alternative hypothesis	H_a: $\mu \neq 50.00$ g.
Choose the significance level of the test, α.	No indication is given in the question of the level of significance at which we should test the null hypothesis, so we choose the 'most commonly' used level, i.e. $\alpha = 0.05$.
Determine the mean, $\bar{x}$, and standard deviation, s, of the values in table 8.4.	$\bar{x} = 50.0695$ g, $s = 0.08950$ g.
Calculate the value of the test statistic, z, using equation (8.4).	$z = \dfrac{\bar{x} - \mu_0}{s/\sqrt{n}} = \dfrac{50.0695 - 50.00}{0.08950/\sqrt{40}} = 4.91.$
Determine the critical value of the test statistic, z_{crit}, based on the chosen significance level.	When the area in one tail of the standard normal distribution is 0.025 ($= \alpha/2$), the magnitude of the z value, found using table 1 in appendix 1, $= 1.96$. Hence $z_{crit} = 1.96$.
Compare $|z|$ and z_{crit}.	$|z| > z_{crit}$.
Make a decision.	As $|z| > z_{crit}$ we *reject* the null hypothesis: There is a significant difference between the sample mean and the hypothesised population mean. Specifically, the probability is less than 0.05 that a sample mean determined from 40 values would differ from the hypothesised mean by 0.0695 g or more.

Table 8.6. *Voltages across bandgap reference diodes.*

1.255	1.259	1.257	1.256	1.260	1.262	1.259	1.257	1.260
1.260	1.261	1.263	1.258	1.261	1.253	1.259	1.257	1.254
1.251	1.257	1.252	1.258	1.253	1.248	1.265	1.255	1.256
1.258	1.265	1.257	1.257	1.264	1.262	1.256	1.258	1.265

8.3.2 Using Excel® to compare sample mean and hypothesised population mean

In section 8.2 we calculated 'by hand' the probability that a particular sample mean would occur for a hypothesised population mean. This process can be carried out using Excel®. The steps in the calculation of the probability are:

(i) Calculate the mean, $\bar{x}$, and the standard deviation, s, and determine the standard error of the mean, $\sigma_{\bar{x}}$, using sample data.

(ii) For a hypothesised mean, μ_0, calculate the value of the z-statistic, z, using $z = (\bar{x} - \mu_0)/\sigma_{\bar{x}}$.

(iii) Use Excel®'s NORMSDIST() function to determine the area in the tail of the distribution[9] between $z = -\infty$ and $z = -|z|$.

(iv) Multiply the area by 2 to obtain the total area in both tails of the distribution.

(v) If a sample is drawn from a population with mean μ_0, then the probability that the sample mean would be at least as far from the mean as $|z|$ is equal to the sum of the areas in the tails of the distribution.

Example 2

Consider the values in table 8.7. Use Excel® to determine the probability that these values have been drawn from a population with a mean of 100.0.

ANSWER

The data in table 8.7 are shown in sheet 8.1. Column F of sheet 8.1 contains the formulae required to determine the sample mean, standard deviation and so on. Sheet 8.2 shows the values returned in column F. The number 0.067 983 returned in cell F6 is the probability of obtaining a sample mean at least as far from the population mean of 100.0 as 101.85. As this probability is greater than 0.05 (the commonly chosen level of significance, α) we cannot reject a null hypothesis that the sample is drawn from a population with mean equal to 100.0.

---

[9] By calculating the area between $-\infty$ and $-|z|$ we are always choosing the tail to the left of $z = 0$.

Table 8.7. *Values for example 2.*

100.5	95.6	103.2	108.4
96.6	98.5	93.5	92.8
102.7	100.2	100.4	100.1
106.3	113.9	98.7	110.3
108.0	110.7	91.1	100.8
97.1	98.1	91.4	99.2
108.7	101.6	101.1	99.4
93.9	104.7	106.5	111.6
107.5	100.0	111.9	101.6

Sheet 8.1. *Determination of p value.*

	A	B	C	D	E	F
1	100.5	95.6	103.2	108.4	hypothesised mean	100.0
2	96.6	98.5	93.5	92.8	sample mean	=AVERAGE(A1:D9)
3	102.7	100.2	100.4	100.1	standard deviation	=STDEV(A1:D9)
4	106.3	113.9	98.7	110.3	standard error of mean	=F3/36^0.5
5	108.0	110.7	91.1	100.8	z-value	=(F2-F1)/F4
6	97.1	98.1	91.4	99.2	probability	=2*NORMSDIST(-ABS(F5))
7	108.7	101.6	101.1	99.4		
8	93.9	104.7	106.5	111.6		
9	107.5	100.0	111.9	101.6		

Sheet 8.2. *Numbers returned in column F.*

	E	F
1	hypothesised mean	100.0
2	sample mean	101.85
3	standard deviation	6.0818
4	standard error of mean	1.013633
5	z-value	1.825118
6	probability	0.067983
7		

Exercise B

Titanium metal is deposited on a glass substrate producing a thin film of nominal thickness 55 nm. The thickness of the film is measured at 30 points, chosen at random, across the film. The values obtained are shown in table 8.8. Determine whether there is a significant difference, at $\alpha=0.05$, between the mean of these values and a hypothesised population mean of 55 nm.

Table 8.8. *Measured thickness (in nm) of titanium film.*

61	53	59	61	60	54	56	56	52	61
57	54	50	60	58	58	55	55	53	58
61	52	53	53	59	61	59	53	54	55

8.3.3 One tailed and two tailed tests of significance

In the test described in section 8.3.2, we are interested in whether the sample mean differs significantly from the hypothesised mean. That is, we are equally concerned whether $\bar{x} - \mu_0$ is positive or negative. Such a test is referred to as 'two tailed' and is appropriate if we have no preconceived notion as to whether to expect $\bar{x} - \mu_0$ to be positive or negative. There are situations in which, prior to carrying out the data analysis, we anticipate $\bar{x} - \mu_0$ to have a specific sign. For example, if we believe we have improved a process in which gold is recovered from sea water, then we would anticipate the new process to have a mean gold recovery greater than the old process. In this case we would express the null and alternative hypotheses as:

$H_0: \mu \leq \mu_0$,
$H_a: \mu > \mu_0$,

where μ_0 is the expected population mean based on the amount of gold recovered using the 'old' process. By contrast, if a surface coating on a lens is designed to reduce the amount of reflected light, then we would write:

$H_0: \mu \geq \mu_0$;
$H_a: \mu < \mu_0$.

Here μ_0 would correspond to the expected reflection coefficient for the lens in the absence of a coating.

Tests in which the alternative hypothesis is $\mu > \mu_0$ or $\mu < \mu_0$ are referred to as one tailed tests. As the name implies, we consider areas only in one tail of the distribution as shown in figure 8.5. If the significance level, $\alpha = 0.05$ is chosen for the test, then the shaded area in figure 8.5 would be equal to 0.05. Any value of z, determined using experimental data, that falls in the shaded region such that $z \geq z_{crit}$ would mean that the null hypothesis is rejected at the $\alpha = 0.05$ level of significance.

In order to determine z_{crit} for a one tailed test, we calculate $(1 - \alpha)$.

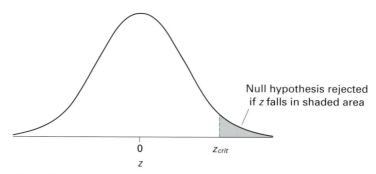

Figure 8.5. Rejection region for a one tailed test.

This gives the area under the standard normal curve between $z = -\infty$ and $z = z_{crit}$. The z value corresponding to the area $(1 - \alpha)$ is found using table 1 in appendix 1.

Example 3

Determine the critical value of the z statistic for a one tailed test when the level of significance, $\alpha = 0.1$.

ANSWER

If $\alpha = 0.1$, then $(1 - \alpha) = 0.9$. Referring to table 1 in appendix 1, if the area under the standard normal curve between $z = -\infty$ and $z = z_{crit}$ is 0.9, then $z_{crit} = 1.28$.

Exercise C

Determine the critical value of the z statistic for a one tailed test when the level of significance α is equal to:

    (i) 0.2,
   (ii) 0.05,
  (iii) 0.01,
  (iv) 0.005.

8.3.4 Type I and type II errors

In making the decision to reject or not reject a null hypothesis we must be aware that we will not always make the right decision. Ideally:

- we reject a null hypothesis when the null hypothesis is *false*, or
- we do not reject a null hypothesis when the null hypothesis is *true*.

We can never be sure that the null hypothesis is true or false, for if we could there would be no need for hypothesis testing! This leads to two undesirable outcomes:

- we could reject a null hypothesis that is *true* (this is called a type I error), or
- we could fail to reject a null hypothesis that is *false* (this is called a type II error[10]).

By choosing a significance level of, say, $\alpha = 0.05$, we limit the probability of a type I error to 0.05, This means that, if we repeat the test on many sets of similar data and the probability of making a type I error is 0.05, we will reject the null hypothesis when it should not have been rejected, on average, only one time in twenty. Quantifying the probability of making a type II error is more difficult but can be done in some situations.[11] Type I errors are usually regarded as more serious than type II errors, as we wish to reduce the probability of rejecting a null hypothesis (usually chosen to be conservative) that is true.

8.4 Comparing $\bar{x}$ with μ_0 when sample sizes are small

In order to compare a hypothesised population mean, μ_0, with the mean, $\bar{x}$, of a small sample, we follow the steps described in section 8.3.1 for large samples. However, we must take into account that when sample sizes are small, the distribution of means is not normal even if the values that make up the sample are normally distributed.

When the population standard deviation, σ, is known then the variable

$$z = \frac{\bar{x} - \mu}{\sigma/\sqrt{n}} \tag{8.5}$$

is normally distributed. In practice, σ is rarely if ever known, and we must estimate σ with the data available. s (given by equation (1.16)) is a good estimate of σ so long as $n \geq 30$ and so equation (8.5) may still be used with s replacing σ. However, a sample size with $n \geq 30$ is a luxury that many experiments cannot afford due to cost or time considerations. If n is less

---

[10] Note type I and type II errors are not the same as the experimental errors discussed in chapter 5.

[11] See Devore (1991) for a discussion of type II errors.

than 30, s can no longer be regarded as a good estimate of σ, and we use the statistic, t, given by

$$t = \frac{\bar{x} - \mu}{s/\sqrt{n}} \tag{8.6}$$

As discussed in chapter 3, the distribution of t has a symmetric 'bell' shape and so is very similar to the normal distribution.[12] An important difference between the normal and t distributions is that there is more area in the tails of the t distribution from, say, $t=2$ to $t=\infty$ than in the normal distribution between $z=2$ to $z=\infty$. We can show this using Excel®'s TDIST() and NORMSDIST() functions.

The cumulative probability between $t=2$ to $t=\infty$ can be obtained for any number of degrees of freedom using Excel®'s TDIST() function.[13] The area in the tail $t=2$ to $t=\infty$ for $\nu=10$ is found by typing =TDIST(2,10,1) into a cell. Excel® returns the number 0.03669. Similarly, the cumulative probability between $z=2$ to $z=\infty$ can be obtained using Excel®'s NORMSDIST() function.[14] Specifically, the area in the tail between $z=2$ and $z=\infty$ is found by typing $=1-$ NORMSDIST(2) into a cell in the spreadsheet. Excel® returns the number 0.02275.

As a result of the increased area in the tail of the t distribution compared to the normal distribution, the critical t value, t_{crit}, is larger for a given level of significance than the corresponding critical value for the standard normal distribution, z_{crit}. As the t distribution depends on the number of degrees of freedom, ν, the critical value of the test statistic depends on ν. Table 2 in appendix 1 gives the critical values of t for various levels of significance, α, for $2 \leq \nu \leq 60$.

Table 8.9. Values of the acceleration due to gravity.

$g\,(\mathrm{m/s^2})$	9.68	9.85	9.73	9.76	9.74

Example 4

Table 8.9 shows the value of the acceleration due to gravity, g, obtained through experiment. Are these values significantly different from a hypothesised value of $g = 9.81$ m/s²? Test the hypothesis at the $\alpha = 0.05$ level of significance.

---

[12] See section 3.9.

[13] See section 3.9.1 for a description of Excel®'s TDIST() function.

[14] See section 3.5.3 for a description of Excel®'s NORMSDIST() function.

ANSWER

H_0: $\mu = 9.81$ m/s^2;

H_a: $\mu \neq 9.81$ m/s^2;

$\bar{x} = 9.752$ m/s^2, $s = 0.06221$ m/s^2, $s/\sqrt{n} = 0.02782$ m/s^2.

Using equation (8.6) with $\mu = \mu_0 = 9.81$ m/s^2 gives

$$t = \frac{9.752 - 9.81}{0.02782} = -2.085$$

The critical value of the test statistic, t_{crit}, with $\alpha = 0.05$ and $\nu = n - 1 = 4$ is found from table 2 in appendix 1 to be $t_{crit} = 2.776$. As $|t| < t_{crit}$, we cannot reject the null hypothesis, i.e. there is no significant difference at $\alpha = 0.05$ between the sample mean and the hypothesised population mean.

Exercise D

The unit cell is the building block of all crystals. An experimenter prepares several crystals of a ceramic material and compares the size of the unit cell to that published by another experimenter. One dimension of the unit cell is the lattice dimension, c. Table 8.10 show eight values of the c dimension obtained by the experimenter. At the $\alpha = 0.05$ level of significance, determine whether the sample mean of the values in table 8.10 differs significantly from the published value of c of 1.1693 nm.

Table 8.10. *c dimension of a ceramic crystal.*

c (nm)	1.1685	1.1672	1.1695	1.1702
	1.1665	1.1675	1.1692	1.1683

8.5 Significance testing for least squares parameters

In chapter 6 we considered fitting the equation $y = a + bx$ to x–y data using the technique of least squares. When a or b is close to zero, we might wonder if we have chosen the best equation to fit to data. For example, if a is not significantly different from 0, we would be justified in omitting a and fitting the equation $y = bx$ to data. Similarly, if b is not significantly different from 0, we could justify removing b from the fit. The consequence of this is that we would find the best estimate of the constant, a, that would fit the data (which would be the mean of the y values).

In order to test whether a or b (or the estimate of other parameters found using least squares) is significantly different from zero we need:

(i)   the best estimate of the parameter found using least squares;
(ii) the standard error in the estimate;
(iii) to choose a significance level for the test.

As the number of x–y pairs in a least squares analysis is routinely less than 30, we assume that the distribution of parameter estimates follows a t distribution. For the intercept of a line through data we hypothesise:

H_0: the population intercept, $\alpha = 0$;
H_a: the population intercept, $\alpha \neq 0$.

The test statistic, t, in this situation is given by

$$t = \frac{a - \alpha_0}{\sigma_a}$$

so that for $\alpha_0 = 0$,

$$t = \frac{a}{\sigma_a} \tag{8.7}$$

where a is the best estimate of the population intercept, α and σ_a is the standard error of a. The critical value of the test statistic, t_{crit}, must be established for a given level of significance and degrees of freedom. If $|t| > t_{crit}$ the null hypothesis is rejected, i.e. the sample intercept, a, is not consistent with a hypothesised intercept of zero.

A similar approach is used to test for the significance of other parameters. For example, for the hypothesised population slope, β_o, we calculate

$$t = \frac{b - \beta_0}{\sigma_b}$$

so that for $\beta_0 = 0$

$$t = \frac{b}{\sigma_b} \tag{8.8}$$

Example 5
Consider the x–y data in table 8.11.

(i)   Using the values in table 8.11, find the intercept, a, and slope, b, of the best straight line through the points using unweighted least squares.
(ii) Calculate the standard errors in a and b.

Table 8.11. *x–y values for example 5.*

x	y
0.1	−0.9
0.2	−2.1
0.3	−3.6
0.4	−4.7
0.5	−5.9
0.6	−6.8
0.7	−7.8
0.8	−9.1

(iii) Determine, at the 0.05 significance level, whether the intercept and slope are significantly different from zero.

ANSWER

Solving for the intercept using unweighted least squares, gives $a = 0.06786$ and the standard error in intercept $\sigma_a = 0.1453$. Similarly, the slope b is -11.51 and the standard error in the slope σ_b is 0.2878. The degrees of freedom, $\nu\, (= n - 2)$ is $8 - 2 = 6$.

To test if the intercept is significantly different from zero, we write the null and alternative hypotheses as:

$H_0: \alpha = 0;$
$H_a: \alpha \neq 0.$

The value of the test statistic $t\, (= a/\sigma_a)$ is $0.06786/0.1453 = 0.4670$. Using the table 2 in appendix 1 gives for a two tailed test[15] $t_{crit} = t_{0.05,6} = 2.447$. As $|t| < t_{crit}$ we cannot reject the null hypothesis, i.e. the intercept is not significantly different from zero.

To test if the slope is significantly different from zero, we write the null and alternative hypotheses as:

$H_0: \beta = 0;$
$H_a: \beta \neq 0.$

The value of the test statistic $t\, (= b/\sigma_b)$ is $-11.51/0.2878 = -40.00$.

$$t_{crit} = t_{0.05,6} = 2.447$$

As $|t| > t_{crit}$ we reject the null hypothesis, i.e. the slope is significantly different from 0.

As the intercept is not significantly different from 0, we have justification for fitting a line to data using least squares in which the line is 'forced' through 0 (i.e. $a = 0$). A convenient way to do this is to use Excel®'s LINEST() function as discussed in section 6.3. Doing this gives the slope, $b = -11.39$ and the standard error in the slope, $\sigma_b = 0.1231$.

---

[15] A short hand way of writing the critical value of t which corresponds to the significance level, α, for ν degrees of freedom is $t_{\alpha,\nu}$.

Exercise E

Table 8.12 shows the variation in voltage across a germanium diode which was measured as the temperature of the diode increased from 250 K to 360 K. Assuming there is a linear relationship between voltage and temperature, find the intercept and slope of the best line through the data in table 8.12. Test whether the intercept and slope differ from zero at the $\alpha=0.05$ significance level.

Table 8.12. *Variation of voltage with temperature for a germanium diode.*

T (K)	V (V)
250	0.440
260	0.550
270	0.469
280	0.486
290	0.508
300	0.494
310	0.450
320	0.451
330	0.434
340	0.385
350	0.458
360	0.451

8.6 Comparison of the means of two samples

As variability occurs in measured values, two samples will seldom have the same mean even if they are drawn from the same population. There are situations in which we would like to compare two means and establish whether they are significantly different.

For example, if a specimen of water from a river is divided and sent to two laboratories for analysis of lead content, it would be reasonable to anticipate that the difference between the means of the values obtained for lead content by each laboratory would not be statistically significant. If a significant difference *is* found then this might be traced to shortcomings in the analysis procedure of one of the laboratories or perhaps inconsistencies in storing or handling the specimens of river water. Other circumstances in which we might wish to compare two sets of data include where:

- the purity of a standard reference material is compared by two independent laboratories;
- the moisture content of a chemical is measured before and after a period of storage;
- the oxygen content of a ceramic is determined before and after heating the ceramic in an oxygen rich environment.

In situations in which we want to know whether there is a significant difference in the mean of two samples in which each sample consists of less than 30 values, we use a t test. An assumption is made that the difference between, say, the mean of sample 1 and the mean of sample 2 follows a t distribution. The t statistic in this situation is given by

$$t = \frac{\bar{x}_1 - \bar{x}_2}{s_{\bar{x}_1 - \bar{x}_2}} \tag{8.9}$$

where $\bar{x}_1$ is the mean of sample 1, $\bar{x}_2$ is the mean of sample 2 and $s_{\bar{x}_1 - \bar{x}_2}$ is the standard error of the difference between the sample means. In situations in which the standard deviations of the samples are not too dissimilar[16]

$$s_{\bar{x}_1 - \bar{x}_2} = s_p \left(\frac{1}{n_1} + \frac{1}{n_2} \right)^{\frac{1}{2}} \tag{8.10}$$

where n_1 and n_2 are the number of values in sample 1 and sample 2 respectively. If the sample sizes are the same, such that $n_1 = n_2 = n$, equation (8.10) becomes

$$s_{\bar{x}_1 - \bar{x}_2} = s_p \left(\frac{2}{n} \right)^{\frac{1}{2}} \tag{8.11}$$

where s_p is the pooled or combined standard deviation and is given by

$$s_p = \left[\frac{s_1^2(n_1 - 1) + s_2^2(n_2 - 1)}{n_1 + n_2 - 2} \right]^{\frac{1}{2}} \tag{8.12}$$

s_1^2 and s_2^2 being the estimated population variances of samples 1 and 2 respectively. The number of degrees of freedom, ν, is $n_1 + n_2 - 2$. If $n_1 = n_2$, then equation (8.12) reduces to

$$s_p = \left(\frac{s_1^2 + s_2^2}{2} \right)^{\frac{1}{2}} \tag{8.13}$$

[16] See McPherson (1990) for a discussion of equation (8.10).

Table 8.13. *Peel off times for two types of adhesive tape.*

Peel off times (s)	
Tape 1	Tape 2
65	176
128	125
87	95
145	255
210	147
85	88

Example 6

In order to compare the adhesive properties of two types of adhesive tape, each tape was pulled from a clean glass slide by a constant force. The time for 5 cm of each tape to peel from the glass is shown in table 8.13. Test at the $\alpha=0.05$ level of significance whether there is any significant difference in the peel off times for tape 1 and tape 2.

ANSWER

To perform the hypothesis test we write:

H$_0$: $\mu_1=\mu_2$ (we hypothesise that the population means for both tapes are the same);

H$_a$: $\mu_1\neq\mu_2$ (two tailed test).

Using the data in table 8.13, we obtain $\bar{x}_1=120.0$ s, $s_1=53.16$ s, $\bar{x}_2=147.7$ s, $s_2=61.92$ s. Using equation (8.13), we have

$$s_p=\left(\frac{2825.6+3834.3}{2}\right)^{\frac{1}{2}}=57.71 \text{ s}$$

Substituting 57.71 s into equation (8.11) (with $n=6$) gives

$$s_{\bar{x}_1-\bar{x}_2}=57.71\times\left(\frac{2}{6}\right)^{\frac{1}{2}}=33.32 \text{ s}$$

The value of the test statistic is found using equation (8.9), i.e.

$$t=\frac{\bar{x}_1-\bar{x}_2}{s_{\bar{x}_1-\bar{x}_2}}=\frac{120.0-147.7}{33.32}=-0.8304$$

Here the number of degrees of freedom, $v=n_1+n_2-2=10$. The critical value of the test statistic, t_{crit}, for a two tailed test at $\alpha=0.05$, found from table 2 in appendix 1, is

$$t_{crit}=t_{0.05,10}=2.228$$

As $|t| < t_{crit}$ we cannot reject the null hypothesis, i.e. based on the data in table 8.13 there is no reason to believe that the peel off time of tape 1 differs from that of tape 2.

Exercise F

As a block slides over a surface, friction acts to oppose the motion of the block. An experiment is devised to determine whether the friction depends on the area of contact between a block and a wooden surface. Table 8.14 shows the coefficient of kinetic friction, μ_k, for two blocks with different contact areas between block and surface. Test at the $\alpha = 0.05$ level of significance whether μ_k is independent of contact area.

Table 8.14. *Coefficients of friction for two contact areas.*

Coefficient of kinetic friction, μ_k	
Block 1 (contact area = 174 cm²)	Block 2 (contact area = 47 cm²)
0.388	0.298
0.379	0.315
0.364	0.303
0.376	0.300
0.386	0.290
0.373	0.287

8.6.1 Excel®'s TTEST()

A convenient way to establish whether there is a significant difference between the means of two samples is to use Excel®'s TTEST() function. The syntax of the function is

TTEST(array1,array2,tails,type)

where array1 and array2 point to cells which contain the values of sample 1 and sample 2 respectively. The argument 'tails' has the value 1 for a one tailed test and the value 2 for a two-tailed test. 'Type' refers to which t test should be performed. For the two sample test considered in section 8.6, type is equal to 2.

The number returned by the function is the probability that the sample means would differ by at least $\bar{x}_1 - \bar{x}_2$, when $\mu_1 = \mu_2$.

Example 7

Is there any significant difference in the peel off times between tape 1 and tape 2 in table 8.13?

ANSWER

We begin by stating the null and alternative hypotheses:

$$H_0: \mu_1 = \mu_2;$$
$$H_a: \mu_1 \neq \mu_2.$$

Sheet 8.3 shows the data entered from table 8.13 along with the TTEST() function. When the Enter key is pressed, the number 0.425681 is returned into cell A9. If the null hypothesis is correct, then the probability that the sample means, $\bar{x}_1$ and $\bar{x}_2$, will differ by at least as much as $\bar{x}_1 - \bar{x}_2$ is ≈ 0.43 (or 43%). This probability is so large that we infer that the values are consistent with the null hypothesis and so we cannot reject that hypothesis.

Sheet 8.3. *Use of Excel®'s* TTEST() *function.*

	A	B	C
1	tape1	tape2	
2	65	176	
3	128	125	
4	87	95	
5	145	255	
6	210	147	
7	85	88	
8			
9	=TTEST(A2:A7,B2:B7,2,2)		

Exercise G

Table 8.15 shows values of the lead content of water drawn from two locations on a river. Use Excel®'s TTEST() function to test the null hypothesis that the mean lead content of both samples is the same at $\alpha = 0.05$.

Table 8.15. *Lead content of river water.*

Lead content	Sample A	42	50	45	48	52	42	53
(ppb)	Sample B	48	58	55	49	56	47	58

8.7 *t* test for paired samples

Another important type of test is that in which the contents of data sets are naturally linked or 'paired'. For example, suppose two chemical processes used to extract copper from ore are to be compared. Assuming that the processes are equally efficient at extracting the copper, then the amount obtained from any particular batch of ore should be independent of the process used, so long as account is taken of random errors.

Other situations in which a paired *t* test might be considered include comparing:

- two methods for analysing the chemical composition of a range of materials;
- heart rate before and after administering a new drug to a group of people;
- the analysis of materials as carried out by two independent laboratories;
- the calibration of hand held multimeters before and after a period of storage.

By calculating the mean of the differences, $\bar{d}$, of the paired values, and the standard deviation of the differences, $s_{\bar{d}}$, the *t* test statistic can be determined using

$$t = \frac{\bar{d} - \delta_0}{s_{\bar{d}}/\sqrt{n}} \tag{8.14}$$

where δ_0 is the hypothesised population mean of the differences between the paired values and n is the number of paired values. As δ_0 is most often taken to be zero, we rewrite equation (8.14) as

$$t = \frac{\bar{d}}{s_{\bar{d}}/\sqrt{n}} \tag{8.15}$$

The number of degrees of freedom, ν, is $n-1$.

Example 8

Table 8.16 shows the amount of copper extracted from ore using two processes. Test at $\alpha = 0.05$ whether there is a significant difference in the yield of copper between the two processes.

Table 8.16. *Copper yields from two processes.*

Batch	Process A yield (%)	Process B yield (%)	d_i = Process A yield – Process B yield (%)
1	32.5	29.6	2.9
2	30.5	31.2	−0.7
3	29.6	29.7	−0.1
4	38.4	37.1	1.3
5	32.8	31.3	1.5
6	34.0	34.5	−0.5
7	32.1	31.0	1.1
8	40.3	38.7	1.6

ANSWER

Though it is possible to calculate the mean yield for process A and compare that with the mean of the yield for process B, the variability between batches (say, due to the batches being obtained from different locations) encourages us to consider a comparison between yields on a batch by batch basis.

To perform the hypothesis test we write:

H_0: $\delta = 0$ (the null hypothesis is that the population mean of the differences between the paired values is zero);

H_a: $\delta \neq 0$.

Using the data in table 8.16, we find $\bar{d} = 0.8875\%$, $s_{\bar{d}} = 1.229\%$. Substituting these numbers into equation (8.15) (and noting the number of pairs, $n = 8$), we have

$t = 2.042$

The number of degrees of freedom, $\nu = n - 1 = 7$. For a two tailed test, the critical value of the test statistic, t_{crit}, for $\nu = 7$ and $\alpha = 0.05$, found from table 2 in appendix 1, is

$t_{crit} = t_{0.05,7} = 2.365$

As $|t| < t_{crit}$ we cannot reject the null hypothesis, i.e. based on the data in table 8.16 there is no difference in the efficiency of the extraction methods at the 0.05 level of significance.

Exercise H

The emfs of a batch of 9 V batteries were measured before and after a storage period of 3 months. Table 8.17 shows the emfs of eight alkaline batteries before and after the storage period. Test at the $\alpha = 0.05$ level of significance whether the emfs of the batteries have changed over the storage period.

Table 8.17. *Emfs of eight alkaline batteries.*

Battery number	Emf at $t=0$ (V)	Emf at $t=3$ months (V)
1	9.49	9.48
2	9.44	9.42
3	9.46	9.46
4	9.47	9.46
5	9.44	9.41
6	9.43	9.40
7	9.39	9.40
8	9.46	9.44

8.7.1 Excel®'s TTEST() for paired samples

Excel®'s TTEST() function may be used to determine the probability of obtaining a mean of differences between paired values, $\bar{d}$, given that the null hypothesis is that the population mean difference between paired values is $\delta_0 = 0$. The syntax of the function is

TTEST(array1,array2,tails,type)

where array1 and array2 point to cells which contain the paired values of sample 1 and sample 2 respectively. The argument 'tails' is equal to 1 for a one tailed test and equal to 2 for a two tailed test. For the paired *t* test, the argument 'type' is equal to 1.

Exercise I

Consider the data given in table 8.16. Use the TTEST() function to determine the probability that the mean difference in yields, $\bar{d}$, would be 0.8875% or larger. Assume that the null and alternative hypotheses given in example 8 still apply.

We can extend the comparison of means to establish whether there is a significant difference between the means of three or more samples. The approach adopted is termed analysis of variance (or ANOVA for short). We will consider ANOVA in section 8.10, but as this technique requires the comparison of variances we will consider this first, and, in particular, the *F* test.

8.8 Comparing variances using the *F* test

The tests of significance we have considered so far have mainly considered population and sample means. Another important population parameter estimated from sample data is the standard deviation, σ. There are situations in which we wish to know if two sets of data have come from populations with the same standard deviation or variance. These situations include where:

- the variability of values returned by two analytical techniques is to be compared;
- the control of a variable (such as temperature) has improved due to the introduction of new technology;
- variability in human reaction time is to be compared before and after the intake of alcohol;
- the electrical noise picked up by a sensitive electronic circuit is to be compared when the circuit is: (a) shielded, (b) unshielded;
- the precision of two instruments is to be compared.

For example, suppose we make ten measurements of the iron concentration in a specimen of stainless steel and calculate the standard deviation of the sample, s_1. Ten measurements of iron concentration are made on another specimen of steel and the standard deviation of the values is s_2. By how much can s_1 differ from s_2 before the difference is regarded as significant? Before we are able to answer that question we consider a probability distribution that describes the probability of obtaining a particular ratio of *variances* of two samples, s_1^2/s_2^2. This is referred to as the *F* distribution.

8.8.1 The *F* distribution

A test that is routinely used to compare the variability of two samples is the '*F* test'. This test is based on the *F* distribution, first introduced by (and named in honour of) R A Fisher, a pioneer in the development of data analysis techniques and experimental design.

  If we take two random samples of sizes n_1 and n_2 from the same normally distributed population, then the ratio

$$F = \frac{s_1^2}{s_2^2}$$

(8.16)

is a statistic which lies between 0 and $+\infty$, s_1^2 is the estimate of the population variance of sample 1, and s_2^2 is the estimate of the population variance of sample 2.

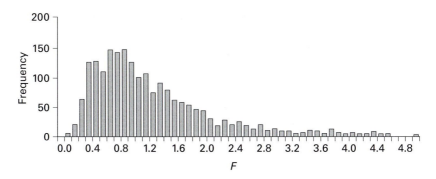

Figure 8.6. Histogram of 1000 simulated F values.

An indication of the shape of the distribution can be obtained by simulating many experiments in which the population variance estimates s_1^2 and s_2^2 are determined for samples containing n_1 and n_2 values respectively, where the samples are drawn from the same population. Figure 8.6 shows a histogram based on such a simulation in which $n_1 = n_2 = 10$ values were generated 1000 times using the normal distribution option of the Random Number Generator in Excel®.[17] Figure 8.6 indicates that, in contrast to the normal and t distributions, the F distribution is not symmetrical. The shape of the distribution depends on the number of degrees of freedom of each sample. The probability density function for the F distribution can be written[18]

$$f(x) = K(\nu_1, \nu_2) \left(\frac{\nu_1}{\nu_2}\right)^{\nu_1/2} \frac{x^{\frac{\nu_1 - 2}{2}}}{\left(1 + \frac{\nu_1 x}{\nu_2}\right)^{\frac{\nu_1 + \nu_2}{2}}} \tag{8.17}$$

where $x = s_1^2/s_2^2$ and $K(\nu_1, \nu_2)$ is a constant which depends on the degrees of freedom, ν_1 and ν_2.

$K(\nu_1, \nu_2)$ is chosen to ensure that the total area under the $f(x)$ curve is equal to 1. Figure 8.7 shows curves of $f(x)$ versus x for $\nu_1 = \nu_2 = 5$, $\nu_1 = \nu_2 = 25$ and $\nu_1 = \nu_2 = 100$.

[17] The population mean, $\mu = 32$, and the population standard deviation, $\sigma = 3$, were arbitrarily chosen for the simulation.

[18] See Graham (1993) for discussion of equation 8.17.

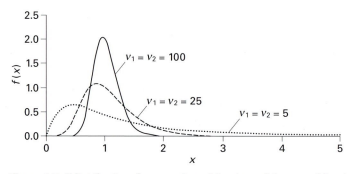

Figure 8.7. *F* distributions for several combinations of degrees of freedom, ν.

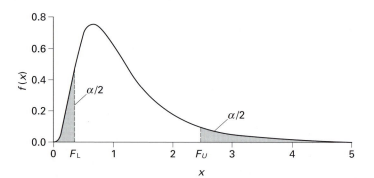

Figure 8.8. The *F* distribution showing the upper and lower critical values for a two tailed test at the α significance level.

8.8.2 The *F* test

In order to determine whether the variances of two sets of data could have come from populations with the same population variance, we begin by stating the null hypothesis that the population variances for the two sets of data are the same, i.e. $\sigma_1^2 = \sigma_2^2$. The alternative hypothesis (for a two tailed test) would be $\sigma_1^2 \neq \sigma_2^2$. The next stage is to determine the critical value for the *F* statistic. If the *F* value determined using the data exceeds the critical value, then we have evidence that the two sets of data do not come from populations that have the same variance.

An added consideration when employing the *F* test is that, owing to the fact that the distribution is not symmetrical, there are two critical *F* values of unequal magnitude as indicated in figure 8.8. This figure shows the upper and lower critical *F* values, F_U and F_L respectively, for a two tailed test at the significance level, α. If $s_1^2 > s_2^2$ and the *F* value determined using the data exceeds F_U, then we reject the null hypothesis (for a given significance level, α). If $s_1^2 < s_2^2$ and the *F* value is less than F_L, then again we reject the null hypothesis. The difficulty of having two critical values can be overcome if, when the ratio s_1^2/s_2^2 is calculated, the larger variance estimate is always placed in the numerator, so that $s_1^2/s_2^2 > 1$. In doing this we need only consider the rejection region in the right hand tail of the *F* distribution. Critical values for *F* (in which the larger of the two variances is placed in the numerator) are given in table 3 in appendix 1 for various probabilities in the right hand tail of the distribution.

Example 9

Two component manufacturers supply capacitors with nominal capacitance of 2.2 μF. Table 8.18 shows the capacitances of six capacitors supplied by each manufacturer. Test at the $\alpha = 0.05$ level of significance whether there is any difference in the variances of capacitances of the components supplied by each manufacturer.

ANSWER

$H_0: \sigma_1^2 = \sigma_2^2$ (the population variance of both samples is the same);
$H_a: \sigma_1^2 \neq \sigma_2^2$ (the population variance differs between samples).

Using the data in table 8.18, the variance of sample 1 is $s_1^2 = 1.575 \times 10^{-2} \ \mu F^2$ and the variance of sample 2 is $s_2^2 = 9.500 \times 10^{-4} \ \mu F^2$, hence $F = s_1^2/s_2^2 = 16.58$. The number of degrees of freedom for each sample is one less than the number of values, i.e. $\nu_1 = \nu_2 = 5$.

The critical value, F_{crit}, for *F* is obtained using table 3 in appendix 1. For $\alpha = 0.05$, the area in the right hand tail of the *F* distribution for a two tailed test is 0.025. For $\nu_1 = \nu_2 = 5$, $F_{crit} = 7.15$. As $F > F_{crit}$ this indicates that there is a significant difference between the variances of sample 1 and sample 2 and so we reject the null hypothesis.

Table 8.18. *Capacitances of two samples.*

Sample 1 capacitance (μF)	Sample 2 capacitance (μF)
2.25	2.23
2.05	2.27
2.27	2.19
2.13	2.20
2.01	2.25
1.97	2.21

Exercise J

In an experiment to measure the purity of reference quality morphine, quantitative nuclear magnetic resonance (NMR) determined the purity as 99.920% with a standard deviation, s, of 0.052% (number of repeat measurements, $n=7$). Using another technique called isotope dilution gas chromatography mass spectrometry (GCMS) the purity of the same material was determined as 99.879% with a standard deviation, s, of 0.035% (again the number of repeat measurements, $n=7$). Determine at the $\alpha=0.05$ level of significance whether there is a significant difference in the variance of the values obtained using each analytical technique.

8.8.3   Excel®'s FINV() function

Excel®'s FINV() function can be used to determine the upper critical value from the F distribution, so avoiding the need to consult tables of critical values. The syntax of the function is

$$FINV(probability, \nu_1, \nu_2)$$

where probability refers to the area in the right hand tail of the F distribution as shown in figure 8.8, ν_1 is the number of degrees of freedom for sample 1 and ν_2 is the number of degrees of freedom for sample 2.

Example 10

Determine the upper critical value of the F distribution when $\nu_1=5$, $\nu_2=7$ and a one tailed test is required to be carried out at the $\alpha=0.05$ level of significance.

ANSWER

Using Excel®, we type into a cell =FINV(0.05,5,7). After pressing the Enter key, Excel® returns the number 3.971522.

Exercise K

What would be the critical F value in example 10 if a two tailed test at $\alpha=0.05$ level of significance were to be carried out?

8.8.4   Robustness of the F test

The F test produces satisfactory results so long as the distribution of the values in each sample is normal. However, if the values that make up a sample are not normally distributed or if the sample contains outliers, then a significance test which is believed to be being carried out at, say, $\alpha = 0.05$ could in fact be either above or below that significance level. That is, the F test is sensitive to the actual distribution of the values. As normality is difficult to establish when sample sizes are small, some workers are reluctant to bestow as much confidence in F tests as in t tests (t tests can be shown to be much less sensitive to violations of normality of data than the F test[19]). A test that works well even when some of the assumptions upon which the test is founded are not valid is referred to as *robust*.

8.9 Comparing expected and observed frequencies using the χ^2 test

Up to this point we have made the assumption that distributions of real data are well described by a theoretical distribution, most notably (and most commonly) the normal distribution. The validity of our analyses would be strengthened if we could show that data are consistent with the distribution used to describe them. While a histogram is a good starting point for examining theoretical and actual distributions, there are tests we can perform to establish quantitatively if data are likely to have been drawn from some hypothesised distribution. One such test is the χ^2 test which employs the χ^2 distribution.

8.9.1 The χ^2 distribution

If data consisting of n values is drawn from a population in which values are normally distributed, then the statistic χ^2 is given by

$$\chi^2 = \sum \left(\frac{x_i - \bar{x}}{\sigma} \right)^2 \tag{8.18}$$

where σ is the population standard deviation and $\bar{x}$ is the mean of the n values. If many samples of the same size are taken from the population, then χ^2 follows a distribution which depends only on the number of degrees of freedom, ν. As an example, figure 8.9 shows the result of a simulation in which values of χ^2 have been generated then displayed in the form of a histogram. Each χ^2 value was determined by first generating five numbers using the normal distribution option of the Random Number

[19] See Moore and McCabe (1989) for a discussion of the robustness of t and F tests.

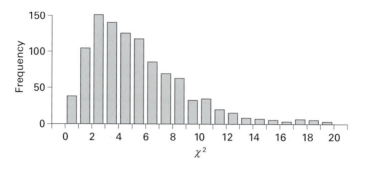

Figure 8.9. Histogram of 1000 simulated χ^2 values.

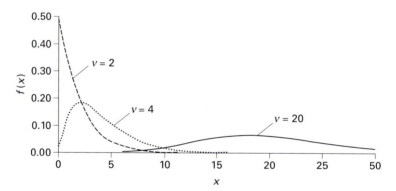

Figure 8.10. χ^2 distribution for $\nu=2$, 4 and 20.

Generator in Excel®. The population mean of the distribution was chosen (arbitrarily) to have a mean of 50 and a standard deviation of 3. Equation (8.18) was used to determine χ^2 for each sample. This process was repeated 1000 times.

The distribution of χ^2 values is non-symmetrical with a single peak with mean $=\nu$ and variance $=2\nu$. The probability density function for the χ^2 distribution with ν degrees of freedom can be written[20]

$$f(x) = K(\nu)x^{\frac{\nu-2}{2}}\exp\left(\frac{-x}{2}\right) \tag{8.19}$$

where $x=\chi^2$ and $K(\nu)$ is a constant which depends on the degrees of freedom. $K(\nu)$ is chosen to ensure that the total area under the $f(x)$ versus x curve is equal to 1. Figure 8.10 shows the distribution function given by equation (8.19) for $\nu=2$, 4 and 20. Note that as ν increases, the distribution becomes more symmetric and in fact tends to the normal distribution when ν is very large.

[20] The χ^2 distribution is discussed by Kennedy and Neville (1986).

8.9.2 The χ^2 test

An important application of the χ^2 distribution is as the basis of a test to establish how well the distribution of sample data gathered as part of an experiment compares with a theoretical distribution of data – this is sometimes called a test of the 'goodness of fit'. To compare a 'real' and a theoretical distribution of data we begin by dividing the range of the data into conveniently sized categories or 'bins' and determine the frequency of values in each category. The categories should be chosen so that the frequency in each category is ≥ 5. This condition occasionally requires that categories containing small frequencies be combined.[21]

We generally refer to the actual frequency in the ith category as O_i (referred to as the 'observed' frequency). If an experiment is carried out many times, we would expect the mean frequency in the ith category to be equal to that predicted by the distribution being tested. We refer to the predicted or expected frequency based on the distribution as E_i. The difference between O_i and E_i is small if the hypothesised distribution describes the data well.

With regard to equation (8.18), it is possible to identify a correspondence between the observed frequency in the ith category, O_i, and the ith observed value of x, x_i. Similarly, the expected frequency can be regarded as corresponding to the mean, $\bar{x}$. The number of counts in any one category will vary from experiment to experiment but we discovered in section 4.3 that so long as events are random and independent then the distribution of counts in each category (should we perform the experiment many times) would follow the Poisson distribution. If this is the case, the standard deviation of the counts is approximately equal to $\sqrt{\bar{x}}$ or, using the symbolism introduced here, $\sigma_i \approx \sqrt{E_i}$.

Replacing x_i by O_i, $\bar{x}$ by E_i and σ_i by $\sqrt{E_i}$ in equation (8.18) gives

$$\chi^2 = \sum \frac{(O_i - E_i)^2}{E_i} \tag{8.20}$$

If there is good agreement between observed and expected frequencies, so that $O_i \approx E_i$, χ^2 as given by equation (8.20) would be small. In fact, so long as χ^2 is approximately equal to the total number of categories, we can show that there is no evidence of significant difference between the actual and predicted frequencies. To put this more formally, if we know the number of degrees of freedom, then we can find the critical values for the χ^2 statistic, χ^2_{crit}, for any chosen level of significance, α. Figure 8.11 shows χ^2_{crit} for a significance level α.

---

[21] See Hamilton (1990) for a discussion on combining categories.

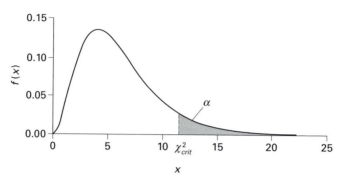

Figure 8.11. χ^2 distribution showing the critical value, χ^2_{crit}, for the significance level, α.

χ^2_{crit} may be found using table 4 in appendix 1. A typical null hypothesis would be that the observed data come from a population that can be described by a particular theoretical distribution, such as the binomial, Poisson or normal distributions. If χ^2, as calculated using equation (8.20), exceeds χ^2_{crit}, we reject the null hypothesis.

8.9.3   Is the fit too good?

It is possible for the fit of a theoretical distribution to experimental data to be too good, i.e. that the observed and expected frequencies in each category match so well that this is unlikely to have happened 'by chance'. The critical value for χ^2 in this situation is found by considering the area corresponding to the level of significance confined to the left hand tail of the χ^2 distribution. In this case if $\chi^2 < \chi^2_{crit}$ we reject the null hypothesis that the observed and expected distributions could match so well 'by chance'. If such a null hypothesis is rejected then it is possible that data gathered in the experiment, or the experimenter's methods, would be regarded with some concern as it is likely to be difficult to explain why $O_i = E_i$ for all i when random error is present.

As we seldom attempt to find out whether the fit between experimental data and hypothesised probability distribution is too good, we use a one tailed test in which the area corresponding to the level of significance is confined to the right hand tail of the χ^2 distribution, as shown in figure 8.11.

8.9.4 Degrees of freedom in χ^2 test

In order to determine χ^2_{crit} we need to know the number of degrees of freedom, ν. Let the frequency in the ith bin be O_i. If there are k categories or bins we can write

$$\sum_{i=1}^{i=k} O_i = n \tag{8.21}$$

where n is the total number of observations. One constraint on the expected number of observations, E_i, in each of the k categories is that

$$\sum_{i=1}^{i=k} E_i = n \tag{8.22}$$

For each constraint that we impose upon the expected frequencies, we reduce the number of degrees of freedom by 1, so that if the only constraint were given by equation (8.22), then[22]

$$\nu = k - 1 \tag{8.23}$$

The number of degrees of freedom is further reduced if the sample data must be used to estimate one or more parameters. As an example, in order to apply the χ^2 test to data that are assumed to be Poisson distributed, we usually need to estimate the mean of the distribution, μ, using the sample data. As the constraint expressed by equation (8.22) still applies, the number of degrees of freedom for a χ^2 test applied to a hypothesised Poisson distribution would be $\nu = k - 2$. Table 8.19 shows the number of degrees of freedom for χ^2 tests applied to various hypothesised distributions.

Table 8.19. *Degrees of freedom for χ^2 tests.*

Null hypothesis	Parameters estimated from sample data	Degrees of freedom, ν
Data follow a uniform distribution	None	$k-1$
Data follow a binomial distribution	The probability of a success on a single trial, p	$k-2$
Data follow a Poisson distribution	Population mean, μ	$k-2$
Data follow a normal distribution	Population mean, μ, and population standard deviation, σ	$k-3$

[22] If there are k categories, then the frequency in each of $k-1$ categories can take on any value. Since the total number of observations is fixed, the kth category is constrained to have a frequency equal to the (total number of observations – sum of frequencies in the $k-1$ categories).

Table 8.20. *Distribution of numbers produced by a uniform random number generator.*

Interval	Frequency
$0.0 \leq x < 0.1$	12
$0.1 \leq x < 0.2$	12
$0.2 \leq x < 0.3$	9
$0.3 \leq x < 0.4$	13
$0.4 \leq x < 0.5$	7
$0.5 \leq x < 0.6$	9
$0.6 \leq x < 0.7$	5
$0.7 \leq x < 0.8$	12
$0.8 \leq x < 0.9$	9
$0.9 \leq x < 1.0$	12

Example 11

Table 8.20 shows the frequency of occurrence of random numbers generated between 0 and 1 by the random number generator on a pocket calculator. The category width has been chosen as 0.1. Use a χ^2 test at the 0.05 level of significance to establish whether the values in table 8.20 are consistent with a number generator that produces random numbers distributed uniformly between 0 and 1.

ANSWER

> H_0: Data have been drawn from a population that consists of numbers which are uniformly distributed between 0 and 1.
>
> H_a: Data have been drawn from a population that consists of numbers which are not uniformly distributed between 0 and 1.

Table 8.20 contains the observed frequencies in each category. If the random number generator produces numbers evenly distributed between 0 and 1, then out of 100 random numbers we would expect, on average, ten numbers to lie between 0 and 0.1, ten to lie between 0.1 and 0.2 and so on. Table 8.21 shows the observed frequencies, expected frequencies and the terms in the summation given in equation (8.20). Summing the last column of table 8.21 gives

$$\chi^2 = \sum \frac{(O_i - E_i)^2}{E_i} = 6.2$$

In order to find χ^2_{crit}, we use table 4 in appendix 1 with $\nu = k - 1 = 10 - 1 = 9$

$$\chi^2_{crit} = \chi^2_{0.05,9} = 16.919$$

As $\chi^2 < \chi^2_{crit}$, we cannot reject the null hypothesis, i.e. the numbers are consistent with having been drawn from a population that is uniformly distributed between 0 and 1.

Table 8.21. *Observed and expected frequencies for example 11.*

Interval	Observed frequency O_i	Expected frequency E_i	$\dfrac{(O_i - E_i)^2}{E_i}$
$0.0 \leq x < 0.1$	12	10	0.4
$0.1 \leq x < 0.2$	12	10	0.4
$0.2 \leq x < 0.3$	9	10	0.1
$0.3 \leq x < 0.4$	13	10	0.9
$0.4 \leq x < 0.5$	7	10	0.9
$0.5 \leq x < 0.6$	9	10	0.1
$0.6 \leq x < 0.7$	5	10	2.5
$0.7 \leq x < 0.8$	12	10	0.4
$0.8 \leq x < 0.9$	9	10	0.1
$0.9 \leq x < 1.0$	12	10	0.4

Table 8.22. *Frequency of occurrence of cosmic rays in 50 one minute intervals.*

Counts	Frequency
0	10
1	13
2	7
3	15
4 or more	5

Exercise L

The number of cosmic rays detected in 50 consecutive one minute intervals is shown in table 8.22.

(i) Assuming that the counts follow a Poisson distribution, determine the expected frequencies for 0, 1, 2 etc. counts. (Note that for the data in table 8.22, the mean number of counts per minute is 1.86.)

(ii) Use a χ^2 test at the 0.05 level of significance to determine whether the distribution of counts is consistent with a Poisson distribution that has a mean

8.9.5 Excel®'s CHIINV() function

Critical values of the χ^2 statistic may be obtained using Excel®'s CHIINV() function instead of using table 4 in appendix 1. The syntax of the function is

CHIINV(probability,ν)

where probability refers to the area in the right hand tail of the distribution as shown by the shaded area in figure 8.11 and ν is the number of degrees of freedom.

Example 12

Calculate the critical value of the χ^2 statistic when the area in the right hand tail of the χ^2 distribution is 0.1 and the number of degrees of freedom, $\nu=3$.

ANSWER

Using Excel®, type into a cell =**CHIINV(0.1,3)**. After pressing the Enter key, Excel® returns the number 6.251394.

Exercise M

Use Excel® to determine the critical value of the χ^2 statistic for:

  (i) $\alpha=0.005$, $\nu=2$;
 (ii) $\alpha=0.01$, $\nu=3$;
(iii) $\alpha=0.05$, $\nu=5$;
 (iv) $\alpha=0.1$, $\nu=10$.

8.10 Analysis of variance

Significance testing can be extended beyond comparing the means of two samples to comparing the means of many samples. One such test is based upon the analysis of variance or 'ANOVA' for short.[23] We consider 'one-way' ANOVA in which a single factor or characteristic may vary from one sample to the next. For example, ANOVA might be used to compare the consistency of analysis of a specimen of blood divided into three parts and sent to three forensic laboratories. In this case the forensic laboratory is the factor that differs from sample to sample.

[23] ANOVA is a versatile and powerful technique. See McPherson (1990) for details of the usual variations of ANOVA.

The null and alternative hypotheses in a one-way ANOVA can be stated quite simply:

H_0: The population means of all samples are equal.
H_a: The population means of all samples are not equal.

Situations in which we might use ANOVA include comparing:

- the effect of three (or more) types of fuel additive on the efficiency of a car engine;
- influence of storage temperature on the adhesive properties of sticky tape;
- the concentration of iron in an ore as determined by several independent laboratories;
- the concentration of carbon dioxide emerging from four volcanic vents.

Although the purpose of ANOVA is to determine whether there is a significant difference between the means of several samples of data, the method actually relies on the calculation of variance of data within samples and compares that with the variance of data determined by considering the variability between samples. If the variances are significantly different then we can infer that the samples are not all drawn from the same population.

We must be careful to use ANOVA in situations in which it is most appropriate. In the physical sciences there is often a known, assumed or hypothesised relationship between variables. If it is possible to express that relationship in the form of an equation, then we may prefer to apply the technique of least squares in order to estimate parameters that appear in the equation rather than to use ANOVA. For example, it makes little sense to make ten repeat measurements of the viscosity of oil at each of five different temperatures and then to use ANOVA to establish whether there is a significant difference between the mean viscosity at each temperature. It is well established that the viscosity of oil is dependent on temperature and a more useful study would focus upon determining a model of temperature dependence of viscosity which best describes the experimental data.

8.10.1 Principle of ANOVA

Analysis using one-way ANOVA relies on using the data in two ways to obtain an estimate of the variance of the population. Firstly, the variance of the values in each sample[24] is estimated. By averaging the estimate of the variance for all samples (assuming that all samples consist of the same number of values) we get the best estimate of the 'within sample variance'. We may write this as $s^2_{within\,samples}$. Suppose, for example, we wish to compare the means of four samples. The estimate of the population variance for each sample is s^2_1, s^2_2, s^2_3 and s^2_4. The best estimate of the within sample variance would be[25]

$$s^2_{within\,samples} = \frac{s^2_1 + s^2_2 + s^2_3 + s^2_4}{4} \tag{8.24}$$

More generally, we write for K samples

$$s^2_{within\,samples} = \frac{s^2_1 + s^2_2 + s^2_3 + \cdots s^2_K}{K}$$

or

$$s^2_{within\,samples} = \frac{1}{K} \sum_{j=1}^{j=K} s^2_j \tag{8.25}$$

Another way to estimate the population variance is to find the variance, $s^2_{\bar{x}}$, of the sample means given by

$$s^2_{\bar{x}} = \sum_{j=1}^{j=K} \frac{(\bar{x}_j - \bar{X})^2}{K-1} \tag{8.26}$$

where $\bar{x}_j$ is the mean of the jth sample and $\bar{X}$ is the mean of all the sample means (sometimes referred to as the 'grand mean').

$s^2_{\bar{x}}$ is related to the estimate of the between sample variance, $s^2_{between\,samples}$, by

$$s^2_{\bar{x}} = \frac{s^2_{between\,samples}}{N} \tag{8.27}$$

where N is the number of values in each sample. The larger the difference between the samples means, the larger will be $s^2_{between\,samples}$. Rearranging equation (8.27) we obtain

$$s^2_{between\,samples} = Ns^2_{\bar{x}} \tag{8.28}$$

If all samples have been drawn from the same population, it should not matter which method is used to estimate the population variance as each

---

[24] Note that some texts and software packages use the word 'group' in place of 'sample' and refer to 'between group variance' and 'within group variance'. In this text we will consistently use the word 'sample'.

[25] For simplicity we assume that each sample consists of the same number of values.

method should give approximately the same variance. However, if the difference between one or more means is large (and it is this we wish to test) then $s^2_{between\ samples}$ will be larger than $s^2_{within\ samples}$. To determine if there is a significant difference between the variances, we use a one tailed F test,[26] where the F statistic is given by

$$F = \frac{s^2_{between\ samples}}{s^2_{within\ samples}} \qquad\qquad (8.29)$$

If all values come from the same population, we expect F to be close to unity. F will be much greater than one if $s^2_{between\ samples} \gg s^2_{within\ samples}$ and this will occur if one or more of the sample means is significantly different from the other means.

In order to demonstrate the application of ANOVA, consider a situation in which an experimenter wishes to determine whether the gas emerging from four volcanic vents is tapping the same source.

8.10.2 Example of ANOVA calculation

Table 8.23 shows the concentration of carbon dioxide for gas emerging from four volcanic vents. We will use ANOVA to test the hypothesis that the gas from each vent comes from a common reservoir. We will carry out the test at $\alpha = 0.05$ level of significance.

The null and alternative hypotheses are:

H_0: The population means of all the samples are the same.
H_a: The population means of all the samples are not the same.

Table 8.23. *Percentage of carbon dioxide from four vents.*

Vent 1 ($\%CO_2$)	Vent 2 ($\%CO_2$)	Vent 3 ($\%CO_2$)	Vent 4 ($\%CO_2$)
21	25	30	31
23	22	25	25
26	28	24	27
28	29	26	28
27	27	27	33
25	25	30	28
24	27	24	32

[26] The F test is described in section 8.8.2.

Table 8.24. *Mean and variance of the carbon dioxide data in table 8.23.*

	Vent 1	Vent 2	Vent 3	Vent 4
Mean, $\bar{x}_j$	24.857	26.143	26.571	29.143
Variance, s_j^2	5.810	5.476	6.619	8.476

The estimate of the population variance of the jth vent is given by

$$s_j^2 = \frac{\sum (x_i - \bar{x}_j)^2}{N-1} \tag{8.30}$$

where N is the number of values in the jth vent and $\bar{x}_j$ is the mean of the values for the jth vent. Table 8.24 shows the sample means and the estimated variances for the data in table 8.23.

Using equation (8.25), the estimate of the within sample population variance is

$$s_{within\ samples}^2 = \frac{5.810 + 5.476 + 6.619 + 8.476}{4} = 6.595$$

The grand mean of the data in table 8.24 is

$$\bar{X} = \frac{1}{K} \sum \bar{x}_j = \frac{24.857 + 26.143 + 26.571 + 29.143}{4} = 26.679$$

Substituting for $\bar{x}_j$ and $\bar{X}$ in equation (8.26) gives

$$s_{\bar{x}}^2 = \frac{(24.857 - 26.679)^2 + (26.143 - 26.679)^2 + (26.571 - 26.679)^2 + (29.143 - 26.679)^2}{4 - 1}$$

$$= 3.230$$

We use equation (8.28) to give the estimate of the between sample variance, i.e.

$$s_{between\ samples}^2 = Ns_{\bar{x}}^2 = 7 \times 3.230 = 22.61$$

The F statistic given by equation (8.29) is

$$F = \frac{s_{between\ samples}^2}{s_{within\ samples}^2} = \frac{22.61}{6.595} = 3.428$$

In order to establish whether the difference in variances is significant we must determine the critical value of the F statistic at whatever significance level is chosen.

When calculating the variance of each sample, the number of degrees of freedom is $N-1$ (one degree of freedom is lost due to the fact that the mean of the sample is used in the calculation of the variance). As K samples contribute to the calculation of the within sample variance, the number of degrees of freedom is

$$\nu_{\text{within samples}} = K(N-1) \tag{8.31}$$

When calculating the between sample variance, we used equation (8.28) in which the number of degrees of freedom is $K-1$, so that

$$\nu_{\text{between samples}} = K-1 \tag{8.32}$$

For the data in table 8.22, $K=4$ and $N=7$. This gives $\nu_{\text{within samples}} = 24$ and $\nu_{\text{between samples}} = 3$. We now require the critical value of the F statistic for $\alpha = 0.05$ when the degrees of freedom in the numerator equals 3, and that in the denominator equals 24. Using table 3 in appendix 1

$$F_{crit} = F_{0.05,3,24} = 3.01$$

Comparing this with $F = 3.428$, as determined using the data, indicates that we should reject the null hypothesis. That is, the population means of all the samples are not the same.

Exercise N

Experimental studies have linked the size of the alpha wave generated by the human brain to the amount of light falling on the retina of the eye. In one study, the size of the alpha signal for nine people is measured at three light levels. Table 8.25 shows the size of the alpha signal at each light level for the nine people. Using ANOVA, determine at the $\alpha = 0.05$ level of significance whether the magnitude of the alpha wave depends on light level.

Table 8.25. *Variation of size of alpha wave with light level.*

Light level	Magnitude of alpha wave (μV)								
High	32	35	40	35	33	37	39	34	37
Medium	36	39	29	33	38	36	32	39	40
Low	39	42	47	39	45	51	43	43	39

8.11 Review

Hypothesis testing assists in data analysis by forcing us to look closely at the data and address in a formal manner the question: 'Is there really something interesting in the data or can apparent relationships or differences between distributions of data be explained by chance?' A hypothesis is formulated, for example that the data are consistent with a normal distribution, and we establish, if the hypothesis is true, what is the probability of obtaining the data set being considered.

While hypothesis testing is powerful we should not forget that the rejection of a null hypothesis should be seen in the context of the whole experiment being performed. A null hypothesis could be rejected when there is a very small and in practice unimportant difference between, say, a sample mean and a hypothesised population mean.[27]

Traditionally, hypothesis testing requires referring to statistical tables at the back of a text book to find critical values of a particular test statistic. This tedious activity is considerably alleviated by the useful built in functions available in Excel®. In the next chapter we will consider more built in data analysis features in Excel®, which allow us, amongst other things, to perform statistical tests efficiently.

Problems

1. (i) Generate a column of 40 normally distributed random numbers with a population mean $\mu = 150$ and a population standard deviation $\sigma = 25$ using the Random Number Generation option in Analysis ToolPak in Excel® (refer to section 9.6 for details of how to use this tool).
    (ii) For the column of numbers, determine the mean, $\bar{x}$.
    (iii) Repeat parts (i) and (ii) for 100 columns each containing 40 random numbers.
    (iv) How many means lie outside the interval given by $\mu \pm 1.96\,\sigma_{\bar{x}}$?
    (v) How many means would you expect to lie outside the interval $\mu \pm 1.96\,\sigma_{\bar{x}}$?

2. The porosity, r, was measured for two samples of the ceramic $YBa_2Cu_3O_7$ prepared at different temperatures. The data are shown in table 8.26. Determine at the $\alpha = 0.05$ level of significance whether there is any difference in the porosity of the two samples.

3. Blood was taken from eight volunteers and sent to two laboratories. The urea concentration in the blood of each volunteer, as determined by both laboratories, is shown in table 8.27. Determine at the $\alpha = 0.05$ level of significance whether there is any difference in the urea concentration determined by the laboratories.

---

[27] This occurs most often when sample sizes are large.

Table 8.26. *Porosity values for two samples of* $YBa_2Cu_3O_7$.

Porosity	Sample 1	0.397	0.385	0.394	0.387	0.362	0.388
r	Sample 2	0.387	0.361	0.377	0.352	0.363	0.387

Table 8.27. *Values of urea concentration for eight volunteers as determined by two laboratories.*

		Volunteer							
		1	2	3	4	5	6	7	8
Urea	Laboratory A	4.1	2.3	8.4	7.4	7.5	3.4	3.9	6.0
concentration									
(mmol/L)	Laboratory B	4.0	2.4	7.9	7.3	7.3	3.0	3.8	5.5

Table 8.28. *Variation of peak area with concentration of isooctane.*

Concentration (moles)	Peak area (arbitrary units)
0.00342	2.10
0.00784	3.51
0.0102	5.11
0.0125	5.93
0.0168	8.06

4. The area under a peak of a chromatogram is measured for various concentrations of isooctane in a mixture of hydrocarbons. Table 8.28 shows calibration data of the peak area as a function of isooctane concentration. Assuming that the peak area is linearly related to concentration, perform unweighted least squares to find the intercept and slope of the best line through the data. Is the intercept significantly different from zero at the $\alpha = 0.05$ level of significance?

5. Two machines, A and B, are used to weigh and pack containers with 50 g of barium carbonate. Table 8.29 shows the mass of nine containers from each machine. Do the data in table 8.29 indicate that there is a difference in the variability in the mass of barium carbonate packed by each machine? Take the level of significance for the hypothesis test to be $\alpha = 0.05$.

Table 8.29. *Mass of barium carbonate in containers packed by two machines.*

Machine A mass (g)	50.0	49.2	49.4	49.8	48.3	50.0	51.3	49.7	49.5
Machine B mass (g)	51.9	48.8	52.0	52.3	51.0	49.6	49.2	49.1	52.4

Table 8.30. Current gain of transistor taken from two batches.

Current gain	Batch A	251	321	617	425	430	512	205	325	415
	Batch B	321	425	502	375	427	522	299	342	420

Table 8.31. *100 values of time of fall of ball (in seconds).*

1.35	1.15	1.46	1.67	1.65	1.76	0.97	1.36	1.63	1.19
1.35	1.35	1.35	1.35	1.35	1.35	1.35	1.35	1.35	1.35
1.27	1.07	1.04	1.21	1.26	0.99	1.30	1.33	1.44	1.34
1.34	1.34	1.68	1.39	1.37	1.31	1.80	1.58	1.89	1.28
1.74	1.09	1.52	1.59	1.79	1.39	1.31	1.55	1.33	1.56
1.12	1.24	1.11	1.34	1.40	1.42	1.35	1.85	1.06	1.26
0.89	1.70	1.15	1.28	1.56	1.50	1.58	1.53	1.14	1.19
1.55	1.47	1.22	1.36	1.44	1.52	1.44	1.23	1.79	1.51
1.42	1.58	1.58	1.28	1.23	1.63	1.17	1.10	1.55	1.54
1.85	1.70	1.67	1.43	1.41	1.50	1.40	1.20	1.06	1.58
1.50	1.53	1.45	1.20	1.66	1.35	1.24	1.25	1.32	1.32

6. After a modification to the process used to manufacture silicon transistors, the supplier claims that the variability in the current gain of the transistors has been reduced. Table 8.30 shows the current gain for transistors from two batches. Batch A corresponds to transistors tested before the new process was introduced and batch B after. Determine at $\alpha = 0.05$ level of significance whether the variability in the current gain has reduced as a result of the new process.

7. The time for a ball to fall 10 m was 'hand timed' using a stopwatch. Table 8.31 shows values of time for 100 successive measurements of elapsed time. Use the chi-squared test to determine whether the data are normally distributed. Carry out the test at the $\alpha = 0.05$ level of significance.

8. Three methods are devised to determine the kinetic energy of an arrow shot from a bow. These methods are based on:

Table 8.32. *Kinetic energy of an arrow as determined by three methods.*

Kinetic energy (joules)		
D45	MH	TE
10.5	9.9	11.0
10.7	10.1	10.9
11.2	11.0	11.3
10.1	10.8	11.4
11.5	10.3	10.9
10.8	10.6	10.6
10.0	10.8	11.1

Table 8.33. *Calcium concentration in blood as measured by three laboratories.*

Amount of calcium in blood (mmol/L)		
Laboratory A	Laboratory B	Laboratory C
2.23	2.35	2.31
2.26	2.28	2.33
2.21	2.29	2.29
2.25	2.28	2.27
2.20	2.27	2.33

(i) distance travelled by the arrow when shot at 45° to the horizontal (D45);

(ii) maximum height attained by the arrow when shot vertically (MH);

(iii) time elapsed for arrow to hit the ground after being shot vertically (TE).

Table 8.32 shows values of kinetic energy of the arrow based on the three methods. Use one-way ANOVA to establish, at the $\alpha = 0.05$ level of significance, whether the determination the kinetic energy depends on the method used.

9. Table 8.33 shows the concentration of calcium in a specimen of human blood as analysed by three laboratories. Use the ANOVA utility in Excel® to determine whether there is any significant difference in the sample means between the three laboratories at $\alpha = 0.05$ (refer to section 9.3 for details on how to use Excel®'s ANOVA utility).

Chapter 9

Data Analysis tools in Excel® and the Analysis ToolPak

9.1 Introduction

The process of analysing experimental data frequently involves many steps such as the tabulation and graphing of data. Numerical analysis of data requires many simple but repetitive calculations such as the summing and averaging of values. Spreadsheet programs are designed to perform these tasks, and in previous chapters we considered how Excel®'s built in functions such as AVERAGE() and CORREL() can assist data analysis. While the functions in Excel® are extremely useful, there is still some effort required to:

- enter data into the functions;
- format numbers returned by the functions so that they are easy to assimilate;
- plot suitable graphs;
- combine functions to perform more advanced analysis.

Excel® contains numerous useful data analysis tools designed around the built in functions which will, as examples, fit an equation to data using least squares or compare the means of many samples using analysis of variance. These tools can be found via the menu bar. The dialog box that appears when a tool is selected allows for the easy input of data. Once the tool is run, results are displayed in a Worksheet with explanatory labels and headings. As an added benefit, some tools offer automatic plotting of data as graphs or charts.

In this chapter we consider several of Excel®'s advanced data analysis tools which form part of the Analysis ToolPak add-in, paying particular

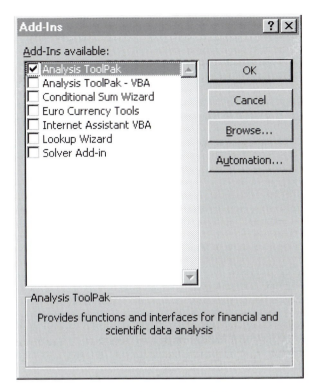

Figure 9.1. Add-in dialog box in Tools pull down menu.

attention to those tools which relate directly to principles and methods described in this book. The Histogram and Descriptive Statistics tools are described in sections 2.8.1 and 2.8.2 respectively and will not be discussed further in this chapter. Tools which relate less closely to the material in this book are described briefly with references given to where more information may be found.[1]

9.2 Activating the Data Analysis tools

To activate the Data Analysis tools, select Tools ➤Data Analysis from the menu bar. If the Data Analysis option does not appear in the Tools pull down menu, then select Tools ➤Add-Ins. The dialog box shown in figure 9.1 should appear. Check the box next to Analysis ToolPak, as shown in figure 9.1, then click OK. Data Analysis should now be added to the Tools pull down menu. It is possible that the Analysis ToolPak utility is not

---

[1] See Orvis (1996) for a useful general reference for this chapter.

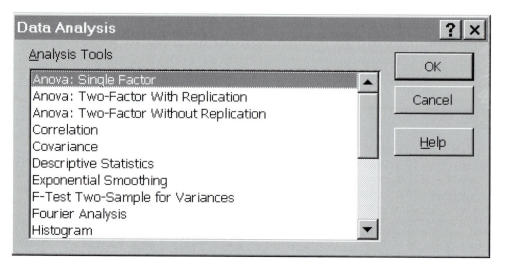

Figure 9.2. Data Analysis dialog box.

resident on your computer, in which case the computer will request you to insert the disc (or equivalent) that holds the Excel® program.

9.2.1 General features

Figure 9.2 shows the dialog box that appears after choosing Tools ➤Data Analysis from the Menu toolbar. The ⑦ or Help buttons in the dialog box can be used to request information on any of the tools in the Analysis ToolPak. Features of the tools in the Analysis ToolPak are:

 (i) If data are changed after using a tool, most tools must be run again to update calculations, i.e. the output is not linked dynamically to the input data. Exceptions to this are the Moving Average and Exponential Smoothing tools.
 (ii) Excel® 'remembers' the cell ranges and numbers typed into each tool's dialog box. This is useful if you wish to run the same tool more than once. When Excel® is exited the dialog boxes within all the tools are cleared.

The Data Analysis tools make extensive use of dialog boxes to allow for the easy entry of data, selection of options such as graph plotting and the input of numbers such as the level of significance, α.

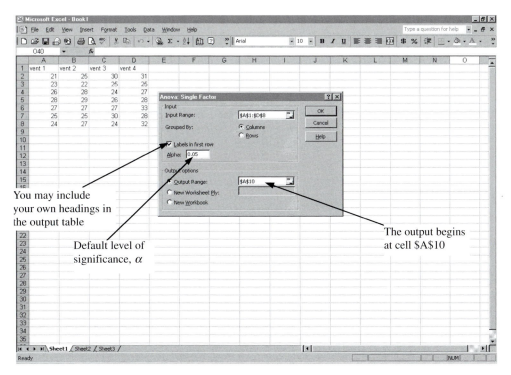

Figure 9.3. Example of the Anova: Single Factor tool.

9.3 Anova: Single Factor

One-way ANOVA (which Excel® refers to as 'single factor' ANOVA) may be used to determine whether there is a significant difference between the means of two or more samples where each sample contains two or more values.[2]

As an example of using the 'Anova: Single Factor' tool in Excel®, consider figure 9.3 which shows data entered into an Excel® Worksheet. The data appearing in the Worksheet are the percentages of carbon dioxide in gas emerging from four volcanic vents as discussed in section 8.10.2. The data to be analysed are contained in cells $A$2 through to $D$8. The range of cells (which includes the headings) is entered in the Input Range box as shown in figure 9.3. If we require the output of the ANOVA to appear on the same sheet as the data, we enter an appropriate cell reference into the

---

[2] Section 8.10 describes the principles of one-way ANOVA.

[3] Alternatively, the output from this (and all the other Data Analysis tools) can be made to appear in a new Worksheet or a new Workbook by clicking on New Worksheet Ply or New Workbook in the dialog box.

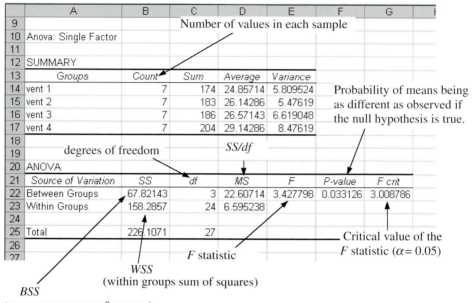

Figure 9.4. Output generated by the Anova: Single Factor tool.

Output Range box.[3] Choosing this cell as $A$10 and pressing OK returns the screen as shown (in part) in figure 9.4. Annotation has been added to figure 9.4 to clarify the labels and headings generated by Excel®. The value of the F statistic (3.428) is greater than the critical value of the F statistic, F_{crit} (3.009). We conclude that there is a significant difference (at the $\alpha = 0.05$ level of significance) between the means of the samples appearing in figure 9.3. The same conclusion was reached when these data were analysed in section 8.10.2. However, by using the ANOVA tool in the Analysis ToolPak, the time to analyse the data is reduced considerably in comparison to the approach adopted in that section. Similar increases in efficiency are found when using other tools in the ToolPak, but one important matter should be borne in mind: if we require numbers returned by formulae to be updated as soon as the contents of cells are modified, it is better not to use the tools in the ToolPak, but to create the spreadsheet 'from scratch' using the built in functions, such as AVERAGE() and STDEV().

9.4 Correlation

The CORREL() function in Excel® was introduced in section 6.6.1 to quantify the extent to which x–y data were correlated. The correlation between any

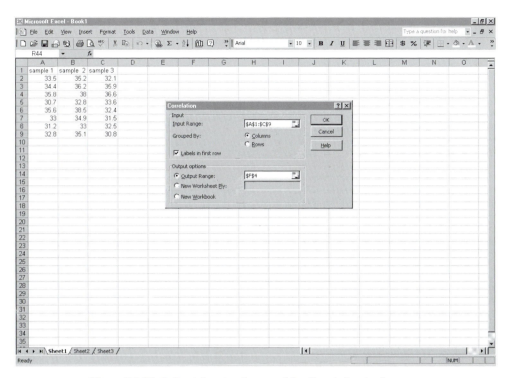

Figure 9.5. Worksheet showing the use of the Correlation tool.

two columns (or rows) containing data can be determined using the Correlation tool in the Analysis ToolPak. Figure 9.5 shows three samples of data entered into a worksheet. Figure 9.6 shows the output of the Correlation tool as a matrix of numbers in which the correlation of every combinations of pairs of columns is given. The correlation matrix in figure 9.6 indicates that values in sample 1 and sample 2 are highly correlated, but the evidence of correlation between samples 1 and 3 and samples 2 and 3 is much less convincing.

9.5 *F* Test Two-Sample for Variances

The *F* test is used to determine whether two samples of data could have been drawn from populations with the same variance as discussed in section 8.8.2. The F-Test tool in Excel® requires that the cell ranges containing the two samples be entered into the dialog box, as shown in figure 9.7. The data in figure 9.7 refer to values of capacitors supplied by two

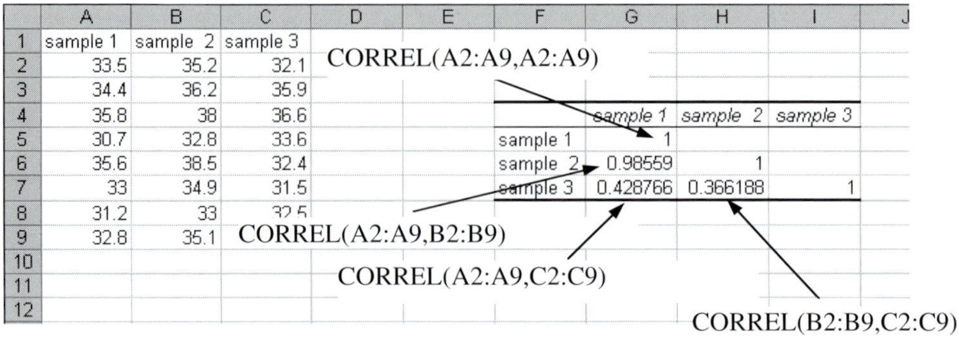

Figure 9.6. Output returned by the Correlation tool.

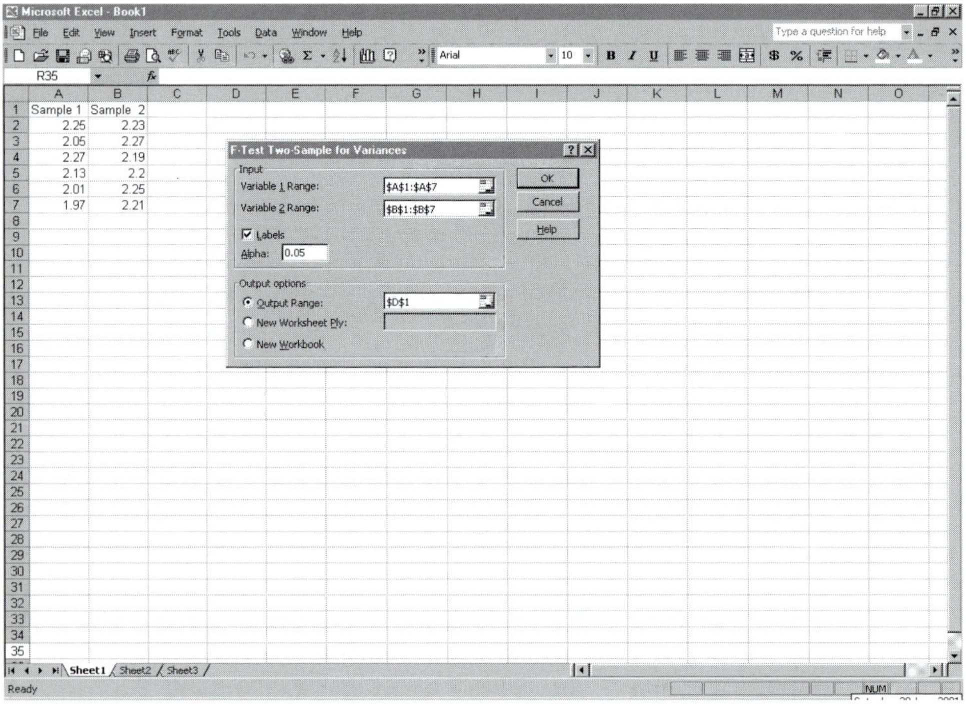

Figure 9.7. Worksheet showing the use of the F-Test tool.

manufacturers as described in example 9 of section 8.8.2. The F-Test tool performs a one tailed test at the chosen significance level, α. When a two tailed F test is required, as in this example, we must enter $\alpha/2$ in the $\underline{A}$lpha box in order to obtain the correct critical value of the test statistic (in which case $F_{crit} = 7.15$). Figure 9.8 shows the output returned by the F-Test tool.

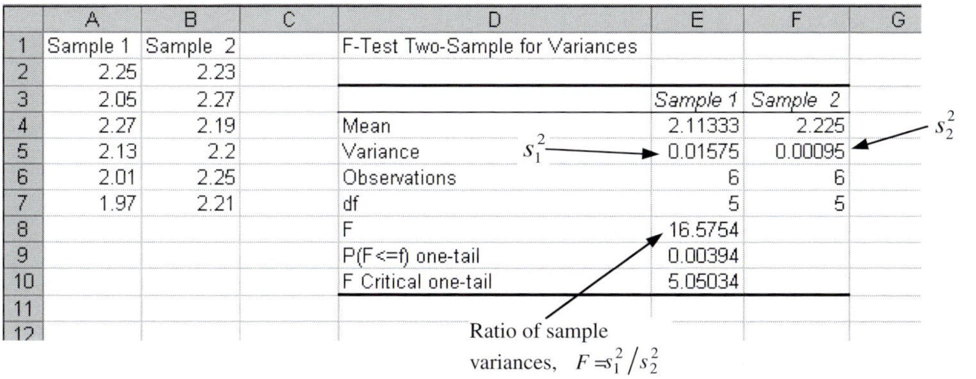

	A	B	C	D	E	F	G
1	Sample 1	Sample 2		F-Test Two-Sample for Variances			
2	2.25	2.23					
3	2.05	2.27			Sample 1	Sample 2	
4	2.27	2.19		Mean	2.11333	2.225	
5	2.13	2.2		Variance	0.01575	0.00095	
6	2.01	2.25		Observations	6	6	
7	1.97	2.21		df	5	5	
8				F	16.5754		
9				P(F<=f) one-tail	0.00394		
10				F Critical one-tail	5.05034		
11							
12							

s_1^2 points to Variance column E; s_2^2 points to column F.

Ratio of sample variances, $F = s_1^2 / s_2^2$

Figure 9.8. Output returned by the F-Test tool.

9.6 Random Number Generation

The Random Number Generation tool in Excel® generates random numbers based on any of seven probability distributions, including the normal, Poisson and binomial distributions. Random numbers are extremely useful for simulating data. For example, the effectiveness of least squares in estimating parameters such as the slope and intercept can be investigated by adding 'noise' with a known distribution to otherwise 'noise free' data. To illustrate this, figure 9.9 shows x–y data in columns A and B of an Excel® worksheet. The relationship between x and y is $y = 3x + 4$. Normally distributed random numbers are generated by choosing the normal distribution in the Random Number Generation dialog box. The default values for the mean and standard deviation of the parent distribution from which the numbers are generated are 0 and 1 respectively, but can be modified if required. The 'Number of Variables' in the Random Number Generation dialog box corresponds to the number of columns containing random numbers, and the 'Number of Random Numbers' corresponds to how many random numbers appear in each column. The random numbers returned by Excel® are shown in figure 9.10. Column D in the worksheet shows the normally distributed 'noise' added to the values in column C.

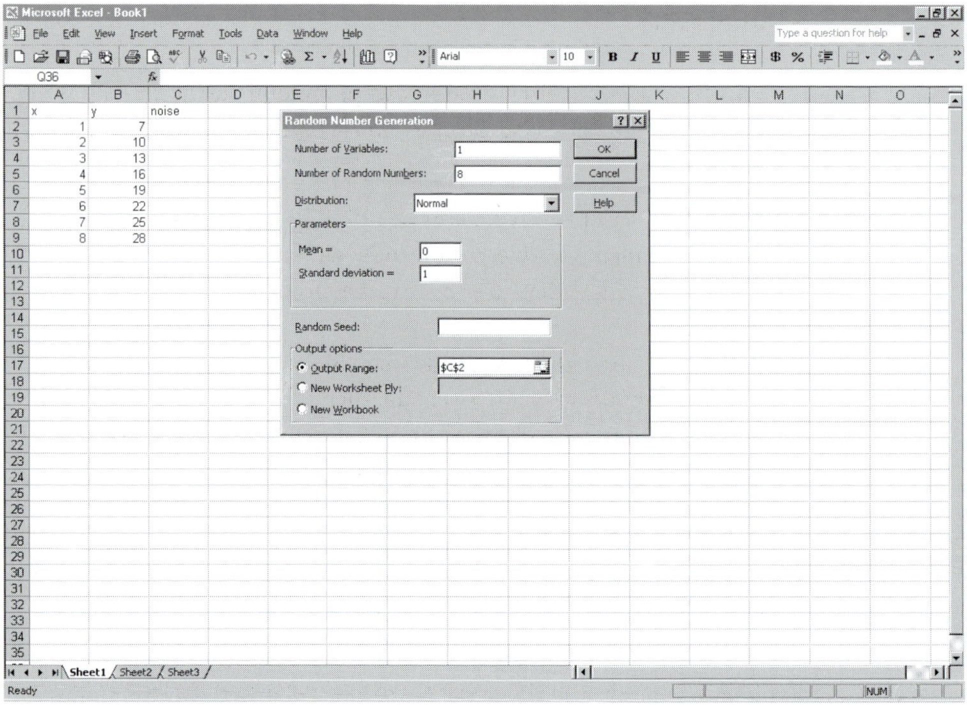

Figure 9.9. Use of the Random Number Generation tool.

	A	B	C	D	E
1	x	y	noise	y+noise	
2	1	7	-0.30023	6.699768	
3	2	10	-1.27768	8.722317	
4	3	13	0.244257	13.24426	
5	4	16	1.276474	17.27647	
6	5	19	1.19835	20.19835	
7	6	22	1.733133	23.73313	
8	7	25	-2.18359	22.81641	
9	8	28	-0.23418	27.76582	
10					

Figure 9.10. Normally distributed 'noise' in the C column. The cells in the D column show the noise added to *y* values appearing in column B.

9.7 Regression

The Regression tool in Excel® estimates parameters using the method of least squares described in chapters 6 and 7. This tool is powerful and is able to:

- estimate parameters when the equation $y = a + bx$ is fitted to data. y is the dependent variable, x is the independent variable. a and b are estimates of intercept and slope respectively, determined using least squares;
- estimate parameters when other equations that are linear in the parameters are to be fitted to data, such as $y = a + bx + cx^2$, where a, b and c are estimates of parameters obtained using least squares;
- determine standard errors in parameter estimates;
- calculate residuals and standardised residuals;
- plot measured and estimated values based on the line of best fit on an x–y graph;
- calculate the p value for each parameter estimate;
- determine confidence intervals for the parameters;
- determine the multiple correlation coefficient, R, the coefficient of multiple determination, R^2, and the adjusted coefficient of multiple determination, R^2_{ADJ} (see section 7.10.1 for more details);
- carry out analysis of variance.

As an example of using the Regression tool, consider the data in figure 9.11 which shows absorbance (y) versus concentration (x) data obtained during an experiment in which standard silver solutions were analysed by flame atomic absorption spectrometry. These data were described in example 2 in section 6.2.4. We use the Regression tool to fit the equation $y = a + bx$ to the data.

Features of the dialog box are described by the annotation on figure 9.11. Figure 9.12 shows the numbers returned by Excel® when applying the Regression tool to the data in figure 9.11. Many useful statistics are shown in figure 9.12. For example, cells B27 and B28 contain the best estimates of intercept and slope respectively. Cells C27 and C28 contain the standard errors in these estimates. Also worthy of special mention are the p values in cells E27 and E28. A p value of 0.472 for the intercept indicates that the intercept is not significantly different from zero. By contrast, the p value for the slope of 5.055×10^{-9} indicates that it is extremely unlikely that the 'true' slope is zero. Note that the Regression tool in Excel® is unable to perform weighted least squares.

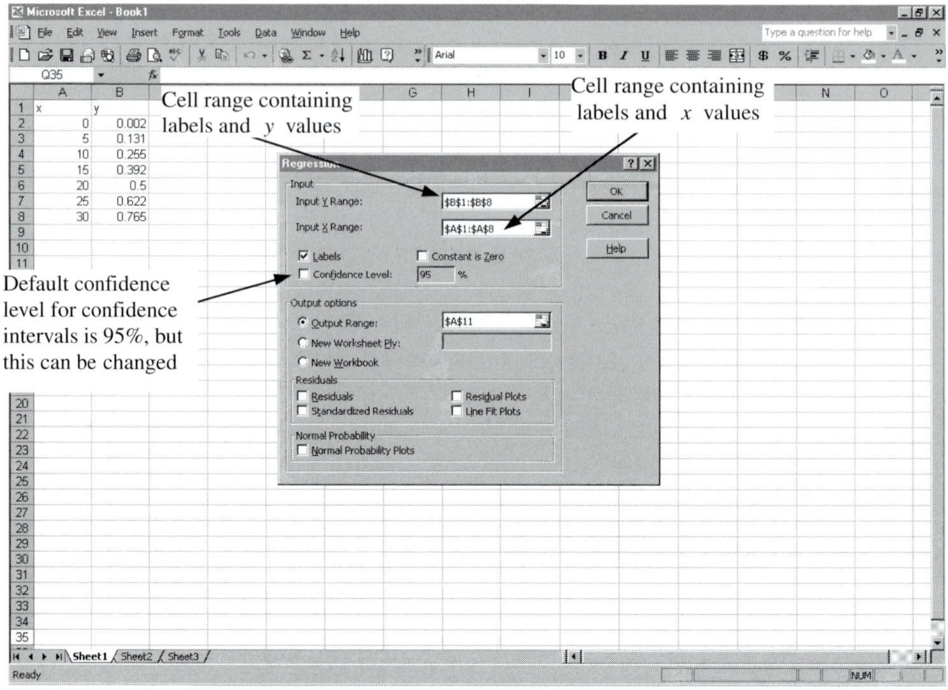

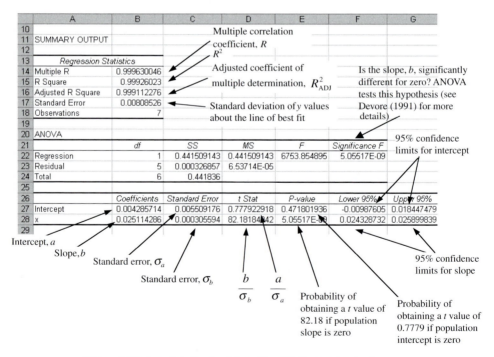

Figure 9.11. Worksheet showing the dialog box for the Regression tool in Excel®.

Figure 9.12. Annotated output from Excel®'s Regression tool.

9.7.1 Advanced linear least squares using Excel®'s Regression tool

It is possible to carry out advanced least squares using the Regression tool in Excel®. As an example, consider fitting the equation $y = a + bx + cx^2$ to data as shown in figure 9.13 (these data are also analysed in example 4 in section 7.6). Each term in the equation that contains the independent variable is allocated its own column in the spreadsheet. For example, column B of the spreadsheet in figure 9.13 contains x and column C contains x^2. To enter the x range (including a label in cell B1) into the dialog box we must highlight both the B and C columns, (or type $B$1:$C$12 into the Input X Range box). By choosing the output to begin at $A$14 on the same Worksheet, numbers are returned as shown in figure 9.14. The three parameters, a, b and c, estimated by the least squares technique appear in cells B30, B31 and B32 respectively, so that (to four significant figures) the equation representing the best line through the points can be written

$$y = 7.847 - 8.862x + 0.6058x^2$$

These estimates are consistent with those obtained in example 4 in section 7.6. Similarly, the standard errors in the estimates which appear in cells C30, C31 and C32 are consistent with those obtained in the same example.

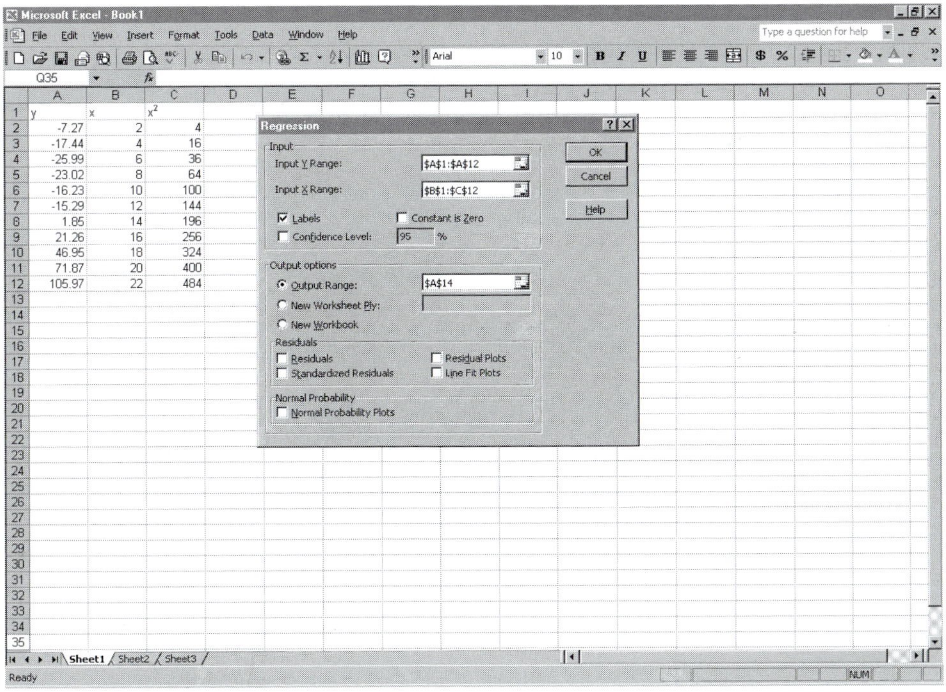

Figure 9.13. Fitting the equation $y = a + bx + cx^2$ to data using Excel®'s Regression tool.

	A	B	C	D	E	F	G
13							
14	SUMMARY OUTPUT						
15							
16	*Regression Statistics*						
17	Multiple R	0.998780654					
18	R Square	0.997562795					
19	Adjusted R Square	0.996953494					
20	Standard Error	2.422563551					
21	Observations	11					
22							
23	ANOVA						
24		*df*	*SS*	*MS*	*F*	*Significance F*	
25	Regression	2	19217.13338	9608.566689	1637.224561	3.52832E-11	
26	Residual	8	46.95051329	5.868814161			
27	Total	10	19264.08389				
28							
29		*Coefficients*	*Standard Error*	*t Stat*	*P-value*	*Lower 95%*	*Upper 95%*
30	Intercept	7.847272727	2.660478446	2.949571998	0.018437751	1.712194461	13.98235099
31	x	-8.861643357	0.509492065	-17.39309394	1.21727E-07	-10.03653493	-7.686751787
32	x2	0.605769231	0.020676239	29.2978443	1.99539E-09	0.558089707	0.653448754

Figure 9.14. Output produced by Excel®'s Regression Tool when fitting $y = a + bx + cx^2$ to data.

9.8 *t* tests

The Analysis ToolPak offers three types of *t* test:

- t-Test: Paired Two Sample for Means;
- t-Test: Two-Sample Assuming Equal Variances;
- t-Test: Two-Sample Assuming Unequal Variances.

The comparison of means of two samples (assuming equal variances) using a *t* test and the paired *t* test are discussed in sections 8.6 and 8.7 respectively. The comparison of means when samples have significantly different variances is not dealt with in this book.[4]

Figure 9.15 shows an Excel® worksheet containing two samples of data and the dialog box that appears when the t-Test tool selected is: Two-Sample Assuming Equal Variances. The data are values of peel off times of two types of adhesive tape, as described in example 6 in section 8.6. Applying the *t* test to data in figure 9.15 returns the numbers shown in figure 9.16.

---

[4] See Devore (1991) for a discussion of the *t* test when samples have unequal variances.

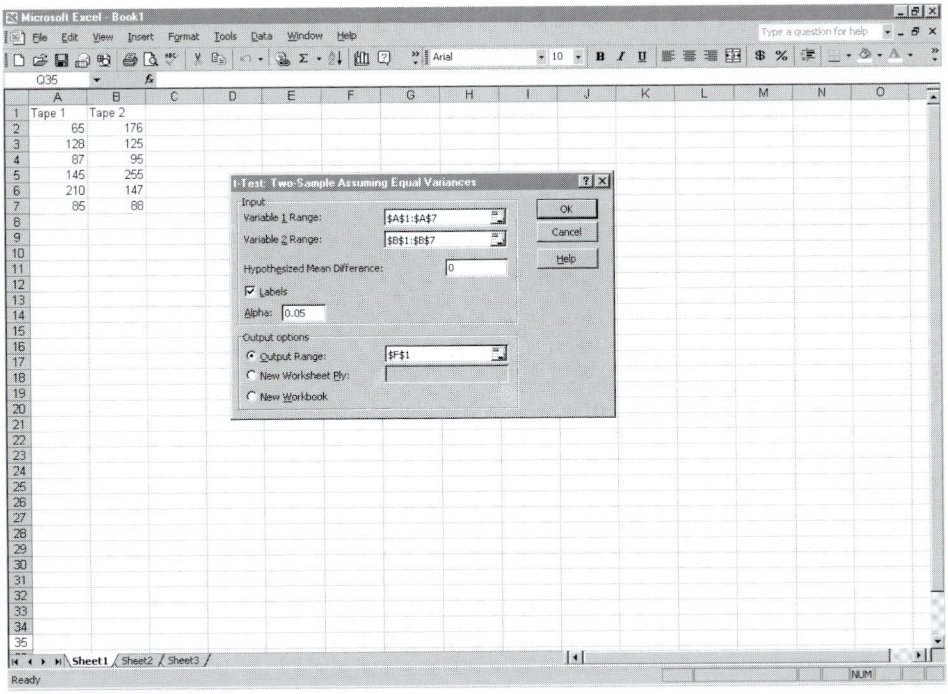

Figure 9.15. Peel off time data and t-Test dialog box.

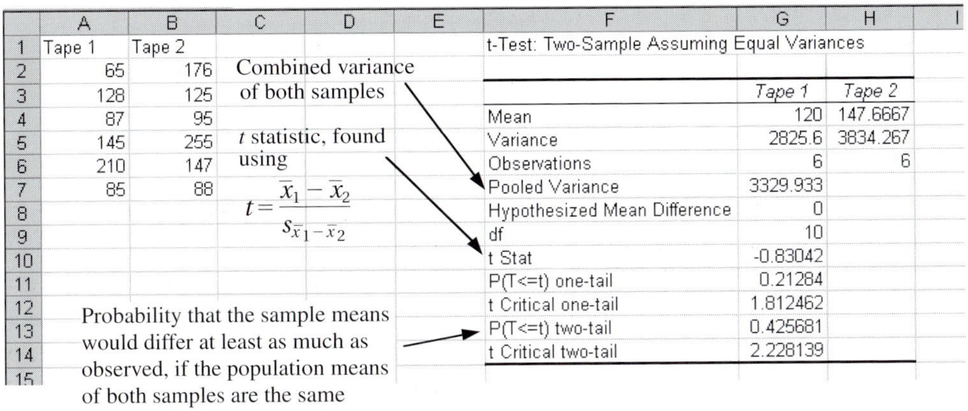

Figure 9.16. Annotated output from Excel®'s t-Test tool.

9.9 Other tools

Excel® possesses several other tools within the Analysis ToolPak used less often by physical scientists. For example, the Moving Average tool is often used to smooth out variations in data observed over time, such as seasonal variations in commodity prices in sales and marketing. The remaining tools are outlined in the following sections, with reference to where more information may be obtained.

9.9.1 Anova: Two-Factor With Replication and Anova: Two-Factor Without Replication

In some situations an experimenter may wish to establish the influence (or lack of it) of two factors on values obtained through experiment. For example, in a study to determine the amount of pollution by lead contamination of an estuary, both the geographical location as well as the time of day at which each sample is taken may affect the concentration of lead. Data categorised by geographical location and time of day can be studied using two-factor (also known as two-way) ANOVA. If only one value of lead is determined at each location at each time point then we would use the Anova: Two-Factor Without Replication tool in Excel®. By contrast, if replicate measurements were made at each time point and location, then we would use the Anova: Two-Factor With Replication tool in Excel®.[5]

9.9.2 Covariance

The covariance of two populations of data is defined as

$$\mathrm{Cov}(x, y) = \frac{1}{n}\sum (x_i - \mu_x)(y_i - \mu_y) \tag{9.1}$$

If x_i is uncorrelated with y_i then the covariance will be close to zero as the product $(x_i - \mu_x)(y_i - \mu_y)$ will vary randomly in size and in sign. The Covariance tool in Excel® determines the value of $\mathrm{Cov}(x, y)$ of two populations of equal size tabulated in columns (or rows).

---

[5] Two factor ANOVA is not dealt with in this text. See Devore (1991) for an introduction to this topic.

9.9.3 Exponential Smoothing

When quantities vary with time it is sometimes useful, especially if the data are noisy, to smooth the data. One way to smooth the data is to add the value at some time point, $t+1$, to a fraction of the value obtained at the prior time point, t. Excel®'s Exponential Smoothing tool does just this. The relationship Excel® uses is

$$(\text{smoothed value})_{t+1} = \alpha \times (\text{actual value})_t + (1-\alpha) \times (\text{smoothed value})_t$$

$$(9.2)$$

α is called the smoothing constant and has a value between 0 and 1. In Excel® $(1-\alpha)$ is referred to as the 'damping factor'. Note that the output of this tool is dynamically linked to the input data, so that any change in the input values causes the output (including any graph created by the tool) to be updated. The Exponential Smoothing tool is discussed by Middleton (1997).

9.9.4 Fourier Analysis

The Fourier analysis tool determines the frequency components of data using the fast Fourier transform (FFT) technique. It is a powerful and widely used technique in signal processing to establish the frequency spectrum of time dependent signals. For a detailed discussion of the use Excel®'s Fourier analysis tool, refer to *Excel® for Engineers and Scientists* by Bloch.

9.9.5 Moving Average

The Moving Average tool smoothes data by replacing the ith value in a column of values, x_i, by $x_{ismooth}$, where

$$x_{ismooth} = \frac{1}{N+1} \sum_{i-N}^{i} x_i \qquad\qquad (9.3)$$

where $N+1$ is the number of values over which $x_{ismooth}$ is calculated. The output of this tool is dynamically linked to the input data, so that any change in the input values causes the output (including any graph created by the tool) to be updated.

9.9.6   Rank and Percentile

The Rank and Percentile tool ranks values in a sample from largest to smallest. Each value is given a rank between 1 and n, where n is the number of values in the sample. The rank is also expressed as a percentage of the data set, such that the first ranked value has a percentile of 100% and that ranked last has a percentile of 0%.

9.9.7   Sampling

The Sampling tool allows for the selection of values from a column or row in a Worksheet. The selection may either be periodic (say every fifth value in a column of values) or random. When selecting at random from a group of values, a value may be absent, appear once or more than once (this is random selection 'with replacement'). Excel® displays the selected values in a column.

9.10   Review

The task of analysing experimental data is considerably assisted by the use of spreadsheets or statistical packages. The Analysis ToolPak in Excel® contains tools usually found in advanced statistical packages that can reduce the time to analyse and display data. For example, the Regression tool, with its capability to perform simple and advanced least squares, as well as plot graphs, is extremely useful for fitting equations to data. As is often the case with such utilities, we need to be careful that the ease with which an analysis may be carried out does not encourage us to suspend our 'critical faculties'. Questions to keep in mind when using the analysis utility are:

- Have we chosen the appropriate analysis tool?
- Have the raw data been entered correctly?
- Are we able to interpret the output?
- Does the output 'make sense'?

These words of caution aside, the Analysis ToolPak is a valuable and powerful aid to the analysis of experimental data.

Appendix 1

Statistical tables

Table 1. *Cumulative distribution function for the standard normal distribution:*
(a) *The table gives the area under the standard normal probability curve between* $z=-\infty$ *and* $z=z_1$.

Example:
If $z_1 = -1.24$, then
$P(-\infty < z < -1.24) = 0.10749$

$P(-\infty \le z \le z_1)$

z_1 0

z_1	0.00	0.01	0.02	0.03	0.04	0.05	0.06	0.07	0.08	0.09
−4.00	0.00003	0.00003	0.00003	0.00003	0.00003	0.00003	0.00002	0.00002	0.00002	0.00002
−3.90	0.00005	0.00005	0.00004	0.00004	0.00004	0.00004	0.00004	0.00004	0.00003	0.00003
−3.80	0.00007	0.00007	0.00007	0.00006	0.00006	0.00006	0.00006	0.00005	0.00005	0.00005
−3.70	0.00011	0.00010	0.00010	0.00010	0.00009	0.00009	0.00008	0.00008	0.00008	0.00008
−3.60	0.00016	0.00015	0.00015	0.00014	0.00014	0.00013	0.00013	0.00012	0.00012	0.00011
−3.50	0.00023	0.00022	0.00022	0.00021	0.00020	0.00019	0.00019	0.00018	0.00017	0.00017
−3.40	0.00034	0.00032	0.00031	0.00030	0.00029	0.00028	0.00027	0.00026	0.00025	0.00024
−3.30	0.00048	0.00047	0.00045	0.00043	0.00042	0.00040	0.00039	0.00038	0.00036	0.00035
−3.20	0.00069	0.00066	0.00064	0.00062	0.00060	0.00058	0.00056	0.00054	0.00052	0.00050
−3.10	0.00097	0.00094	0.00090	0.00087	0.00084	0.00082	0.00079	0.00076	0.00074	0.00071
−3.00	0.00135	0.00131	0.00126	0.00122	0.00118	0.00114	0.00111	0.00107	0.00104	0.00100
−2.90	0.00187	0.00181	0.00175	0.00169	0.00164	0.00159	0.00154	0.00149	0.00144	0.00139
−2.80	0.00256	0.00248	0.00240	0.00233	0.00226	0.00219	0.00212	0.00205	0.00199	0.00193
−2.70	0.00347	0.00336	0.00326	0.00317	0.00307	0.00298	0.00289	0.00280	0.00272	0.00264
−2.60	0.00466	0.00453	0.00440	0.00427	0.00415	0.00402	0.00391	0.00379	0.00368	0.00357
−2.50	0.00621	0.00604	0.00587	0.00570	0.00554	0.00539	0.00523	0.00508	0.00494	0.00480

Table 1(a). (*cont.*)

z_I	0.00	0.01	0.02	0.03	0.04	0.05	0.06	0.07	0.08	0.09
−2.40	0.00820	0.00798	0.00776	0.00755	0.00734	0.00714	0.00695	0.00676	0.00657	0.00639
−2.30	0.01072	0.01044	0.01017	0.00990	0.00964	0.00939	0.00914	0.00889	0.00866	0.00842
−2.20	0.01390	0.01355	0.01321	0.01287	0.01255	0.01222	0.01191	0.01160	0.01130	0.01101
−2.10	0.01786	0.01743	0.01700	0.01659	0.01618	0.01578	0.01539	0.01500	0.01463	0.01426
−2.00	0.02275	0.02222	0.02169	0.02118	0.02068	0.02018	0.01970	0.01923	0.01876	0.01831
−1.90	0.02872	0.02807	0.02743	0.02680	0.02619	0.02559	0.02500	0.02442	0.02385	0.02330
−1.80	0.03593	0.03515	0.03438	0.03362	0.03288	0.03216	0.03144	0.03074	0.03005	0.02938
−1.70	0.04457	0.04363	0.04272	0.04182	0.04093	0.04006	0.03920	0.03836	0.03754	0.03673
−1.60	0.05480	0.05370	0.05262	0.05155	0.05050	0.04947	0.04846	0.04746	0.04648	0.04551
−1.50	0.06681	0.06552	0.06426	0.06301	0.06178	0.06057	0.05938	0.05821	0.05705	0.05592
−1.40	0.08076	0.07927	0.07780	0.07636	0.07493	0.07353	0.07215	0.07078	0.06944	0.06811
−1.30	0.09680	0.09510	0.09342	0.09176	0.09012	0.08851	0.08692	0.08534	0.08379	0.08226
−1.20	0.11507	0.11314	0.11123	0.10935	0.10749	0.10565	0.10383	0.10204	0.10027	0.09853
−1.10	0.13567	0.13350	0.13136	0.12924	0.12714	0.12507	0.12302	0.12100	0.11900	0.11702
−1.00	0.15866	0.15625	0.15386	0.15151	0.14917	0.14686	0.14457	0.14231	0.14007	0.13786
−0.90	0.18406	0.18141	0.17879	0.17619	0.17361	0.17106	0.16853	0.16602	0.16354	0.16109
−0.80	0.21186	0.20897	0.20611	0.20327	0.20045	0.19766	0.19489	0.19215	0.18943	0.18673
−0.70	0.24196	0.23885	0.23576	0.23270	0.22965	0.22663	0.22363	0.22065	0.21770	0.21476
−0.60	0.27425	0.27093	0.26763	0.26435	0.26109	0.25785	0.25463	0.25143	0.24825	0.24510
−0.50	0.30854	0.30503	0.30153	0.29806	0.29460	0.29116	0.28774	0.28434	0.28096	0.27760
−0.40	0.34458	0.34090	0.33724	0.33360	0.32997	0.32636	0.32276	0.31918	0.31561	0.31207
−0.30	0.38209	0.37828	0.37448	0.37070	0.36693	0.36317	0.35942	0.35569	0.35197	0.34827
−0.20	0.42074	0.41683	0.41294	0.40905	0.40517	0.40129	0.39743	0.39358	0.38974	0.38591
−0.10	0.46017	0.45620	0.45224	0.44828	0.44433	0.44038	0.43644	0.43251	0.42858	0.42465
−0.00	0.50000	0.49601	0.49202	0.48803	0.48405	0.48006	0.47608	0.47210	0.46812	0.46414

Table 1. *Cumulative distribution function for the standard normal distribution (continued):*
(b) *The table gives the area under the standard normal probability curve between $z=-\infty$ and $z=z_1$.*

Example:
If $z_1 = 2.56$, then
$P(-\infty < z < 2.56) = 0.99477$

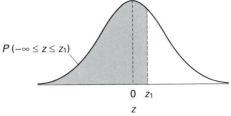

$P(-\infty \le z \le z_1)$

0   z_1

z

z_1	0.00	0.01	0.02	0.03	0.04	0.05	0.06	0.07	0.08	0.09
0.00	0.50000	0.50399	0.50798	0.51197	0.51595	0.51994	0.52392	0.52790	0.53188	0.53586
0.10	0.53983	0.54380	0.54776	0.55172	0.55567	0.55962	0.56356	0.56749	0.57142	0.57535
0.20	0.57926	0.58317	0.58706	0.59095	0.59483	0.59871	0.60257	0.60642	0.61026	0.61409
0.30	0.61791	0.62172	0.62552	0.62930	0.63307	0.63683	0.64058	0.64431	0.64803	0.65173
0.40	0.65542	0.65910	0.66276	0.66640	0.67003	0.67364	0.67724	0.68082	0.68439	0.68793
0.50	0.69146	0.69497	0.69847	0.70194	0.70540	0.70884	0.71226	0.71566	0.71904	0.72240
0.60	0.72575	0.72907	0.73237	0.73565	0.73891	0.74215	0.74537	0.74857	0.75175	0.75490
0.70	0.75804	0.76115	0.76424	0.76730	0.77035	0.77337	0.77637	0.77935	0.78230	0.78524
0.80	0.78814	0.79103	0.79389	0.79673	0.79955	0.80234	0.80511	0.80785	0.81057	0.81327
0.90	0.81594	0.81859	0.82121	0.82381	0.82639	0.82894	0.83147	0.83398	0.83646	0.83891
1.00	0.84134	0.84375	0.84614	0.84849	0.85083	0.85314	0.85543	0.85769	0.85993	0.86214
1.10	0.86433	0.86650	0.86864	0.87076	0.87286	0.87493	0.87698	0.87900	0.88100	0.88298
1.20	0.88493	0.88686	0.88877	0.89065	0.89251	0.89435	0.89617	0.89796	0.89973	0.90147
1.30	0.90320	0.90490	0.90658	0.90824	0.90988	0.91149	0.91308	0.91466	0.91621	0.91774
1.40	0.91924	0.92073	0.92220	0.92364	0.92507	0.92647	0.92785	0.92922	0.93056	0.93189
1.50	0.93319	0.93448	0.93574	0.93699	0.93822	0.93943	0.94062	0.94179	0.94295	0.94408
1.60	0.94520	0.94630	0.94738	0.94845	0.94950	0.95053	0.95154	0.95254	0.95352	0.95449
1.70	0.95543	0.95637	0.95728	0.95818	0.95907	0.95994	0.96080	0.96164	0.96246	0.96327
1.80	0.96407	0.96485	0.96562	0.96638	0.96712	0.96784	0.96856	0.96926	0.96995	0.97062
1.90	0.97128	0.97193	0.97257	0.97320	0.97381	0.97441	0.97500	0.97558	0.97615	0.97670
2.00	0.97725	0.97778	0.97831	0.97882	0.97932	0.97982	0.98030	0.98077	0.98124	0.98169
2.10	0.98214	0.98257	0.98300	0.98341	0.98382	0.98422	0.98461	0.98500	0.98537	0.98574
2.20	0.98610	0.98645	0.98679	0.98713	0.98745	0.98778	0.98809	0.98840	0.98870	0.98899
2.30	0.98928	0.98956	0.98983	0.99010	0.99036	0.99061	0.99086	0.99111	0.99134	0.99158
2.40	0.99180	0.99202	0.99224	0.99245	0.99266	0.99286	0.99305	0.99324	0.99343	0.99361
2.50	0.99379	0.99396	0.99413	0.99430	0.99446	0.99461	0.99477	0.99492	0.99506	0.99520
2.60	0.99534	0.99547	0.99560	0.99573	0.99585	0.99598	0.99609	0.99621	0.99632	0.99643
2.70	0.99653	0.99664	0.99674	0.99683	0.99693	0.99702	0.99711	0.99720	0.99728	0.99736
2.80	0.99744	0.99752	0.99760	0.99767	0.99774	0.99781	0.99788	0.99795	0.99801	0.99807
2.90	0.99813	0.99819	0.99825	0.99831	0.99836	0.99841	0.99846	0.99851	0.99856	0.99861

Table 1. (b) (*cont.*)

z_1	0.00	0.01	0.02	0.03	0.04	0.05	0.06	0.07	0.08	0.09
3.00	0.99865	0.99869	0.99874	0.99878	0.99882	0.99886	0.99889	0.99893	0.99896	0.99900
3.10	0.99903	0.99906	0.99910	0.99913	0.99916	0.99918	0.99921	0.99924	0.99926	0.99929
3.20	0.99931	0.99934	0.99936	0.99938	0.99940	0.99942	0.99944	0.99946	0.99948	0.99950
3.30	0.99952	0.99953	0.99955	0.99957	0.99958	0.99960	0.99961	0.99962	0.99964	0.99965
3.40	0.99966	0.99968	0.99969	0.99970	0.99971	0.99972	0.99973	0.99974	0.99975	0.99976
3.50	0.99977	0.99978	0.99978	0.99979	0.99980	0.99981	0.99981	0.99982	0.99983	0.99983
3.60	0.99984	0.99985	0.99985	0.99986	0.99986	0.99987	0.99987	0.99988	0.99988	0.99989
3.70	0.99989	0.99990	0.99990	0.99990	0.99991	0.99991	0.99992	0.99992	0.99992	0.99992
3.80	0.99993	0.99993	0.99993	0.99994	0.99994	0.99994	0.99994	0.99995	0.99995	0.99995
3.90	0.99995	0.99995	0.99996	0.99996	0.99996	0.99996	0.99996	0.99996	0.99997	0.99997
4.00	0.99997	0.99997	0.99997	0.99997	0.99997	0.99997	0.99998	0.99998	0.99998	0.99998

Table 2. *Critical values for the t distribution.*

Example: For a 95% confidence interval, with $\nu = 6$, then

$$t_{X\%,\nu} = t_{95\%,6} = 2.447$$

Shaded area = $X\%$/100

$X\%$ confidence interval

ν	50%	68%	90%	95%	99%	99.5%	Confidence interval ($X\%$)
	0.5	0.32	0.1	0.05	0.01	0.005	α (Two tailed test)
	0.25	0.16	0.05	0.025	0.005	0.0025	α (One tailed test)
1	1.000	1.819	6.314	12.706	63.656	127.321	
2	0.816	1.312	2.920	4.303	9.925	14.089	
3	0.765	1.189	2.353	3.182	5.841	7.453	
4	0.741	1.134	2.132	2.776	4.604	5.598	
5	0.727	1.104	2.015	2.571	4.032	4.773	
6	0.718	1.084	1.943	2.447	3.707	4.317	
7	0.711	1.070	1.895	2.365	3.499	4.029	
8	0.706	1.060	1.860	2.306	3.355	3.833	
9	0.703	1.053	1.833	2.262	3.250	3.690	
10	0.700	1.046	1.812	2.228	3.169	3.581	
11	0.697	1.041	1.796	2.201	3.106	3.497	
12	0.695	1.037	1.782	2.179	3.055	3.428	
13	0.694	1.034	1.771	2.160	3.012	3.372	
14	0.692	1.031	1.761	2.145	2.977	3.326	
15	0.691	1.029	1.753	2.131	2.947	3.286	
16	0.690	1.026	1.746	2.120	2.921	3.252	
17	0.689	1.024	1.740	2.110	2.898	3.222	
18	0.688	1.023	1.734	2.101	2.878	3.197	
19	0.688	1.021	1.729	2.093	2.861	3.174	
20	0.687	1.020	1.725	2.086	2.845	3.153	
24	0.685	1.015	1.711	2.064	2.797	3.091	
25	0.684	1.015	1.708	2.060	2.787	3.078	
29	0.683	1.012	1.699	2.045	2.756	3.038	
30	0.683	1.011	1.697	2.042	2.750	3.030	
49	0.680	1.005	1.677	2.010	2.680	2.940	
50	0.679	1.004	1.676	2.009	2.678	2.937	
99	0.677	0.999	1.660	1.984	2.626	2.871	
100	0.677	0.999	1.660	1.984	2.626	2.871	
999	0.675	0.995	1.646	1.962	2.581	2.813	
1000	0.675	0.995	1.646	1.962	2.581	2.813	

Table 3. *Critical values of the F distribution.*

Critical values for the F distribution for various probabilities, p, in the right hand tail of the distribution with ν_1 degrees of freedom in the numerator, and ν_2 degrees of freedom in the denominator.

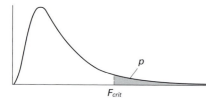

Example:

if $p = 0.1$, $\nu_1 = 6$, and $\nu_2 = 8$, then $F_{crit} = F_{0.1,6,8} = 2.67$.

						Numerator degrees of freedom, ν_1								
	p	1	2	3	4	5	6	7	8	10	12	15	20	50
1	0.1	39.86	49.50	53.59	55.83	57.24	58.20	58.91	59.44	60.19	60.71	61.22	61.74	62.69
	0.05	161.45	199.50	215.71	224.58	230.16	233.99	236.77	238.88	241.88	243.90	245.95	248.02	251.77
	0.025	647.79	799.48	864.15	899.60	921.83	937.11	948.20	956.64	968.63	976.72	984.87	993.08	1008.10
2	0.1	8.53	9.00	9.16	9.24	9.29	9.33	9.35	9.37	9.39	9.41	9.42	9.44	9.47
	0.05	18.51	19.00	19.16	19.25	19.30	19.33	19.35	19.37	19.40	19.41	19.43	19.45	19.48
	0.025	38.51	39.00	39.17	39.25	39.30	39.33	39.36	39.37	39.40	39.41	39.43	39.45	39.48
	0.01	98.50	99.00	99.16	99.25	99.30	99.33	99.36	99.38	99.40	99.42	99.43	99.45	99.48
	0.005	198.50	199.01	199.16	199.24	199.30	199.33	199.36	199.38	199.39	199.42	199.43	199.45	199.48
3	0.1	5.54	5.46	5.39	5.34	5.31	5.28	5.27	5.25	5.23	5.22	5.20	5.18	5.15
	0.05	10.13	9.55	9.28	9.12	9.01	8.94	8.89	8.85	8.79	8.74	8.70	8.66	8.58
	0.025	17.44	16.04	15.44	15.10	14.88	14.73	14.62	14.54	14.42	14.34	14.25	14.17	14.01
	0.01	34.12	30.82	29.46	28.71	28.24	27.91	27.67	27.49	27.23	27.05	26.87	26.69	26.35
	0.005	55.55	49.80	47.47	46.20	45.39	44.84	44.43	44.13	43.68	43.39	43.08	42.78	42.21
4	0.1	4.54	4.32	4.19	4.11	4.05	4.01	3.98	3.95	3.92	3.90	3.87	3.84	3.80
	0.05	7.71	6.94	6.59	6.39	6.26	6.16	6.09	6.04	5.96	5.91	5.86	5.80	5.70
	0.025	12.22	10.65	9.98	9.60	9.36	9.20	9.07	8.98	8.84	8.75	8.66	8.56	8.38
	0.01	21.20	18.00	16.69	15.98	15.52	15.21	14.98	14.80	14.55	14.37	14.20	14.02	13.69
	0.005	31.33	26.28	24.26	23.15	22.46	21.98	21.62	21.35	20.97	20.70	20.44	20.17	19.67
5	0.1	4.06	3.78	3.62	3.52	3.45	3.40	3.37	3.34	3.30	3.27	3.24	3.21	3.15
	0.05	6.61	5.79	5.41	5.19	5.05	4.95	4.88	4.82	4.74	4.68	4.62	4.56	4.44
	0.025	10.01	8.43	7.76	7.39	7.15	6.98	6.85	6.76	6.62	6.52	6.43	6.33	6.14
	0.01	16.26	13.27	12.06	11.39	10.97	10.67	10.46	10.29	10.05	9.89	9.72	9.55	9.24
	0.005	22.78	18.31	16.53	15.56	14.94	14.51	14.20	13.96	13.62	13.38	13.15	12.90	12.45
6	0.1	3.78	3.46	3.29	3.18	3.11	3.05	3.01	2.98	2.94	2.90	2.87	2.84	2.77
	0.05	5.99	5.14	4.76	4.53	4.39	4.28	4.21	4.15	4.06	4.00	3.94	3.87	3.75
	0.025	8.81	7.26	6.60	6.23	5.99	5.82	5.70	5.60	5.46	5.37	5.27	5.17	4.98
	0.01	13.75	10.92	9.78	9.15	8.75	8.47	8.26	8.10	7.87	7.72	7.56	7.40	7.09
	0.005	18.63	14.54	12.92	12.03	11.46	11.07	10.79	10.57	10.25	10.03	9.81	9.59	9.17

Denominator degrees of freedom, ν_2

Table 3 (continued).

Numerator degrees of freedom, ν_1

	p	1	2	3	4	5	6	7	8	10	12	15	20	50
7	0.1	3.59	3.26	3.07	2.96	2.88	2.83	2.78	2.75	2.70	2.67	2.63	2.59	2.52
	0.05	5.59	4.74	4.35	4.12	3.97	3.87	3.79	3.73	3.64	3.57	3.51	3.44	3.32
	0.025	8.07	6.54	5.89	5.52	5.29	5.12	4.99	4.90	4.76	4.67	4.57	4.47	4.28
	0.01	12.25	9.55	8.45	7.85	7.46	7.19	6.99	6.84	6.62	6.47	6.31	6.16	5.86
	0.005	16.24	12.40	10.88	10.05	9.52	9.16	8.89	8.68	8.38	8.18	7.97	7.75	7.35
8	0.1	3.46	3.11	2.92	2.81	2.73	2.67	2.62	2.59	2.54	2.50	2.46	2.42	2.35
	0.05	5.32	4.46	4.07	3.84	3.69	3.58	3.50	3.44	3.35	3.28	3.22	3.15	3.02
	0.025	7.57	6.06	5.42	5.05	4.82	4.65	4.53	4.43	4.30	4.20	4.10	4.00	3.81
	0.01	11.26	8.65	7.59	7.01	6.63	6.37	6.18	6.03	5.81	5.67	5.52	5.36	5.07
	0.005	14.69	11.04	9.60	8.81	8.30	7.95	7.69	7.50	7.21	7.01	6.81	6.61	6.22
10	0.1	3.29	2.92	2.73	2.61	2.52	2.46	2.41	2.38	2.32	2.28	2.24	2.20	2.12
	0.05	4.96	4.10	3.71	3.48	3.33	3.22	3.14	3.07	2.98	2.91	2.85	2.77	2.64
	0.025	6.94	5.46	4.83	4.47	4.24	4.07	3.95	3.85	3.72	3.62	3.52	3.42	3.22
	0.01	10.04	7.56	6.55	5.99	5.64	5.39	5.20	5.06	4.85	4.71	4.56	4.41	4.12
	0.005	12.83	9.43	8.08	7.34	6.87	6.54	6.30	6.12	5.85	5.66	5.47	5.27	4.90
12	0.1	3.18	2.81	2.61	2.48	2.39	2.33	2.28	2.24	2.19	2.15	2.10	2.06	1.97
	0.05	4.75	3.89	3.49	3.26	3.11	3.00	2.91	2.85	2.75	2.69	2.62	2.54	2.40
	0.025	6.55	5.10	4.47	4.12	3.89	3.73	3.61	3.51	3.37	3.28	3.18	3.07	2.87
	0.01	9.33	6.93	5.95	5.41	5.06	4.82	4.64	4.50	4.30	4.16	4.01	3.86	3.57
	0.005	11.75	8.51	7.23	6.52	6.07	5.76	5.52	5.35	5.09	4.91	4.72	4.53	4.17
14	0.1	3.10	2.73	2.52	2.39	2.31	2.24	2.19	2.15	2.10	2.05	2.01	1.96	1.87
	0.05	4.60	3.74	3.34	3.11	2.96	2.85	2.76	2.70	2.60	2.53	2.46	2.39	2.24
	0.025	6.30	4.86	4.24	3.89	3.66	3.50	3.38	3.29	3.15	3.05	2.95	2.84	2.64
	0.01	8.86	6.51	5.56	5.04	4.69	4.46	4.28	4.14	3.94	3.80	3.66	3.51	3.22
	0.005	11.06	7.92	6.68	6.00	5.56	5.26	5.03	4.86	4.60	4.43	4.25	4.06	3.70
16	0.1	3.05	2.67	2.46	2.33	2.24	2.18	2.13	2.09	2.03	1.99	1.94	1.89	1.79
	0.05	4.49	3.63	3.24	3.01	2.85	2.74	2.66	2.59	2.49	2.42	2.35	2.28	2.12
	0.025	6.12	4.69	4.08	3.73	3.50	3.34	3.22	3.12	2.99	2.89	2.79	2.68	2.47
	0.01	8.53	6.23	5.29	4.77	4.44	4.20	4.03	3.89	3.69	3.55	3.41	3.26	2.97
	0.005	10.58	7.51	6.30	5.64	5.21	4.91	4.69	4.52	4.27	4.10	3.92	3.73	3.37
18	0.1	3.01	2.62	2.42	2.29	2.20	2.13	2.08	2.04	1.98	1.93	1.89	1.84	1.74
	0.05	4.41	3.55	3.16	2.93	2.77	2.66	2.58	2.51	2.41	2.34	2.27	2.19	2.04
	0.025	5.98	4.56	3.95	3.61	3.38	3.22	3.10	3.01	2.87	2.77	2.67	2.56	2.35
	0.01	8.29	6.01	5.09	4.58	4.25	4.01	3.84	3.71	3.51	3.37	3.23	3.08	2.78
	0.005	10.22	7.21	6.03	5.37	4.96	4.66	4.44	4.28	4.03	3.86	3.68	3.50	3.14
20	0.1	2.97	2.59	2.38	2.25	2.16	2.09	2.04	2.00	1.94	1.89	1.84	1.79	1.69
	0.05	4.35	3.49	3.10	2.87	2.71	2.60	2.51	2.45	2.35	2.28	2.20	2.12	1.97
	0.025	5.87	4.46	3.86	3.51	3.29	3.13	3.01	2.91	2.77	2.68	2.57	2.46	2.25
	0.01	8.10	5.85	4.94	4.43	4.10	3.87	3.70	3.56	3.37	3.23	3.09	2.94	2.64
	0.005	9.94	6.99	5.82	5.17	4.76	4.47	4.26	4.09	3.85	3.68	3.50	3.32	2.96

Denominator degrees of freedom, ν_2

Table 3 (continued).

Numerator degrees of freedom, v_1

	p	1	2	3	4	5	6	7	8	10	12	15	20	50
22	0.1	2.95	2.56	2.35	2.22	2.13	2.06	2.01	1.97	1.90	1.86	1.81	1.76	1.65
	0.05	4.30	3.44	3.05	2.82	2.66	2.55	2.46	2.40	2.30	2.23	2.15	2.07	1.91
	0.025	5.79	4.38	3.78	3.44	3.22	3.05	2.93	2.84	2.70	2.60	2.50	2.39	2.17
	0.01	7.95	5.72	4.82	4.31	3.99	3.76	3.59	3.45	3.26	3.12	2.98	2.83	2.53
	0.005	9.73	6.81	5.65	5.02	4.61	4.32	4.11	3.94	3.70	3.54	3.36	3.18	2.82
24	0.1	2.93	2.54	2.33	2.19	2.10	2.04	1.98	1.94	1.88	1.83	1.78	1.73	1.62
	0.05	4.26	3.40	3.01	2.78	2.62	2.51	2.42	2.36	2.25	2.18	2.11	2.03	1.86
	0.025	5.72	4.32	3.72	3.38	3.15	2.99	2.87	2.78	2.64	2.54	2.44	2.33	2.11
	0.01	7.82	5.61	4.72	4.22	3.90	3.67	3.50	3.36	3.17	3.03	2.89	2.74	2.44
	0.005	9.55	6.66	5.52	4.89	4.49	4.20	3.99	3.83	3.59	3.42	3.25	3.06	2.70
26	0.1	2.91	2.52	2.31	2.17	2.08	2.01	1.96	1.92	1.86	1.81	1.76	1.71	1.59
	0.05	4.23	3.37	2.98	2.74	2.59	2.47	2.39	2.32	2.22	2.15	2.07	1.99	1.82
	0.025	5.66	4.27	3.67	3.33	3.10	2.94	2.82	2.73	2.59	2.49	2.39	2.28	2.05
	0.01	7.72	5.53	4.64	4.14	3.82	3.59	3.42	3.29	3.09	2.96	2.81	2.66	2.36
	0.005	9.41	6.54	5.41	4.79	4.38	4.10	3.89	3.73	3.49	3.33	3.15	2.97	2.61
28	0.1	2.89	2.50	2.29	2.16	2.06	2.00	1.94	1.90	1.84	1.79	1.74	1.69	1.57
	0.05	4.20	3.34	2.95	2.71	2.56	2.45	2.36	2.29	2.19	2.12	2.04	1.96	1.79
	0.025	5.61	4.22	3.63	3.29	3.06	2.90	2.78	2.69	2.55	2.45	2.34	2.23	2.01
	0.01	7.64	5.45	4.57	4.07	3.75	3.53	3.36	3.23	3.03	2.90	2.75	2.60	2.30
	0.005	9.28	6.44	5.32	4.70	4.30	4.02	3.81	3.65	3.41	3.25	3.07	2.89	2.53
30	0.1	2.88	2.49	2.28	2.14	2.05	1.98	1.93	1.88	1.82	1.77	1.72	1.67	1.55
	0.05	4.17	3.32	2.92	2.69	2.53	2.42	2.33	2.27	2.16	2.09	2.01	1.93	1.76
	0.025	5.57	4.18	3.59	3.25	3.03	2.87	2.75	2.65	2.51	2.41	2.31	2.20	1.97
	0.01	7.56	5.39	4.51	4.02	3.70	3.47	3.30	3.17	2.98	2.84	2.70	2.55	2.25
	0.005	9.18	6.35	5.24	4.62	4.23	3.95	3.74	3.58	3.34	3.18	3.01	2.82	2.46
35	0.1	2.85	2.46	2.25	2.11	2.02	1.95	1.90	1.85	1.79	1.74	1.69	1.63	1.51
	0.05	4.12	3.27	2.87	2.64	2.49	2.37	2.29	2.22	2.11	2.04	1.96	1.88	1.70
	0.025	5.48	4.11	3.52	3.18	2.96	2.80	2.68	2.58	2.44	2.34	2.23	2.12	1.89
	0.01	7.42	5.27	4.40	3.91	3.59	3.37	3.20	3.07	2.88	2.74	2.60	2.44	2.14
	0.005	8.98	6.19	5.09	4.48	4.09	3.81	3.61	3.45	3.21	3.05	2.88	2.69	2.33
40	0.1	2.84	2.44	2.23	2.09	2.00	1.93	1.87	1.83	1.76	1.71	1.66	1.61	1.48
	0.05	4.08	3.23	2.84	2.61	2.45	2.34	2.25	2.18	2.08	2.00	1.92	1.84	1.66
	0.025	5.42	4.05	3.46	3.13	2.90	2.74	2.62	2.53	2.39	2.29	2.18	2.07	1.83
	0.01	7.31	5.18	4.31	3.83	3.51	3.29	3.12	2.99	2.80	2.66	2.52	2.37	2.06
	0.005	8.83	6.07	4.98	4.37	3.99	3.71	3.51	3.35	3.12	2.95	2.78	2.60	2.23
50	0.1	2.81	2.41	2.20	2.06	1.97	1.90	1.84	1.80	1.73	1.68	1.63	1.57	1.44
	0.05	4.03	3.18	2.79	2.56	2.40	2.29	2.20	2.13	2.03	1.95	1.87	1.78	1.60
	0.025	5.34	3.97	3.39	3.05	2.83	2.67	2.55	2.46	2.32	2.22	2.11	1.99	1.75
	0.01	7.17	5.06	4.20	3.72	3.41	3.19	3.02	2.89	2.70	2.56	2.42	2.27	1.95
	0.005	8.63	5.90	4.83	4.23	3.85	3.58	3.38	3.22	2.99	2.82	2.65	2.47	2.10

Denominator degrees of freedom, v_2

Table 4. *Critical values of the χ^2 distribution for various probabilities in the right hand tail of the distribution and degrees of freedom, ν.*

Example:

If $p = 0.025$ and $\nu = 10$, then

$\chi^2_{crit} = \chi^2_{0.025,10} = 20.483$

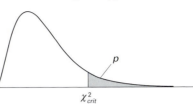

χ^2_{crit}

Degrees of freedom, ν	Probability, p				
	0.1	0.05	0.025	0.01	0.005
1	2.706	3.841	5.024	6.635	7.879
2	4.605	5.991	7.378	9.210	10.597
3	6.251	7.815	9.348	11.345	12.838
4	7.779	9.488	11.143	13.277	14.860
5	9.236	11.070	12.832	15.086	16.750
6	10.645	12.592	14.449	16.812	18.548
7	12.017	14.067	16.013	18.475	20.278
8	13.362	15.507	17.535	20.090	21.955
9	14.684	16.919	19.023	21.666	23.589
10	15.987	18.307	20.483	23.209	25.188
11	17.275	19.675	21.920	24.725	26.757
12	18.549	21.026	23.337	26.217	28.300
13	19.812	22.362	24.736	27.688	29.819
14	21.064	23.685	26.119	29.141	31.319
15	22.307	24.996	27.488	30.578	32.801
16	23.542	26.296	28.845	32.000	34.267
17	24.769	27.587	30.191	33.409	35.718
18	25.989	28.869	31.526	34.805	37.156
19	27.204	30.144	32.852	36.191	38.582
20	28.412	31.410	34.170	37.566	39.997
25	34.382	37.652	40.646	44.314	46.928
30	40.256	43.773	46.979	50.892	53.672
35	46.059	49.802	53.203	57.342	60.275
40	51.805	55.758	59.342	63.691	66.766
45	57.505	61.656	65.410	69.957	73.166
50	63.167	67.505	71.420	76.154	79.490
60	74.397	79.082	83.298	88.379	91.952
70	85.527	90.531	95.023	100.425	104.215
80	96.578	101.879	106.629	112.329	116.321
90	107.565	113.145	118.136	124.116	128.299
100	118.498	124.342	129.561	135.807	140.170

Appendix **2**

Propagation of uncertainties

If y depends on x and z, then errors in values of x and z contribute to the error in the value of y. We are able to quote an uncertainty in y if we know the uncertainties in x and z. We derive an equation for the propagation of uncertainties under the assumption that errors in values are independent, i.e. the error in x is not correlated with the error in z. We start with the expression for the variance in y, given by

$$\sigma_y^2 = \frac{\sum (y_i - \mu_y)^2}{n} \tag{A2.1}$$

where μ_y is the population mean or true value of the y quantity and n is the number of values. If we write $\delta y_i = (y_i - \mu_y)$, then δy_i is the deviation of the ith value from the population mean. Another way to regard δy_i is that if repeat measurements are made of a quantity, then δy_i is the experimental error (as discussed in section 5.3). Replacing $(y_i - \mu_y)$ in equation (A2.1) by δy_i gives

$$\sigma_y^2 = \frac{\sum (\delta y_i)^2}{n} \tag{A2.2}$$

Assuming $y = f(x,z)$, we use the approximation (valid for $\delta x_i \ll x_i$ and $\delta z_i \ll z_i$)

$$\delta y_i = \frac{\partial y}{\partial x} \delta x_i + \frac{\partial y}{\partial z} \delta z_i \tag{A2.3}$$

Substituting equation (A2.3) into equation (A2.2) gives

$$\sigma_y^2 = \frac{\sum \left(\frac{\partial y}{\partial x} \delta x_i + \frac{\partial y}{\partial z} \delta z_i \right)^2}{n} \tag{A2.4}$$

Expanding the brackets in equation (A2.4) gives

$$\sigma_y^2 = \frac{\sum\left[\left(\frac{\partial y}{\partial x}\right)\delta x_i\right]^2}{n} + \frac{\sum\left[\left(\frac{\partial y}{\partial z}\right)\delta z_i\right]^2}{n} + \frac{\sum 2\frac{\partial y}{\partial x}\frac{\partial y}{\partial z}\delta x_i\,\delta z_i}{n} \tag{A2.5}$$

where δx_i and δz_i are the deviations (or errors) in the ith values of x and z respectively. $\partial y/\partial x$ and $\partial y/\partial z$ are determined at $x=\bar{x}$ and $z=\bar{z}$.

Concentrating on the last term in equation (A2.5) for a moment, δx_i and δz_i take on both positive and negative values. If δx_i and δz_i are random and mutually independent, then the summation of positive and negative terms should give a sum that tends to zero, particularly for large data sets. We therefore argue that the last term in equation (A2.5) is negligible compared to the other terms and omit it.

The first two terms in equation (A2.5) can be written

$$\sigma_y^2 = \left(\frac{\partial y}{\partial x}\right)^2 \frac{\sum(\delta x_i)^2}{n} + \left(\frac{\partial y}{\partial z}\right)^2 \frac{\sum(\delta z_i)^2}{n} \tag{A2.6}$$

but

$$\frac{\sum(\delta x_i)^2}{n} = \sigma_x^2 \qquad \frac{\sum(\delta z_i)^2}{n} = \sigma_z^2$$

so that equation (A2.6) becomes

$$\sigma_y^2 = \left(\frac{\partial y}{\partial x}\right)^2 \sigma_x^2 + \left(\frac{\partial y}{\partial z}\right)^2 \sigma_z^2 \tag{A2.7}$$

or

$$\sigma_y = \left[\left(\frac{\partial y}{\partial x}\sigma_x\right)^2 + \left(\frac{\partial y}{\partial z}\sigma_z\right)^2\right]^{\frac{1}{2}} \tag{A2.8}$$

Equation (A2.8) can be extended to any number of variables that possess error.

If we determine the means $\bar{x}$ and $\bar{z}$ and their respective standard errors, $\sigma_{\bar{x}}$ and $\sigma_{\bar{z}}$, we can adapt equation (A2.8) to give the standard error in $\bar{y}$, i.e.

$$\sigma_{\bar{y}} = \left[\left(\frac{\partial y}{\partial x}\sigma_{\bar{x}}\right)^2 + \left(\frac{\partial y}{\partial z}\sigma_{\bar{z}}\right)^2\right]^{\frac{1}{2}} \tag{A2.9}$$

where $\partial y/\partial x$ and $\partial y/\partial z$ are determined at $x=\bar{x}$ and $z=\bar{z}$.

If, more generally, we write the uncertainties in y, x and z as u_y, u_x and u_z respectively, equation (A2.9) is rewritten

$$u_y = \left[\left(\frac{\partial y}{\partial x}u_x\right)^2 + \left(\frac{\partial y}{\partial z}u_z\right)^2\right]^{\frac{1}{2}} \tag{A2.10}$$

Appendix 3

Least squares and the principle of maximum likelihood

A3.1 Mean and weighted mean

We can use the principle of maximum likelihood to show that the best estimate of the population mean is the sample mean as given by equation (1.6).
Using this approach we can also obtain the best estimate of the mean when
the standard deviation differs from value to value. In this case we derive an
expression for what is usually termed the *weighted* mean. The argument is as
follows.[1]

A distribution of values that occurs when repeat measurements are made
is most likely to come from a population with a mean, μ, rather than from any
other population. If the normal distribution is valid for the data, then the probability, P_i, of observing the value x_i is

$$P_i \propto \exp\left\{ -\frac{1}{2}\left[\frac{(x_i - \mu)}{\sigma_i} \right]^2 \right\} \tag{A3.1}$$

where μ is the population mean and σ_i is the standard deviation of the ith value.

If we write the probability of observing n values when the population
mean is μ, as $P(\mu)$, then so long as probabilities are independent

$$P(\mu) = P_1 \times P_2 \times P_3 \times P_4 \cdots \times P_n \tag{A3.2}$$

which may be written more succinctly as

$$P(\mu) = \Pi P_i$$

[1] This follows the approach of Bevington and Robinson (1992).

392

Substituting equation (A3.1) into (A3.2) gives

$$P(\mu) \propto \exp\left\{ -\frac{1}{2}\left[\frac{(x_1 - \mu)}{\sigma_1}\right]^2\right\} \times \exp\left\{ -\frac{1}{2}\left[\frac{(x_2 - \mu)}{\sigma_2}\right]^2\right\} \dots$$

$$\times \exp\left\{ -\frac{1}{2}\left[\frac{(x_n - \mu)}{\sigma_n}\right]^2\right\} \tag{A3.3}$$

Equation (A3.3) may be rewritten[2]

$$P(\mu) \propto \exp\left\{ -\frac{1}{2}\sum\left[\frac{(x_i - \mu)}{\sigma_i}\right]^2\right\} \tag{A3.4}$$

For any *estimate*, X, of the population mean, we can calculate the probability, $P(X)$, of making a particular set of n measurements as

$$P(X) = \Pi P_i$$

where P_i is now given by

$$P_i \propto \exp\left\{ -\frac{1}{2}\left[\frac{(x_i - X)}{\sigma_i}\right]^2\right\} \tag{A3.5}$$

It follows that

$$P(X) \propto \exp\left\{ -\frac{1}{2}\sum\left[\frac{(x_i - X)}{\sigma_i}\right]^2\right\} \tag{A3.6}$$

We assume that the values x_i are more likely to have come from the distribution given by equation (A3.1) than any other distribution. It follows that the probability given by equation (A3.4) is the maximum attainable by equation (A3.6). If we find the value of X that maximises $P(X)$, we will have found the best estimate of the population mean.

We introduce χ^2, which we may regard as a 'weighted sum of squares' (and which is the χ^2 statistic discussed in chapter 8), where

$$\chi^2 = \sum\left[\frac{(x_i - X)}{\sigma_i}\right]^2 \tag{A3.7}$$

In order to maximise $P(X)$, it is necessary to find the value of X that minimises χ^2. This is done by differentiating χ^2 with respect to X, then setting the resulting equation equal to zero, i.e.

$$\frac{\partial \chi^2}{\partial X} = -2\sum\frac{1}{\sigma_i^2}(x_i - X) = 0$$

or

---

[2] Recalling $\exp(a) \times \exp(b) = \exp(a + b)$.

$$\sum \frac{x_i}{\sigma_i^2} = X \sum \frac{1}{\sigma_i^2}$$

It follows that

$$X = \frac{\sum \dfrac{x_i}{\sigma_i^2}}{\sum \dfrac{1}{\sigma_i^2}} \tag{A3.8}$$

In preference to using X to represent the weighted mean we will use (for consistency with the way the mean is usually written) $\bar{x}_w$, so that

$$\bar{x}_w = \frac{\sum \dfrac{x_i}{\sigma_i^2}}{\sum \dfrac{1}{\sigma_i^2}} \tag{A3.9}$$

Equation (A3.9) gives the weighted mean. If the standard deviation for every value is the same such that $\sigma_1 = \sigma_2 = \sigma$ then equation (A3.9) becomes

$$\bar{x}_w = \frac{\dfrac{1}{\sigma^2} \sum x_i}{\dfrac{1}{\sigma^2} \sum 1} = \frac{\sum x_i}{n} = \bar{x} \tag{A3.10}$$

Appendix A4.1 considers the uncertainty (as expressed by the standard error) in the weighted mean.

A3.2 Best estimates of slope and intercept

Chapter 6 introduced the equations needed to calculate the 'best' straight line of the form $y = a + bx$ through a set of x–y data, where a is the intercept on the y axis at $x = 0$ and b is the slope of the line. Let us consider how the equations for a and b are derived.

Assume that sample observations are extracted from a population with parameters α and β, and that the true relationship between y and x is

$$y = \alpha + \beta x \tag{A3.11}$$

We can never know the exact values of α and β, but we can find best estimates of these by using the principle of maximum likelihood.

For any given value of $x = x_i$, we can calculate the probability, P_i, of making a particular measurement of $y = y_i$. Assuming that the y observations are normally distributed and taking the true value of y at $x = x_i$ to be $y(x_i)$,

$$P_i \propto \exp \left\{ -\frac{1}{2} \left[\frac{y_i - y(x_i)}{\sigma_i} \right]^2 \right\} \tag{A3.12}$$

Here $y(x_i) = \alpha + \beta x_i$ and σ_i is the standard deviation of the observed y values.

The probability of making any observed set of n independent measurements, $P(\alpha,\beta)$, is the product of the individual probabilities

$$P(\alpha,\beta) = \Pi P_i = P_1 \times P_2 \times P_3 \times P_4 \cdots \times P_n$$

$$\propto \exp\left\{-\tfrac{1}{2}\sum\left[\frac{y_i - y(x_i)}{\sigma_i}\right]^2\right\} \tag{A3.13}$$

Similarly, if estimated values of the parameters α and β are a and b respectively, we can calculate the probability, $P(a,b)$, of making a particular set of n measurements:

$$P(a,b) = \Pi P_i$$

where P_i is now given by

$$P_i \propto \exp\left\{-\tfrac{1}{2}\left[\frac{y_i - \hat{y}_i}{\sigma_i}\right]^2\right\} \tag{A3.14}$$

Here $\hat{y}_i = a + bx_i$. Therefore,

$$P(a,b) \propto \exp\left\{-\tfrac{1}{2}\sum\left[\frac{y_i - \hat{y}_i}{\sigma_i}\right]^2\right\} \tag{A3.15}$$

We assume that the observed set of values is more likely to have come from the parent distribution given by equation (A3.12) rather than any other distribution and therefore the probability given by equation (A3.13) is the maximum probability attainable by equation (A3.15). The best estimates of α and β are those which maximise the probability given by equation (A3.15). Maximising the exponential term means minimising the sum that appears within the exponential. Writing the summation in equation (A3.15) as χ^2, we have

$$\chi^2 = \sum\left[\frac{y_i - \hat{y}_i}{\sigma_i}\right]^2 \tag{A3.16}$$

Equation (A3.16) is at the heart of analysis by least squares. When χ^2 is minimised then we will have found the best estimates for the parameters which appear in the equation relating x to y.

If $\hat{y}_i = a + bx_i$, then equation (A3.16) can be rewritten as

$$\chi^2 = \sum\left[\frac{y_i - a - bx_i}{\sigma_i}\right]^2 \tag{A3.17}$$

To find values for a and b which will minimise χ^2, we partially differentiate equation (A3.17) with respect to a and b in turn, set the resulting equations to zero, then solve for a and b.

Note, in general, the equation relating x to y may be more complicated than simply $y = a + bx$, for example, $y = a + bx + cx^2$, where a, b and c are parameters. In such a situation we must differentiate with respect to each of the parameters in turn, set the equations to zero and solve for a, b and c.

Returning to equation (A3.17) and differentiating, we get

$$\frac{\partial \chi^2}{\partial a} = -2 \sum \frac{1}{\sigma_i^2}(y_i - a - bx_i) = 0 \tag{A3.18}$$

$$\frac{\partial \chi^2}{\partial b} = -2 \sum \frac{x_i}{\sigma_i^2}(y_i - a - bx_i) = 0 \tag{A3.19}$$

where σ_i is the standard deviation associated with the ith value of y, y_i. In many cases, σ_i is a constant and can be replaced by σ. For example, when using an instrument such as a voltmeter we might assess that all voltages have an uncertainty of ± 10 mV. When σ_i is replaced by σ, we can write equations (A3.18) and (A3.19) as

$$\frac{-2}{\sigma^2} \sum (y_i - a - bx) = 0 \tag{A3.20}$$

$$\frac{-2}{\sigma^2} \sum x_i(y_i - a - bx_i) = 0 \tag{A3.21}$$

Equations (A3.20) and (A3.21) may be rearranged to give

$$na + b \sum x_i = \sum y_i \tag{A3.22}$$

and

$$a \sum x_i + b \sum x_i^2 = \sum x_i y_i \tag{A3.23}$$

Manipulating equations (A3.22) and (A3.23) gives

$$a = \frac{\sum x_i^2 \sum y_i - \sum x_i \sum x_i y_i}{n \sum x_i^2 - \left(\sum x_i\right)^2} \tag{A3.24}$$

$$b = \frac{n \sum x_i y_i - \sum x_i \sum y_i}{n \sum x_i^2 - \left(\sum x_i\right)^2} \tag{A3.25}$$

A3.3   The line of best fit passes through $\bar{x}$, $\bar{y}$

If we divide all the terms in equation (A3.22) by the number of values, n, we obtain

$$a + b \frac{\sum x_i}{n} = \frac{\sum y_i}{n}$$

which can be written

$$\bar{y} = a + b\bar{x} \tag{A3.26}$$

Equation (A3.26) indicates that a 'line of best fit' found using least squares passes through the point given by $\bar{x}, \bar{y}$.

A3.4 Weighting the fit

If σ_i is not constant (for example, in a nuclear counting experiment the standard deviation in the number of counts recorded, N, is not constant but is equal to $\sqrt{N}$), what do we do? We return to equations (A3.18) and (A3.19) and solve for a and b:

$$b\sum \frac{x_i}{\sigma_i^2} + a\sum \frac{1}{\sigma_i^2} = \sum \frac{y_i}{\sigma_i^2} \tag{A3.27}$$

$$b\sum \frac{x_i^2}{\sigma_i^2} + a\sum \frac{x_i}{\sigma_i^2} = \sum \frac{x_i y_i}{\sigma_i^2} \tag{A3.28}$$

Solving for a and b gives

$$a = \frac{\sum \dfrac{x_i^2}{\sigma_i^2} \sum \dfrac{y_i}{\sigma_i^2} - \sum \dfrac{x_i}{\sigma_i^2} \sum \dfrac{x_i y_i}{\sigma_i^2}}{\Delta} \tag{A3.29}$$

$$b = \frac{\sum \dfrac{1}{\sigma_i^2} \sum \dfrac{x_i y_i}{\sigma_i^2} - \sum \dfrac{x_i}{\sigma_i^2} \sum \dfrac{y_i}{\sigma_i^2}}{\Delta} \tag{A3.30}$$

where

$$\Delta = \sum \frac{1}{\sigma_i^2} \sum \frac{x_i^2}{\sigma_i^2} - \left(\sum \frac{x_i}{\sigma_i^2}\right)^2 \tag{A3.31}$$

Expressions for the standard errors in a and b are considered in appendix 4.

Appendix 4

Standard errors in mean, intercept and slope

A4.1 Standard error in the mean and weighted mean

In appendix 3 we showed that the weighted mean (i.e. the mean when the standard deviation in x values is not constant) may be written

$$\bar{x}_w = \frac{\sum \dfrac{x_i}{\sigma_i^2}}{\sum \dfrac{1}{\sigma_i^2}} \tag{A4.1}$$

where x_i is the ith value, and σ_i is the standard deviation in the ith value. The 'unweighted' mean is written as usual as

$$\bar{x} = \frac{\sum x_i}{n} \tag{A4.2}$$

where n is the number of values.

Now we consider the standard error in weighted and unweighted means. Assuming errors in values of x are independent, we write

$$\sigma_{\bar{x}_w}^2 = \left(\frac{\partial \bar{x}_w}{\partial x_1} \sigma_1\right)^2 + \left(\frac{\partial \bar{x}_w}{\partial x_2} \sigma_2\right)^2 + \cdots \left(\frac{\partial \bar{x}_w}{\partial x_n} \sigma_n\right)^2 \tag{A4.3}$$

or

$$\sigma_{\bar{x}_w}^2 = \sum \left(\frac{\partial \bar{x}_w}{\partial x_i} \sigma_i\right)^2 \tag{A4.4}$$

From equation (A4.1),

$$\frac{\partial \bar{x}_w}{\partial x_i} = \frac{\dfrac{1}{\sigma_i^2}}{\sum \dfrac{1}{\sigma_i^2}} \tag{A4.5}$$

Substituting equation (A4.5) into (A4.4) gives

$$\sigma_{\bar{x}_w}^2 = \sum \left(\frac{\frac{1}{\sigma_i^2}}{\sum \frac{1}{\sigma_i^2}} \right)^2 \sigma_i^2 = \frac{\sum \frac{1}{\sigma_i^2}}{\left(\sum \frac{1}{\sigma_i^2} \right)^2} = \frac{1}{\sum \frac{1}{\sigma_i^2}}$$

so that

$$\sigma_{\bar{x}_w} = \left(\frac{1}{\sum \frac{1}{\sigma_i^2}} \right)^{\frac{1}{2}} \tag{A4.6}$$

Equation (A4.6) gives the standard error for the *weighted* mean.

In situations in which $\sigma_1 = \sigma_2 = \sigma_i = \sigma$, equation (A4.6) reduces to

$$\sigma_{\bar{x}} = \left(\frac{1}{\frac{1}{\sigma^2} \sum 1} \right)^{\frac{1}{2}} = \left(\frac{\sigma^2}{n} \right)^{\frac{1}{2}}$$

(note that $\sum 1 = n$) i.e.

$$\sigma_{\bar{x}} = \frac{\sigma}{\sqrt{n}} \tag{A4.7}$$

A4.2 Standard error in intercept a and slope b for a straight line

The best line through linearly related data is written

$$y = a + bx \tag{A4.8}$$

In order to estimate the standard error in a and b, we consider the propagation of uncertainties that arises due to the fact that each value of y has some uncertainty. The intercept, a, may be regarded as a function of the y values, i.e.

$$a = f(y_1, y_2, y_3, y_4, y_5, \ldots y_n)$$

where n is the number of data points. Similarly, the slope, b, may be written

$$b = f(y_1, y_2, y_3, y_4, y_5, \ldots y_n)$$

(a and b can also be taken to be functions of x as well as y, but as we assume negligible error in x values, they do not contribute to the standard error in a and b.)

The standard error in a is represented by σ_a and is found from

$$\sigma_a^2 = \sum \sigma_i^2 \left(\frac{\partial a}{\partial y_i}\right)^2 \tag{A4.9}$$

where σ_i is the standard deviation in the ith value; similarly, for σ_b

$$\sigma_b^2 = \sum \sigma_i^2 \left(\frac{\partial b}{\partial y_i}\right)^2 \tag{A4.10}$$

In an unweighted fit we take all the σ_is to be the same and replace them by σ, where σ is the population standard deviation in the y values and, as we cannot know this value, we adopt the usual approximation:

$$s^2 \approx \sigma^2$$

where[1]

$$s^2 = \frac{1}{n-2} \sum (\Delta y_i)^2 \tag{A4.11}$$

n is the number of data points and $\Delta y_i = y_i - a - bx_i$.

Now we must deal with the partial derivatives appearing in equations (A4.9) and (A4.10). For an unweighted fit, we have from section A3.2,

$$a = \frac{\sum x_i^2 \sum y_i - \sum x_i \sum x_i y_i}{\Delta} \tag{A4.12}$$

$$b = \frac{n \sum x_i y_i - \sum x_i \sum y_i}{\Delta} \tag{A4.13}$$

where

$$\Delta = n \sum x_i^2 - \left(\sum x_i\right)^2 \tag{A4.14}$$

Consider the jth value of y and its contribution to the uncertainty in a and b. We have

$$\sigma_b^2 = \sum_{j=1}^{j=n} \sigma_{b(y_j)}^2 \tag{A4.15}$$

where

$$\sigma_{b(y_j)}^2 = \sigma^2 \left(\frac{\partial b}{\partial y_j}\right)^2 \tag{A4.16}$$

---

[1] The sum of squares of the residuals is divided by $(n-2)$ which is the number of degrees of freedom. The number of degrees of freedom is 2 less than the number of data points because a degree of freedom is 'lost' for every parameter that is calculated using the sample observations. Here there are two such parameters, a and b.

A similar equation can be written for σ_a^2 (just replace b by a in equations (A4.15) and (A4.16)).

Return to equation (A4.13) and differentiate[2] b with respect to y_j

$$\frac{\partial b}{\partial y_j} = \frac{1}{\Delta}\left(nx_j - \sum x_i\right) \tag{A4.17}$$

Substituting into (A4.16) gives

$$\sigma_{b(y_j)}^2 = \frac{\sigma^2}{\Delta^2}\left(nx_j - \sum x_i\right)^2 \tag{A4.18}$$

Substituting equation (A4.18) into equation (A4.15) gives

$$\sigma_b^2 = \frac{\sigma^2}{\Delta^2}\left[\sum_{j=1}^{j=n}\left((n^2x_j^2 - 2nx_j\sum x_i + \left(\sum x_i\right)^2\right)\right] \tag{A4.19}$$

$$= \frac{\sigma^2}{\Delta^2}\left[n^2\sum x_i^2 - 2n\left(\sum x_i\right)^2 + n\left(\sum x_i\right)^2\right] \tag{A4.20}$$

$$= \frac{\sigma^2 n}{\Delta^2}\left[n\sum x_i^2 - \left(\sum x_i\right)^2\right] \tag{A4.21}$$

As $\Delta = n\sum x_i^2 - (\sum x_i)^2$, we have

$$\sigma_b^2 = \frac{\sigma^2 n}{\Delta} \tag{A4.22}$$

or

$$\sigma_b = \frac{\sigma n^{\frac{1}{2}}}{\Delta^{\frac{1}{2}}} \tag{A4.23}$$

Using the same arguments to find σ_a^2

$$\frac{\partial a}{\partial y_j} = \frac{1}{\Delta}\left(x_i^2 - x_j\sum x_i\right) \tag{A4.24}$$

so that

$$\sigma_a^2 = \frac{\sigma^2\sum x_i^2}{\Delta} \tag{A4.25}$$

[2] In order to be convinced that equation (A4.17) is correct, return to equation (A4.13) and write the summation out more fully, i.e.

$$b = \frac{1}{\Delta}n(x_1y_1 + x_2y_2 + x_3y_3\cdots) - (x_1 + x_2 + x_3\cdots)(y_1 + y_2 + y_3\cdots)$$

then differentiate with respect to y_1, then y_2 and so on.

or

$$\sigma_a = \frac{\sigma\left(\sum x_i^2\right)^{\frac{1}{2}}}{\Delta^{\frac{1}{2}}} \tag{A4.26}$$

If the standard deviations in the y values are not equal, we introduce σ_i^2 explicitly into the equations for the standard error in a and b. The standard errors in a and b are now given by

$$\sigma_a = \left(\frac{\sum \dfrac{x_i^2}{\sigma_i^2}}{\Delta}\right)^{\frac{1}{2}} \tag{A4.27}$$

$$\sigma_b = \left(\frac{\sum \dfrac{1}{\sigma_i^2}}{\Delta}\right)^{\frac{1}{2}} \tag{A4.28}$$

where

$$\Delta = \sum \frac{1}{\sigma_i^2} \sum \frac{x_i^2}{\sigma_i^2} - \left(\sum \frac{x_i}{\sigma_i^2}\right)^2 \tag{A4.29}$$

Appendix 5

Introduction to matrices for least squares analysis

Applying least squares to the problem of finding the best straight line represented by the equation, $y = a + bx$ through data creates two equations which must be solved for a and b (see equations (A3.22) and (A3.23)). The equations can be solved by the method of 'elimination and substitution' but this approach becomes increasingly cumbersome when equations to be fitted to data contain three or more parameters that must be estimated such as

$$y = a + bx + cx \ln x + dx^2 \ln x \tag{A5.1}$$

Fitting equation (A5.1) to data using least squares creates four equations to be solved for a, b, c and d. The preferred method for dealing with fitting of equations to data where the equations consist of several parameters and/or independent variables is to use matrices. Matrices provide an efficient means of solving linear equations as well as offering compact and elegant notation. In this appendix we consider matrices and some of their basic properties, especially those useful in parameter estimation by linear least squares.[1]

A matrix consists of a rectangular array of elements, for example:

$$\mathbf{A} = \begin{bmatrix} a_{11} & a_{12} & a_{13} \\ a_{21} & a_{22} & a_{23} \\ a_{31} & a_{32} & a_{33} \end{bmatrix} \qquad \mathbf{B} = \begin{bmatrix} b_1 \\ b_2 \\ b_3 \end{bmatrix}$$

By convention, a matrix is represented by a bold symbol such as $\mathbf{A}$. The elements of the matrix $\mathbf{A}$ are represented by a_{11}, a_{12}, a_{13} etc.

If a matrix consists of more than one row and more than one column (such as matrix $\mathbf{A}$) it is customary to use a double subscript to identify a particular element within the matrix. A general element of a matrix is designated a_{ij},

---

[1] Neter, Kutner, Nachtsheim and Wasserman (1996) deals with the application of matrices to least squares.

where i refers to the ith row and j refers to the jth column. Matrix **B** above consists of a single column of elements where each element can be identified unambiguously by a single subscript.

In general, a matrix consists of m rows and n columns and is usually referred to as a matrix of *dimension* $m \times n$ (note that the number of rows is specified first, then the number of columns). If $m = n$, as it does for matrix **A**, then the matrix is said to be 'square'. By contrast, **B** is a 3×1 matrix. A matrix consisting of a single column of elements is sometimes referred to as a column vector, or simply as a vector.

Many operations such as addition, subtraction and multiplication can be defined for matrices. For data analysis using least squares, it is often required to multiply two matrices together.

Matrix multiplication

Consider matrices **A** and **B** where

$$\mathbf{A} = \begin{bmatrix} 2 & 4 \\ 7 & 5 \end{bmatrix} \qquad \mathbf{B} = \begin{bmatrix} 1 & 5 \\ 3 & 9 \end{bmatrix}$$

If $\mathbf{AB} = \mathbf{P}$, where **A** can be written most generally as

$$\mathbf{A} = \begin{bmatrix} a_{11} & a_{12} \\ a_{21} & a_{22} \end{bmatrix}$$

and

$$\mathbf{B} = \begin{bmatrix} b_{11} & b_{12} \\ b_{21} & b_{22} \end{bmatrix}$$

then the elements of **P**

$$\begin{bmatrix} p_{11} & p_{12} \\ p_{21} & p_{22} \end{bmatrix}$$

can be found by multiplying each row of **A** by each column of **B**, in the following manner:

$$\begin{aligned} p_{11} &= a_{11} \times b_{11} + a_{12} \times b_{21} = 2 \times 1 + 4 \times 3 = 14 \\ p_{12} &= a_{11} \times b_{12} + a_{12} \times b_{22} = 2 \times 5 + 4 \times 9 = 46 \\ p_{21} &= a_{21} \times b_{11} + a_{22} \times b_{21} = 7 \times 1 + 5 \times 3 = 22 \\ p_{22} &= a_{21} \times b_{12} + a_{22} \times b_{22} = 7 \times 5 + 5 \times 9 = 80 \end{aligned}$$

so that

$$\mathbf{P} = \begin{bmatrix} 14 & 46 \\ 22 & 80 \end{bmatrix}$$

If we reverse the order of **A** and **B**, we find that

$$\mathbf{BA} = \begin{bmatrix} 37 & 29 \\ 69 & 57 \end{bmatrix}$$

In this example (and generally), $\mathbf{AB} \neq \mathbf{BA}$, so that the order in which the matrices are multiplied is important.

If matrix $\mathbf{A}$ consists of r_1 rows and c_1 columns and matrix $\mathbf{B}$ consists of r_2 rows and c_2 columns, then the product $\mathbf{AB}$ can only be formed if $c_1 = r_2$. If $c_1 = r_2$, then $\mathbf{AB}$ is a matrix with r_1 rows and c_2 columns:

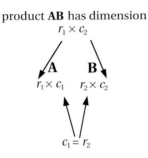

For example

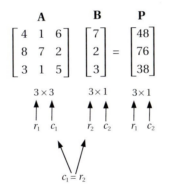

In contrast,

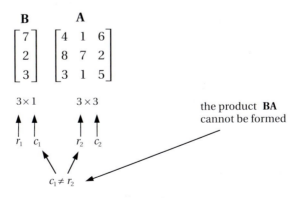

The identity matrix, I

Identity matrices play a role equivalent to the number 1 in conventional algebra (and are sometimes termed 'unit' matrices). When multiplying any number or variable by unity, the number or variable remains unchanged. Similarly, if you multiply a matrix, **A**, by its identity matrix, **I**, we find $\mathbf{AI} = \mathbf{A}$.

As an example, if we multiply the square matrix

$$\mathbf{I} = \begin{bmatrix} 1 & 0 & 0 \\ 0 & 1 & 0 \\ 0 & 0 & 1 \end{bmatrix}$$

by any other 3×3 matrix such as

$$\mathbf{A} = \begin{bmatrix} 2 & 1 & -4 \\ 1 & 8 & 2 \\ 3 & 4 & 2 \end{bmatrix}$$

we find that $\mathbf{IA} = \mathbf{AI} = \mathbf{A}$. The matrix

$$\begin{bmatrix} 1 & 0 & 0 \\ 0 & 1 & 0 \\ 0 & 0 & 1 \end{bmatrix}$$

is the identity matrix of dimension 3.

Inverse matrix, $\mathbf{A}^{-1}$

If we multiply the matrix

$$\mathbf{A} = \begin{bmatrix} 2 & 1 & -4 \\ 1 & 8 & 2 \\ 3 & 4 & 2 \end{bmatrix}$$

by the matrix

$$\mathbf{A}^{-1} = \begin{bmatrix} 0.08 & -0.18 & 0.34 \\ 0.04 & 0.16 & -0.18 \\ -0.20 & -0.05 & 0.15 \end{bmatrix}$$

we obtain the identity matrix, **I**. That is

$$\begin{bmatrix} 2 & 1 & -4 \\ 1 & 8 & 2 \\ 3 & 4 & 2 \end{bmatrix} \begin{bmatrix} 0.08 & -0.18 & 0.34 \\ 0.04 & 0.16 & -0.18 \\ -0.20 & -0.05 & 0.15 \end{bmatrix} = \begin{bmatrix} 1 & 0 & 0 \\ 0 & 1 & 0 \\ 0 & 0 & 1 \end{bmatrix}$$

$\mathbf{A}^{-1}$ is referred to as the inverse matrix and in general

$$\mathbf{AA}^{-1} = \mathbf{A}^{-1}\mathbf{A} = \mathbf{I}$$

It is not always possible to determine an inverse matrix. For example if two columns in a matrix have equal elements, such as

$$A = \begin{bmatrix} -4 & 1 & -4 \\ 2 & 8 & 2 \\ 2 & 4 & 2 \end{bmatrix}$$

then it is not possible to determine A^{-1} and A is said to be 'singular'. If A^{-1} can be found then A is said to be 'non-singular'. Matrix inversion is a challenging operation to perform 'by hand' even for small matrices. A computer package with matrix manipulation routines is almost mandatory if matrices larger than 3×3 are to be inverted.

Using the inverse matrix to solve for parameter estimates in least squares

In chapters 6 and 7 we discovered that the application of the least squares technique leads to two or more simultaneous equations that must be solved to find best estimates for the parameters that appear in an equation that is to be fitted to data. The equations can be written in matrix form (see equation (7.11)),

$$AB = P \tag{A5.2}$$

where it is the elements of matrix B which are the best estimates of the parameters. To isolate these elements we multiply both sides of equation (A5.2) by A^{-1}. This gives

$$A^{-1}AB = A^{-1}P \tag{A5.3}$$

but $A^{-1}A = I$, so equation (A5.3) becomes

$$IB = A^{-1}P \tag{A5.4}$$

Now $IB = B$, so we have

$$B = A^{-1}P \tag{A5.5}$$

Example

Suppose that after applying the method of least squares to experimental data, we obtain the following equations which must be solved for a, b, c and d:

$$1.75a + 18.3b + 42.8c - 25.9d = 49.3$$
$$3.26a - 19.8b + 17.4c - 32.2d = 65.3$$
$$18.6a + 14.7b + 12.2c + 14.3d = -18.1$$
$$65.7a - 15.3b - 18.9c + 25.3d = 19.1$$

We can write this in matrix form as

$$
\begin{array}{ccc}
\mathbf{A} & \mathbf{B} & \mathbf{P}
\end{array}
$$

$$
\begin{bmatrix}
1.75 & 18.3 & 42.8 & -25.9 \\
3.26 & -19.8 & 17.4 & -32.2 \\
18.6 & 14.7 & 12.2 & 14.3 \\
65.7 & -15.3 & -18.9 & 25.3
\end{bmatrix}
\begin{bmatrix}
a \\ b \\ c \\ d
\end{bmatrix}
\begin{bmatrix}
49.3 \\ 65.3 \\ -18.1 \\ 19.1
\end{bmatrix}
$$

The elements of $\mathbf{B}$ are found from $\mathbf{B} = \mathbf{A}^{-1}\mathbf{P}$. An efficient way to obtain $\mathbf{A}^{-1}$ is to use the MINVERSE() function in Excel® as described in section 7.4.1. Using this function we find

$$
\mathbf{A}^{-1} =
\begin{bmatrix}
0.028909 & -0.01895 & -0.03514 & 0.025339 \\
0.072522 & -0.07832 & -0.0971 & 0.029445 \\
-0.05058 & 0.064565 & 0.114316 & -0.03422 \\
-0.069 & 0.050075 & 0.117925 & -0.03403
\end{bmatrix}
$$

$$
\mathbf{B} = \mathbf{A}^{-1}\mathbf{P} =
\begin{bmatrix}
1.307834 \\
0.780933 \\
-1.00025 \\
-2.91625
\end{bmatrix}
$$

It follows that, to four significant figures

$$a = 1.308 \qquad b = 0.7809 \qquad c = -1.000 \qquad d = -2.916$$

As a final step, it is useful to verify that $\mathbf{AB} = \mathbf{P}$.

Appendix 6

Useful formulae

If events are mutually exclusive	$P(\text{A or B}) = P(\text{A}) + P(\text{B})$
If events are independent	$P(\text{A and B}) = P(\text{A}) \times P(\text{B})$
Standard normal variable, z	$z = \dfrac{x - \mu}{\sigma}$
t variable	$t = \left(\dfrac{\bar{x} - \mu}{s/\sqrt{n}} \right)$
Mean	$\bar{x} = \dfrac{\sum x_i}{n}$
Weighted mean	$\bar{x}_w = \dfrac{\sum \dfrac{x_i}{\sigma_i^2}}{\sum \dfrac{1}{\sigma_i^2}}$
Standard deviation	$\sigma \approx s = \left[\dfrac{\sum (x_i - \bar{x})^2}{n-1} \right]^{\frac{1}{2}}$
Standard error of mean	$\sigma_{\bar{x}} = \dfrac{\sigma}{\sqrt{n}} \approx \dfrac{s}{\sqrt{n}}$
Standard error of weighted mean	$\sigma_{\bar{x}_w} = \left(\dfrac{1}{\sum \dfrac{1}{\sigma_i^2}} \right)^{\frac{1}{2}}$
Level of significance, α, for $X\%$ confidence level	$\alpha = \dfrac{(100\% - X\%)}{100\%}$

Fractional uncertainty

$$\frac{u}{|\bar{x}|}$$

Percentage uncertainty

$$\frac{u}{|\bar{x}|} \times 100\%$$

Uncertainty in y (y is function of x only)

$$u_y = \left|\frac{dy}{dx}\right| u_x$$

Maximum uncertainty in y for uncertainty in x and z

$$u_y = \left|\frac{\partial y}{\partial x}\right| u_x + \left|\frac{\partial y}{\partial z}\right| u_z$$

Uncertainty in y for uncorrelated uncertainty in x and z

$$u_y = \left[\left(\frac{\partial y}{\partial x} u_x\right)^2 + \left(\frac{\partial y}{\partial z} u_z\right)^2\right]^{\frac{1}{2}}$$

Uncertainty due to random and systematic errors

$$u = \sqrt{u_r^2 + u_s^2}$$

Equation of a straight line

$$y = a + bx$$

Residual

$$\Delta y_i = y_i - \hat{y}_i$$

Standardised residual

$$\Delta y_{is} = \frac{\Delta y_i}{\sigma_i}$$

Unweighted sum of squares of residuals

$$SSR = \sum (y_i - \hat{y}_i)^2$$

Weighted sum of squares of residuals

$$\chi^2 = \sum \left(\frac{y_i - \hat{y}_i}{\sigma_i}\right)^2$$

Slope (unweighted least squares)

$$b = \frac{n\sum x_i y_i - \sum x_i \sum y_i}{n\sum x_i^2 - \left(\sum x_i\right)^2}$$

Intercept (unweighted least squares)

$$a = \frac{\sum x_i^2 \sum y_i - \sum x_i \sum x_i y_i}{n\sum x_i^2 - \left(\sum x_i\right)^2}$$

Standard deviation in y values (unweighted least squares)

$$\sigma = \left[\frac{1}{n-2}\sum (y_i - \hat{y}_i)^2\right]^{\frac{1}{2}}$$

Standard error in slope (unweighted least squares)

$$\sigma_b = \frac{\sigma n^{\frac{1}{2}}}{\left[n\sum x_i^2 - \left(\sum x_i\right)^2\right]^{\frac{1}{2}}}$$

Standard error in intercept (unweighted least squares)

$$\sigma_a = \frac{\sigma\left(\sum x_i^2\right)^{\frac{1}{2}}}{\left[n\sum x_i^2 - \left(\sum x_i\right)^2\right]^{\frac{1}{2}}}$$

Unweighted correlation coefficient

$$r = \frac{n\sum x_i y_i - \sum x_i \sum y_i}{\left[n\sum x_i^2 - \left(\sum x_i\right)^2\right]^{\frac{1}{2}}\left[n\sum y_i^2 - \left(\sum y_i\right)^2\right]^{\frac{1}{2}}}$$

Estimate of y for $x = x_0$

$$\hat{y}_0 = \bar{y} + b(x_0 - \bar{x})$$

Standard error in estimate of y for $x = x_0$

$$\sigma_{\hat{y}_0} = \sigma\left[\frac{1}{n} + \frac{n(x_0 - \bar{x})^2}{n\sum x_i^2 - \left(\sum x_i\right)^2}\right]^{\frac{1}{2}}$$

Estimate of x for mean y, $\bar{y}_0$

$$\hat{x}_0 = \frac{\bar{y}_0 - a}{b}$$

Standard error in estimate of x for mean y, $\bar{y}_0$

$$\sigma_{\hat{x}_0} = \frac{\sigma}{b}\left\{\frac{1}{m} + \frac{1}{n} + \frac{n(\bar{y}_0 - \bar{y})^2}{b^2\left[\sum x_i^2 - \left(\sum x_i\right)^2\right]}\right\}^{\frac{1}{2}}$$

Slope (weighted least squares)

$$b = \frac{\sum\frac{1}{\sigma_i^2}\sum\frac{x_i y_i}{\sigma_i^2} - \sum\frac{x_i}{\sigma_i^2}\sum\frac{y_i}{\sigma_i^2}}{\Delta}$$

Intercept (weighted least squares)

$$a = \frac{\sum\frac{x_i^2}{\sigma_i^2}\sum\frac{y_i}{\sigma_i^2} - \sum\frac{x_i}{\sigma_i^2}\sum\frac{x_i y_i}{\sigma_i^2}}{\Delta}$$

Denominator for weighted least squares equations

$$\Delta = \sum\frac{1}{\sigma_i^2}\sum\frac{x_i^2}{\sigma_i^2} - \left(\sum\frac{x_i}{\sigma_i^2}\right)^2$$

Standard error in intercept (weighted least squares, absolute σ_i known)

$$\sigma_a = \left(\frac{\sum\frac{x_i^2}{\sigma_i^2}}{\Delta}\right)^{\frac{1}{2}}$$

Standard error in slope (weighted least squares, absolute σ_i known)

$$\sigma_b = \left(\frac{\sum\frac{1}{\sigma_i^2}}{\Delta}\right)^{\frac{1}{2}}$$

Standard error in intercept (weighted least squares, relative σ_i known)

$$\sigma_a = \sigma_w\left(\frac{\sum\frac{1}{\sigma_i^2}\sum\frac{x_i^2}{\sigma_i^2}}{n\Delta}\right)^{\frac{1}{2}}$$

Standard error in slope (weighted least squares, relative σ_i known)

$$\sigma_b = \frac{\sigma_w\sum\frac{1}{\sigma_i^2}}{(n\Delta)^{\frac{1}{2}}}$$

Standard deviation in y values (weighted least squares)

$$\sigma_w = \frac{\left(\dfrac{n}{n-2}\right)^{\frac{1}{2}}}{\sum \dfrac{1}{\sigma_i^2}} \left[\sum \frac{1}{\sigma_i^2} \sum \frac{y_i^2}{\sigma_i^2} - \left(\sum \frac{y_i}{\sigma_i^2}\right)^2 - \frac{\left(\sum \dfrac{1}{\sigma_i^2} \sum \dfrac{x_i y_i}{\sigma_i^2} - \sum \dfrac{x_i}{\sigma_i^2} \sum \dfrac{y_i}{\sigma_i^2}\right)^2}{\Delta} \right]^{\frac{1}{2}}$$

Weighted correlation coefficient

$$r_w = \frac{\sum \dfrac{1}{\sigma_i^2} \sum \dfrac{x_i y_i}{\sigma_i^2} - \sum \dfrac{x_i}{\sigma_i^2} \sum \dfrac{y_i}{\sigma_i^2}}{\left[\sum \dfrac{1}{\sigma_i^2} \sum \dfrac{x_i^2}{\sigma_i^2} - \left(\sum \dfrac{x_i}{\sigma_i^2}\right)^2\right]^{\frac{1}{2}} \left[\sum \dfrac{1}{\sigma_i^2} \sum \dfrac{y_i^2}{\sigma_i^2} - \left(\sum \dfrac{y_i}{\sigma_i^2}\right)^2\right]^{\frac{1}{2}}}$$

Answers to exercises and problems

Chapter 1

Exercise A

$kg \cdot m^2 \cdot s^{-2} \cdot A^{-2}$.

Exercise B

1. (i) 13.8 zJ; (ii) 0.36 µs; (iii) 43.258 kW; (iv) 780 Mm/s.
2. (i) 6.50×10^{-10} m; (ii) 3.7×10^{-11} C; (iii) 1.915×10^6 W; (iv) 1.25×10^{-4} s.

Exercise C

1. (i) three; (ii) three; (iii) two; (iv) four; (v) one; (vi) four.
2.

Part	Two significant figures	Three significant figures	Four significant figures
(i)	7.8×10^5 m/s²	7.76×10^5 m/s²	7.757×10^5 m/s²
(ii)	1.3×10^{-3} s	1.27×10^{-3} s	1.266×10^{-3} s
(iii)	-1.1×10^2 °C	-1.05×10^2 °C	-1.054×10^2 °C
(iv)	1.4×10^{-5} H	1.40×10^{-5} H	1.400×10^{-5} H
(v)	1.2×10^4 J	1.24×10^4 J	1.240×10^4 J
(vi)	1.0×10^{-7} m	1.02×10^{-7} m	1.016×10^{-7} m

Exercise D

(i) Using the guidelines in section 1.4.1, the number of intervals, $N = \sqrt{n} = \sqrt{52} \approx 7$. Dividing the range by 7 and rounding up gives an interval width of 0.1 g. Now we can construct a grouped frequency distribution:

Interval (g)	Frequency
$49.8 < x \leq 49.9$	4
$49.9 < x \leq 50.0$	3
$50.0 < x \leq 50.1$	22
$50.1 < x \leq 50.2$	18
$50.2 < x \leq 50.3$	4
$50.3 < x \leq 50.4$	0
$50.4 < x \leq 50.5$	1

Exercise E

A graph with semi-logarithmic scales is most appropriate.

Exercise F

$\bar{x} = 102.04$ pF, median $= 101.25$ pF.

Exercise G

1. Expanding equation (1.10) gives

$$\sigma = \left[\frac{\sum x_i^2 - 2\bar{x}\sum x_i + \sum(\bar{x})^2}{n} \right]^{\frac{1}{2}}$$

now $\sum x_i = n\bar{x}$, and $\sum(\bar{x})^2 = n(\bar{x})^2$, so that

$$\sigma = \left[\frac{\sum x_i^2}{n} - \frac{2n(\bar{x})^2}{n} + \frac{n(\bar{x})^2}{n} \right]^{\frac{1}{2}}$$

hence equation (1.11) follows.

2. (i) range $= 0.22$ cm; (ii) $\bar{x} = 4.163$ cm; (iii) median $= 4.15$ cm; (iv) variance $= 3.9 \times 10^{-3}$ cm^2; (v) standard deviation $= 0.063$ cm.

Exercise H

$\bar{x} = 2.187$ s, standard deviation, $s = 0.75$ s.

Exercise I

(i) $s = 0.052$ s (using equation (1.16));
(ii) $s = 0.047$ s (using equation (1.19));
(iii) percentage difference $\approx 10\%$.

Problems

1. (i) J/(kg·K); (ii) N/m^2; (iii) W/(m·K).
2. (i) m^2·s^{-2}·K^{-1}; (ii) kg·m^{-1}·s^{-2}; (iii) kg·m·s^{-3}·K^{-1}.

3. $kg^{-1} \cdot m^{-3} \cdot s^4 \cdot A^2$.

4. (i) 5.7×10^{-5} s; (ii) 1.4×10^4 K; (iii) 1.4×10^3 m/s; (iv) 1.0×10^5 Pa; (v) $1.5 \times 10^{-3} \, \Omega$.

5. (ii) Median lead content $= 51.5$ ppb.

6. (ii) Mean retention time $= 6.116$ s; (iii) standard deviation, $s = 0.081$s.

7. (ii) When the resistance is 5000 Ω, the humidity is approximately 66%.

8. (i) $\bar{x} = 0.4915 \, \mu mol/mL$, standard deviation, $s = 0.019 \, \mu mol/mL$.

9. (i) $\bar{x} = 120.4$ m, median $= 120$ m; (ii) range $= 32$ m; (iii) $s = 9.5$ m, $s^2 = 90 \, m^2$.

Chapter 2

In the interests of brevity, answers to exercises and end of chapter problems for this chapter show only relevant extracts from a Worksheet.

Exercise A

1.

t(s)	V(volts)	I(amps)	Q(coulombs)
0	3.98	3.32E-07	1.8706E-06
5	1.58	1.32E-07	7.426E-07
10	0.61	5.08E-08	2.867E-07
15	0.24	2E-08	1.128E-07
20	0.094	7.83E-09	4.418E-08
25	0.035	2.92E-09	1.645E-08
30	0.016	1.33E-09	7.52E-09
35	0.0063	5.25E-10	2.961E-09
40	0.0031	2.58E-10	1.457E-09
45	0.0017	1.42E-10	7.99E-10
50	0.0011	9.17E-11	5.17E-10
55	0.0007	5.83E-11	3.29E-10
60	0.0006	5E-11	2.82E-10

2.

t(s)	V(volts)	I(amps)	Q(coulombs)	$I^{0.5}(amps)^{0.5}$
0	3.98	3.32E-07	1.87E-06	0.000576
5	1.58	1.32E-07	7.43E-07	0.000363
10	0.61	5.08E-08	2.87E-07	0.000225
15	0.24	2E-08	1.13E-07	0.000141
20	0.094	7.83E-09	4.42E-08	8.85E-05
25	0.035	2.92E-09	1.65E-08	5.4E-05
30	0.016	1.33E-09	7.52E-09	3.65E-05
35	0.0063	5.25E-10	2.96E-09	2.29E-05
40	0.0031	2.58E-10	1.46E-09	1.61E-05
45	0.0017	1.42E-10	7.99E-10	1.19E-05
50	0.0011	9.17E-11	5.17E-10	9.57E-06
55	0.0007	5.83E-11	3.29E-10	7.64E-06
60	0.0006	5E-11	2.82E-10	7.07E-06

Exercise B

20	6.5	130
30	7.2	216
40	8.5	340

Exercise C

(i) When $g=9.81$, column B reads: (ii) when $g=1.6$, column B reads:

t (s)		
0.638551		
0.903047		
1.106003		
1.277102		
1.427843		

t (s)		
1.581139		
2.236068		
2.738613		
3.162278		
3.535534		

Exercise D

T (K)	Q (J/s)			
1000	3515.4		SB (W/(m²·K⁴))	5.67E-08
2000	56246.4		A (m²)	0.062
3000	284747.4			
4000	899942.4			
5000	2197125			
6000	4555958			

Exercise E

1.

r_m (m)	6.35E-03
r_n (m)	6.72E-03
m	52
n	86
l (m)	6.02E-07
$R_{numerator}$	4.84E-06
$R_{denominator}$	2.05E-05
R (m)	2.36E-01

2.

d (m)	v_d (m/s)		
0.1	339.3848	v (m/s)	344
0.2	341.6924	f (Hz)	5
0.3	342.4616		
0.4	342.8462		
0.5	343.077		
0.6	343.2308		
0.7	343.3407		
0.8	343.4231		
0.9	343.4872		
1	343.5385		

Problems

1. Note that each term appearing within the square root in equation (2.12) is evaluated in a separate column of the spreadsheet.

λ (m)	1st_term	2nd_term	v (m/s)			
0.01	0.015613	4.59E-02	0.247952		γ(N/m)	7.30E-02
0.02	0.031226	2.29E-02	0.232723		ρ(kg/m^3)	1.00E+03
0.03	0.046839	1.53E-02	0.249256		g (m/s^2)	9.81
0.04	0.062452	1.15E-02	0.271881			
0.05	0.078065	9.17E-03	0.295362			
0.06	0.093679	7.64E-03	0.318313			
0.07	0.109292	6.55E-03	0.340359			
0.08	0.124905	5.73E-03	0.361439			
0.09	0.140518	5.10E-03	0.381594			

2. Parentheses required in the denominator of the formula in cell B2.

v (m/s)	m (kg)		
2.90E+08	1.39E-29	mo	9.10E-31
2.91E+08	1.54E-29	c_	3.00E+08
2.92E+08	1.73E-29		
2.93E+08	1.97E-29		
2.94E+08	2.30E-29		
2.95E+08	2.75E-29		
2.96E+08	3.44E-29		
2.97E+08	4.57E-29		
2.98E+08	6.85E-29		
2.99E+08	1.37E-28		

3. (i)

mean (%)	66.19
standard deviation (%)	5.482193
maximum (%)	79
minimum (%)	54
range (%)	25

4. (ii)

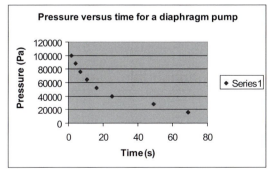

5.

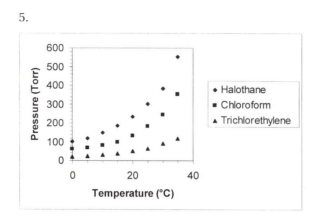

6. (ii)

H (%)	$\dfrac{\rho}{\rho_P}$				
	f=4	f=3	f=2.5	f=2	f=1.5
0	1	1	1	1	1
5	1.210526	1.157895	1.131579	1.105263	1.078947
10	1.444444	1.333333	1.277778	1.222222	1.166667
15	1.705882	1.529412	1.441176	1.352941	1.264706
20	2	1.75	1.625	1.5	1.375
25	2.333333	2	1.833333	1.666667	1.5
30	2.714286	2.285714	2.071429	1.857143	1.642857
35	3.153846	2.615385	2.346154	2.076923	1.807692
40	3.666667	3	2.666667	2.333333	2
45	4.272727	3.454545	3.045455	2.636364	2.227273
50	5	4	3.5	3	2.5
55	5.888889	4.666667	4.055556	3.444444	2.833333
60	7	5.5	4.75	4	3.25
65	8.428571	6.571429	5.642857	4.714286	3.785714
70	10.33333	8	6.833333	5.666667	4.5
75	13	10	8.5	7	5.5
80	17	13	11	9	7

7.

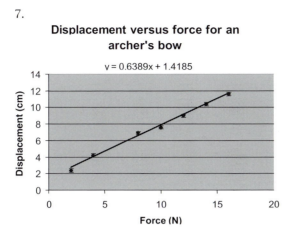

Displacement versus force for an archer's bow

$y = 0.6389x + 1.4185$

Chapter 3

Exercise A

1. (ii) $A = \dfrac{1}{8}$; (iii) 0.4375; (iv) 437.5.

2. (ii) 0.4512; (iii) 0.5488.

Exercise B

x(g)	7.2	7.4	7.6	7.8	8.0	8.2	8.4	8.6	8.8	9.0
cdf	3.401×10^{-6}	2.327×10^{-4}	6.210×10^{-3}	6.681×10^{-2}	0.3085	0.6915	0.9332	0.9938	0.9998	1.000

Exercise C

0.0476.

Exercise D

z	0.0	0.1	0.2	0.3	0.4	0.5	0.6	0.7	0.8	0.9	1.0
cdf	0.5000	0.5398	0.5793	0.6179	0.6554	0.6915	0.7257	0.7580	0.7881	0.8159	0.8413

Exercise E

1. (i) 0.2192; (ii) 7.9 (round to 8).

2. (i) 0.02275; (ii) 0.9545; (iii) 0.0214; (iv) 0.02275.

3. 0.01242.

4. (i)(a) 0.1587; (b) 0.1359; (c) 0.00135; (ii) 13.5.

Exercise F

50% confidence interval: 17.99 °C to 18.83 °C;
68% confidence interval: 17.79 °C to 19.03 °C;
90% confidence interval: 17.38 °C to 19.44 °C;
95% confidence interval: 17.18 °C to 19.64 °C;
99% confidence interval: 16.79 °C to 20.03 °C.

Exercise G

0.84162

Exercise H

99% confidence interval: 893 kg/m^3 to 917 kg/m^3

Exercise I

Using equation (3.21), $\sigma_{\bar{x}} = 0.014$ eV. Using equation (3.27), $\sigma_{\bar{x}} = 0.014$ eV.

Exercise J

90% confidence interval: 4.639 mm to 4.721 mm

Exercise K

(i)

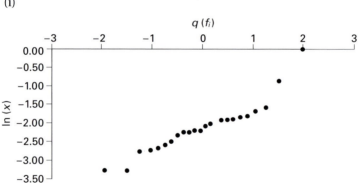

(ii) Comparing $\ln(x)$ versus $q(f_i)$ with the graph given in figure 3.26 indicates that the normality has been much improved by the logarithmic transformation.

Exercise L

5.

Problems

1. (ii) $A=\frac{3}{2}$;

(iii)   (a) 0.145; (b) 0.055.

4. Mean = 916.7 MPa, standard deviation = 20 MPa.

5. $\bar{x}=0.64086$ V, $s=0.0027$ V. Number of diodes with voltage in excess of 0.6400 V would be 125.

6. (i) (b)

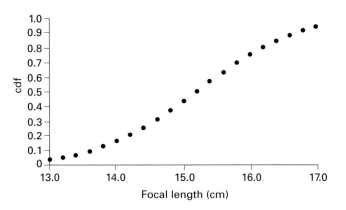

(ii)   (a) 0.4004; (b) 0.1857.

7.

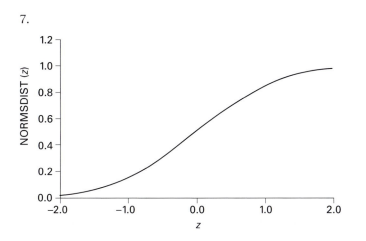

8. 90% confidence interval for the population mean: 2080 Hz to 2132 Hz.

9. 95% confidence interval for the population mean: 131.5 μPa to 144.5 μPa.

10. (i) 0.147584; (ii) 0.02275; (iii) overleaf; (iv) when $\nu=61$.

10. (iii)

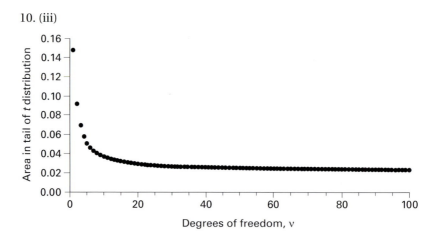

Degrees of freedom, ν

11. (i) Graph of x_i versus $q(f_i)$ is not linear suggesting data are not normally distributed.

(ii) Graph of x_i versus $q(f_i)$ for transformed data is linear indicating that the original data are lognormally distributed.

12. (iii) Normal quantile plot indicates that distribution is consistent with a lognormal distribution.

Chapter 4

Exercise A

(i) $P(r=0) = 0.8904$;

(ii) $P(r=1) = 0.1035$;

(iii) $P(r>1) = 0.0061$.

Exercise B

(i) $P(r=20) = 0.09306$;

(ii) $P(r\leq20) = 0.7622$;

(iii) $P(r<20) = 0.6692$;

(iv) $P(r>20) = 0.2378$;

(v) $P(r\geq20) = 0.3308$.

Exercise C

(i) $\mu = 8$;

(ii) $\sigma = 2.8$;

(iii) $P(r\geq2) = 0.9972$.

Exercise D

(i)   Using binomial distribution, $P(290 \leq r \leq 320) = 0.6857$. Using Normal approximation, $P(289.5 \leq x \leq 320.5) = 0.6871$.

(ii) Using binomial distribution, $P(r > 320) = 0.07923$. Using Normal approximation, $P(x > 320.5) = 0.07857$.

Exercise E

(i) $P(r=0) = 0.6065$;

(ii) $P(r \leq 3) = 0.9982$;

(iii) $P(2 \leq r \leq 4) = 0.0900$.

Exercise F

(i) $P(r=0) = 0.3535$;

(ii) $P(r=1) = 0.3676$;

(iii) $P(r>2) = 0.0878$.

Exercise G

(i) $P(r=0) = 0.2209$;

(ii) $P(r=1) = 0.3336$;

(iii) $P(r=3) = 0.1267$;

(iv) $P(2 \leq r \leq 4) = 0.4265$;

(v) $P(r>6) = 0.000962$.

Exercise H

(i)   (Using normal approximation to the Poisson distribution) $P(179.5 \leq r \leq 230.5) = 0.9109$.

Problems

1. $C_{10,5} = 252$, $C_{15,2} = 105$, $C_{42,24} = 3.537 \times 10^{11}$, $C_{580,290} = 1.311 \times 10^{173}$.

2. (i) $P(r=1) = 0.0872$; (ii) $P(r>1) = 0.004323$.
Number of screens expected to have more than one faulty transistor $= 22$.

3. (i)   Number with three functioning ammeters $= 18$ (rounded).

(ii) Number with two functioning ammeters $= 6$ (rounded).

(iii) Number with less than two functioning ammeters $= 1$ (rounded).

4. (i) $P(r \geq 20) = 0.8835$;

(ii) $n = 32$.

5. (i) Estimate of population mean = 123 counts (rounded).

   (ii) Estimate of population standard deviation = 11 counts (rounded).

6. (i) 1.3167 flaws per metre.

   (ii)

r	N(rounded)
0	16
1	21
2	14
3	6
4	2
5	1

7. (i) We require $P(0) = 0.2$. Using equation (4.9), $0.2 = \exp(-\mu)$, so that $\mu = 1.609$.

   (ii) $P(r > 4) = 0.02421$.

8. 0.9817.

Chapter 5

Exercise A

(i) Mean light intensity = 345.2 lx.

(ii) Standard deviation $s = 23$ lx.

(iii) 99% confidence interval = (345 ± 15) lx.

Exercise B

Fractional uncertainty = 0.05.

Absolute uncertainty = 73 °C.

Exercise C

1. $f = 1812$ Hz, $u_f = 39$ Hz.

2. $V = 38.8$ mm³, $u_v = 8.3$ mm³.

Exercise D

$p = 1669$ Pa, $u_p = 64$ Pa.

Exercise E

1. $n = 1.487$, $u_n = 0.046$.

2. (i) $f = 104.6$ mm, $u_f = 1.0$ mm; (ii) $m = 5.004$, $u_m = 0.061$.

Exercise F

(i)   Mean$=221.8$ s, standard deviation, $s=7.9$ s.
(ii) Value furthest from mean is 235 s.
(iii) Number expected to be at least as far from the mean as the 'suspect' value is 0.465. Note this is very close to 0.5 and it would be sensible, if possible, to acquire more data rather than eliminate the 'outlier'.
(iv) New mean$=218.5$ s, new standard deviation, $s=3.1$ s.

Exercise G

(i)   Upper limit for standard deviation $s=0.0031$ mm. Upper limit for standard error of mean $=0.0014$ mm.
(ii) 95% confidence interval $=(1.2200\pm0.0038)$ mm.

Exercise H

Uncertainty in capacitance $=3.5$ nF.

Exercise I

(i)   Gain $=1730$, uncertainty in gain $=12$.
(ii) Assumptions:
    (a)   There is no uncertainty in $R_g(0)$ and α.
    (b) α is constant over the temperature range 21 °C to 25 °C.
    (c)   The gain setting resistor is at the same temperature as the room.

Exercise J

$R_{int}=267$ MΩ (beware of premature rounding in this problem).

Exercise K

17.27 s.

Exercise L

(i)   Mean $=853.75$ mm.
(ii) Standard deviation $s=2.0$ mm.
(iii) Standard error $=0.70$ mm.
(iv) 95% confidence interval $=(853.8\pm1.7)$ mm.
(v)   95% confidence interval $=(853.8\pm0.85)$ mm.
(vi) 95% confidence interval $=(853.8\pm1.9)$ mm.

Exercise M

Weighted mean $=1.103$ s.

Exercise N

Standard error of weighted mean = 0.11 s.

Problems

1. (i)   Fractional uncertainty = 0.034;
    (ii)   Percentage uncertainty = 3.4%.
2. (i)   Mean rebound height = 186.0 mm.
    (ii)   Standard error in rebound height = 1.2 mm.
    (iii) 95% confidence interval for the rebound height = (186.0 ± 3.9) mm.
3. (i)   Mean film thickness = 328.33 nm.
    (ii)   Standard error in mean = 12 nm.
    (iii) 99% confidence interval = (328 ± 47) nm.
4. $n = 1.466$, $u_n = 0.027$.
5. $r = (0.386 \pm 0.032)$.
6. $\theta_c = (5.52 \pm 0.10) \times 10^{-3}$ Rad.
7. $R = (125.57 \pm 0.59)$ Ω.
8. (i) $c = (0.6710 \pm 0.0084)$.
    (ii) $c = (0.6710 \pm 0.0073)$.
9. $H = (-6.54 \pm 0.99)$ W.
10. (i)   Mean length = 47.83 cm.
    (ii)   Value furthest from mean is 42.7 cm.
    (iii) Yes, reject outlier.
    (iv)   New mean = 48.4 cm.
11. (i) (166.0 ± 2.0) mV.
    (ii) (166.00 ± 0.93) mV.
    (iii) (166.0 ± 2.2) mV.
12. (i)   Mean mass = 0.9656 g.
    (ii)   Standard error of mean = 0.0032 g.
    (iii) 95% confidence interval = (0.9656 ± 0.0072) g.
13. (i) $\bar{x} = 33.38$ mL, $s = 0.28$ mL.
    (ii)   Possible outlier is 33.9 mL. Applying Chauvenet's criterion indicates that outlier should be removed.
    (iii) New $\bar{x} = 33.28$ mL, new $s = 0.13$ mL.
14. Weighted mean = 1.0650 g/cm^3, standard error in weighted mean = 0.0099 g/cm^3.

Chapter 6

Exercise A

$a = 332.1$ m/s, $b = 0.6496$ m/(s·°C), $SSR = 15.41$ m^2/s^2.

Exercise B

$\sigma_a = 1.5$, $\sigma_b = 0.23$.

Exercise C

1. The 99% confidence interval for α is $(4 \pm 22) \times 10^{-3}$. The 99% confidence interval for β is $(2.51 \pm 0.12) \times 10^{-2}$ mL/ng.

2.
(i) $a = 10.24$ Ω, $b = 4.324 \times 10^{-2}$ Ω/°C.
(ii) $\sigma_a = 0.066$ Ω, $\sigma_b = 1.4 \times 10^{-3}$ Ω/°C.
(iii) The 95% confidence intervals for $A = (10.24 \pm 0.14)$ Ω and for $B = (4.32 \pm 0.32) \times 10^{-2}$ Ω/°C.

Exercise D

(i) $a = 3.981$, $b = 16.48$, $\sigma_a = 1.1$, $\sigma_b = 1.9$.

Exercise E

(i)   Plot P versus h; (ii) $a = P_A$, $b = \rho g$.

Exercise F

(i) $a = 1.20046$ m, $b = 1.07818 \times 10^{-5}$ m/°C, $\sigma_a = 5.0 \times 10^{-4}$ m, $\sigma_b = 8.0 \times 10^{-7}$ m/°C.
(ii) $\alpha = 8.981 \times 10^{-6}$ °C^{-1}.
(iii) $\sigma_\alpha = 6.7 \times 10^{-7}$ °C^{-1}.

Exercise G

The 99% confidence interval for $\mu_{y|x_0}$ when $x_0 = 15$ is 4.5 ± 7.1.

Exercise H

The 95% prediction interval for y at $x_0 = 12$ is 12 ± 12.

Exercise I

(i) $\hat{y} = 4677.89 + 14415.11 x_i$.
(ii) $x_0 = 3.640$ ppm, $\sigma_{x_0} = 0.17$ ppm.

Exercise J

(i)   'Usual' least squares (error in values of V), $k_0 = 0.6842$ V, $k_1 = -2.391 \times 10^{-3}$ V/°C.

(ii)   When errors are in θ values, $k_0 = 0.6847$ V, $k_1 = -2.401 \times 10^{-3}$ V/°C.

Exercise K

(i) $r = -0.8636$.

(ii) $a = 31.794$ °C, $b = -0.5854$ °C/cm³.

(iv)   A plot of data indicates that the assumption of linearity is not valid.

Exercise L

$r = 0.9667$.

Exercise M

(i) $r = 0.7262$.

(ii)   Value of r is not significant.

Exercise N

(ii) $a = 0.4173$ s, $b = 0.4779$ s/kg.

(iii)

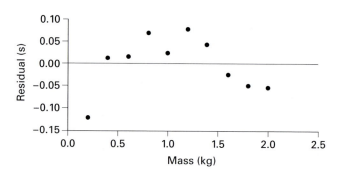

(iv)   Yes, probably wrong equation fitted to data (i.e. period is not linearly related to mass).

Exercise O

Number of data expected to be at least as far from the best line as the outlier is 0.392. Based on Chauvenet's criterion, the outlier should be rejected and the intercept and slope recalculated.

Exercise P

1. (i) Plot $\ln R$ versus T. Intercept $= \ln A$, slope $= -B$.
   (ii) Plot v versus t. Intercept $= u$, slope $= g$.
   (iii) Plot H versus T. Intercept $= -CT_0$, slope $= C$.
   (iv) Plot T_w versus R^2. Intercept $= T_c$, slope $= -k$.
   (v) Plot T versus $\sqrt{m}$. Intercept $= 0$, slope $= 2\pi/\sqrt{k}$.
   (vi) Plot $1/I$ versus R. Intercept $= r/E$, slope $= -1/E$.
   (vii) Plot $1/v$ versus $1/u$. Intercept $= 1/f$, slope $= -1$.
   (viii) Plot $\ln N$ versus $\ln C$. Intercept $= \ln k$, slope $= 1/n$.
   (ix) Plot $1/(n^2-1)$ versus $1/\lambda^2$. Intercept $= 1/A$, slope $= -B/A$.
   (x) Plot tD^2 versus D. Intercept $= A$, slope $= AB$.
2. $k = 2.311 \times 10^{-6} \, \text{V}^{-1} \cdot \text{pF}^{-2}$, $\phi = 1.064 \, \text{V}$.

Exercise Q

(i) $y = 7.483$, $u_y = 0.13$.
(ii) $y = 3136$, $u_y = 220$.
(iii) $y = 0.01786$, $u_y = 6.4 \times 10^{-4}$.
(iv) $y = 3.189 \times 10^{-4}$, $u_y = 2.3 \times 10^{-5}$.
(v) $y = 1.748$, $u_y = 0.016$.

Exercise S

(i) To linearise the equation, take the natural logarithms of both sides of the equation to give $\ln C = \ln A - \lambda d^2$. This is of the form $y = a + bx$, where $y = \ln C$, $a = \ln A$, $b = -\lambda$ and $x = d^2$.
(ii) $A = 1.998 \times 10^4$ counts, $\lambda = 6.087 \times 10^{-4} \, \text{mm}^{-2}$, $\sigma_A = 1.3 \times 10^2$ counts, $\sigma_\lambda = 4.7 \times 10^{-6} \, \text{mm}^{-2}$.

Exercise T

$a = 2.311$, $b = -19.70 \, \text{V}^{-1}$, $\sigma_a = 0.063$, $\sigma_b = 0.82 \, \text{V}^{-1}$.

Problems

1. (ii) $a = 9.803 \, \text{m/s}^2$, $b = -2.915 \times 10^{-6} \, \text{s}^{-2}$.
   (iii) $SSR = 0.0063 \, (\text{m/s}^2)^2$, $\sigma = 0.028 \, \text{m/s}^2$.
   (iv) $\sigma_a = 0.019 \, \text{m/s}^2$, $\sigma_b = 3.1 \times 10^{-7} \, \text{s}^{-2}$.
   (v) $r = -0.9579$.
   (vi) $a = 9.7927 \, \text{m/s}^2$, $b = -2.867 \times 10^{-6} \, \text{s}^{-2}$, $SSR = 0.0004533 \, (\text{m/s}^2)^2$, $\sigma = 0.0075 \, \text{m/s}^2$, $\sigma_a = 0.0051 \, \text{m/s}^2$, $\sigma_b = 8.3 \times 10^{-8} \, \text{s}^{-2}$, $r = -0.9967$.
2. $k = 27.09 \, \text{MPa}$, $\sigma_k = 0.75 \, \text{MPa}$

3. (ii) $K_1 = 2.556 \times 10^{-10}\,\text{m}/(\text{s}\cdot\mu\text{A}^4)$, $K_2 = 5.583 \times 10^4\,\mu\text{A}^2$.
    (iii) No discernible pattern in residuals – weighting seems appropriate.
4. (i)   Taking natural logarithms of both sides of equation (6.78) gives

    $$\ln Y = \ln k + (1/n)\ln C$$

    This is the form $y = a + bx$, where $y = \ln Y$, $a = \ln k$, $b = 1/n$ and $x = \ln C$.
    (ii) $k = 2.638$, $n = 2.346$, $\sigma_k = 0.036$, $\sigma_n = 0.028$.
    (iii) There is an indication that as $\ln C$ increases, so do the standard-ised residuals. A weighted fit is probably appropriate, but more points are needed to confirm this.
    (iv)   When $C = 0.085$ mol/L, $Y = (0.923 \pm 0.017)$ mol.
5. (i) $a = 0.9591$, $b = 0.9039$.
    (ii) $I_{max} = 1.863$, $I_{min} = 0.9591$.
    (iii) Note $I_{max} = a + b$. As a and b are correlated, replace a with $\bar{y} - b\bar{x}$ before proceeding to calculate $\sigma_{I_{max}}$. The calculation gives $\sigma_{I_{max}} = 0.038$.
6. (i)   Plot $1/X$ versus $1/P$. The intercept $a = -B/A$, and the slope $b = 1/A$.
    (ii) $a = 2228\,\text{m}^2/\text{kg}$, $b = 1338\,\text{N/kg}$, $\sigma_a = 31\,\text{m}^2/\text{kg}$, $\sigma_b = 15\,\text{N/kg}$.
    (iii) $A = 7.473 \times 10^{-4}\,\text{kg/N}$, $B = -1.665\,\text{m}^2/\text{N}$, $\sigma_A = 8.6 \times 10^{-6}\,\text{kg/N}$, $\sigma_B = 0.040\,\text{m}^2/\text{N}$.
7. (i) $b = 5.833\,\text{m}^{-1}\cdot\text{Pa}^{-1}$, $\sigma_b = 0.55\,\text{m}^{-1}\cdot\text{Pa}^{-1}$.
    (ii) $d = 1.748 \times 10^{-10}\,\text{m}$, $\sigma_d = 8.3 \times 10^{-12}\,\text{m}$.
11. (ii) $r = 0.9220$.
    (iii) Here we have six points. Using table 6.20, the probability of having $r > 0.9$ when data are uncorrelated is 0.014, therefore we have evidence that the correlation is significant.

Chapter 7

Exercise A

Begin by writing the equation of 'best fit' as $V_i = a + bT_i + cT_i \ln T_i$, where a, b and c are best estimates of α, β and γ respectively. The matrix equation to be solved for a, b and c is

$$
\begin{bmatrix}
n & \sum T_i & \sum T_i \ln T_i \\
\sum T_i & \sum T_i^2 & \sum T_i^2 \ln T_i \\
\sum T_i \ln T_i & \sum T_i^2 \ln T_i & \sum (T_i \ln T_i)^2
\end{bmatrix}
\begin{bmatrix}
a \\ b \\ c
\end{bmatrix}
=
\begin{bmatrix}
\sum V_i \\
\sum V_i T_i \\
\sum V_i T_i \ln T_i
\end{bmatrix}
$$

Exercise B

(i)
$$
\begin{vmatrix}
-0.0598 & -0.00427 & 0.134406 \\
0.099696 & -0.14207 & 0.097605 \\
0.024255 & 0.21152 & -0.19437
\end{vmatrix}
$$

(ii)
$$
\begin{vmatrix}
-0.02049 & 0.000461 & 0.025877 & -0.00427 \\
0.02333 & -0.00381 & -0.03049 & 0.02004 \\
0.02103 & -0.00537 & -0.00979 & -0.00087 \\
-0.01745 & 0.016652 & 0.014903 & -0.0125
\end{vmatrix}
$$

(iii)
$$
\begin{vmatrix}
-0.06572 & -0.00915 & 0.019853 & -0.0018 & 0.102176 \\
-0.17055 & -0.10107 & -0.57789 & 0.651764 & -0.01951 \\
0.080788 & 0.050332 & 0.368402 & -0.2621 & -0.11095 \\
-0.11792 & -0.3062 & -0.79712 & 0.88727 & 0.06521 \\
0.320455 & 0.414455 & 0.995139 & -1.21103 & -0.04705
\end{vmatrix}
$$

Exercise C

(i)
$$
\begin{vmatrix}
9436.22 \\
5547.23 \\
7173.82
\end{vmatrix}
$$

(ii)
$$
\begin{vmatrix}
5529 \\
6140 \\
4428 \\
6961
\end{vmatrix}
$$

Exercise D

Best estimate of A, $a = -11.82\ \Omega$.

Best estimate of B, $b = 0.4244\ \Omega/\text{K}$.

Best estimate of C, $c = -5.928 \times 10^{-5}\ \Omega/\text{K}^2$.

Exercise E

$a = 15.36$, $b = 2.408$, $c = 1.876$.

Exercise F

Best estimate of A, $a = -6.776\ \mu\text{V}$.

Best estimate of B, $b = 4.922 \times 10^{-3}\ \mu\text{V}/\text{K}$.

Best estimate of C, $c = 9.121 \times 10^{-5}\ \mu\text{V}/\text{K}^2$.

Best estimate of D, $d = -6.747 \times 10^{-8}\ \mu\text{V}/\text{K}^3$.

$\sigma_a = 0.083\ \mu\text{V}$.

$\sigma_b = 1.6 \times 10^{-3}\ \mu\text{V}/\text{K}$.

$\sigma_c = 9.0 \times 10^{-6}\ \mu\text{V}/\text{K}^2$.

$\sigma_d = 1.6 \times 10^{-8}\ \mu\text{V}/\text{K}^3$.

Exercise G

$\alpha = (0.010 \pm 0.011)$ N, $\beta = (3.9 \pm 3.4)$ N/m, $\gamma = (427 \pm 24)$ N/m^2.

Exercise H

(i) $a = 13.27$, $b = 3.628$, $c = 1.426$.

(ii) $\sigma_a = 1.1$, $\sigma_b = 0.55$, $\sigma_c = 0.095$.

Exercise I

$R^2 = 0.9988$.

Exercise J

(i) $a = 5.102 \times 10^{-3}$, $b = 1.028 \times 10^{-2}$, $c = 8.019 \times 10^{-3}$.

(ii) $\sigma_a = 7.7 \times 10^{-4}$, $\sigma_b = 2.7 \times 10^{-3}$, $\sigma_c = 2.0 \times 10^{-4}$.

(iii) $R^2 = 0.9997$.

(iv) $\sigma = 1.2 \times 10^{-4}$.

Problems

1.

(ii)
$$\begin{bmatrix} n & \sum x_i & \sum \exp x_i \\ \sum x_i & \sum x_i^2 & \sum x_i \exp x_i \\ \sum \exp x_i & \sum x_i \exp x_i & \sum (\exp x_i)^2 \end{bmatrix} \begin{bmatrix} a \\ b \\ c \end{bmatrix} = \begin{bmatrix} \sum y_i \\ \sum y_i x_i \\ \sum y_i \exp x_i \end{bmatrix}$$

(iii) $a = 6.076$, $b = -0.9074$, $c = 1.492 \times 10^{-4}$.

2.

(i)
$$\begin{bmatrix} n & \sum \dfrac{1}{x_i} & \sum x_i \\ \sum \dfrac{1}{x_i} & \sum \dfrac{1}{x_i^2} & n \\ \sum x_i & n & \sum x_i^2 \end{bmatrix} \begin{bmatrix} a \\ b \\ c \end{bmatrix} = \begin{bmatrix} \sum y_i \\ \sum \dfrac{y_i}{x_i} \\ \sum x_i y_i \end{bmatrix}$$

(ii) Writing best estimates of A, B and C as a, b and c, respectively, we have $a = 1.740$ mm, $b = 26.87$ mm·mL/minute, $c = 0.02366$ mm·minute/mL.

(iii) $\sigma_a = 0.13$ mm, $\sigma_b = 0.89$ mm·mL/minute, $\sigma_c = 1.7 \times 10^{-3}$ mm·minute/ mL.

3. Writing best estimates of s_0, u and g as a, b, and c respectively, we have $a = 134.2$ m, $b = 46.27$ m/s, $c = -10.06$ m/s^2, $\sigma_a = 3.4$ m, $\sigma_b = 2.6$ m/s, $\sigma_c = 0.42$ m/s^2.

4. Writing best estimates of A, B, C and D as a, b, c and d, respectively, we have

$a=0.9230$, $b=-67.57$ cm^3, $c=1977$ cm^6 and $d=2.387\times10^4$ cm^9; $\sigma_a=0.017$, $\sigma_b=5.2$ cm^3, $\sigma_c=430$ cm^6 and $\sigma_d=9.7\times10^3$ cm^9.

5. For two parameters, AIC$=149.3$, $R^2_{ADJ}=0.9707$. For three parameters, AIC$=151.3$, $R^2_{ADJ}=0.9675$.

Both indicators of goodness of fit support the equation $y=a+bx$ being the better fit to data.

6. (i) Writing best estimates of A, B and C as a, b and c respectively, we have $a=29.90$ J·mol^{-1}·K^{-1}, $b=4.304\times10^{-3}$ J·mol^{-1}·K^{-2}, $c=-1.632\times10^5$ J·mol^{-1}·K.

   (ii) $\sigma_a=0.22$ J·mol^{-1}·K^{-1}, $\sigma_b=2.4\times10^{-4}$ J·mol^{-1}·K^{-2}, $\sigma_c=1.8\times10^4$ J·mol^{-1}·K.

   (iii) $A=(29.90\pm0.47)$ J·mol^{-1}·K^{-1}, $B=(4.30\pm0.53)\times10^{-3}$ J·mol^{-1}·K^{-2}, $C=(-1.63\pm0.40)\times10^5$ J·mol^{-1}·K.

7. Indicators of goodness of fit including AIC and residuals should indicate that equation (7.62) is a better fit to data than equation (7.63).

Chapter 8

Exercise A

H_0: $\mu=1.260$ V; H_a: $\mu\neq1.260$ V. For data, $z=-3.0$. $z_{crit}=1.96$. As $|z|>z_{crit}$, reject null hypothesis at $\alpha=0.05$.

Exercise B

p value$=0.024$, therefore at $\alpha=0.05$, there is a significant difference between the hypothesised population mean and the sample mean.

Exercise C

(i)   When $\alpha=0.2$, $z_{crit}=0.84$.
(ii) When $\alpha=0.05$, $z_{crit}=1.64$.
(iii) When $\alpha=0.01$, $z_{crit}=2.33$.
(iv) When $\alpha=0.005$, $z_{crit}=2.58$.

Exercise D

$t=-2.119$, $t_{crit}=2.365$. As $|t|<t_{crit}$, we cannot reject the null hypothesis, i.e. the mean of the values in table 8.10 is not significantly different from the published value of c.

Exercise E

H_0: population intercept$=0$. Carry out two tailed test at $\alpha=0.05$, $t=6.917$, $t_{crit}=2.228$, therefore reject null hypothesis.

H_0: population slope $=0$, $|t|=2.011$, $t_{crit}=2.228$, therefore cannot reject null hypothesis.

Exercise F

$t=14.49$ and $t_{crit}=2.228$, therefore reject the null hypothesis, i.e. the means of the coefficient of kinetic friction for the two contacts areas are significantly different at the $\alpha=0.05$ level of significance.

Exercise G

Using Excel®'s TTEST() function, the p value is 0.046. As this is less than $\alpha=0.05$ we reject the null hypothesis, i.e. there is a significant difference (at $\alpha=0.05$) between the lead content at the two locations.

Exercise H

Carry out t test for paired samples.
$t=2.762$. For a two tailed test at $\alpha=0.05$ and with number of degrees of freedom $=7$, $t_{crit}=2.365$. As $t>t_{crit}$ reject null hypothesis, i.e. the emfs of the batteries have changed over the period of storage.

Exercise I

$p=0.08037$. As $p>0.05$, we would not reject the null hypothesis at $\alpha=0.05$.

Exercise J

$F=2.207$, $F_{crit}=5.82$. As $F<F_{crit}$, we cannot reject null hypothesis at $\alpha=0.05$.

Exercise K

$F_{crit}=5.285$.

Exercise L

(i)

Count	Observed frequency	Expected frequency
0	10	7.78
1	13	14.48
2	7	13.46
3	15	8.35
≥ 4	5	5.93

(ii) $\chi^2=9.331$, $\chi^2_{crit}=7.815$. Reject null hypothesis at $\alpha=0.05$.

Exercise M

(i)   10.60.

(ii)   11.34.

(iii) 11.07.

(iv)   15.99.

Exercise N

$F=12.82$, $F_{crit}=3.40$, as $F>F_{crit}$, the ANOVA indicates that, at $\alpha=0.05$, the magnitude of the alpha wave does depends on light level.

Problems

1. (ii) For my random numbers, I found $\bar{x} =152.7297$.

   (iv) For my 100 columns of random numbers, I found four means to lie outside interval $\mu\pm1.96\,\sigma_{\bar{x}}$.

   (v) $\mu\pm1.96\sigma_{\bar{x}}$ is the 95% confidence interval for the sample mean, so expect 5% of sample means to lie outside this interval, i.e. five means.

2. Two tailed t test required (samples not paired). $t=1.831$, $t_{crit}=2.228$. As $t<t_{crit}$ we cannot reject the hypothesis that both samples have the same population mean (at $\alpha=0.05$).

3. Paired sample t test required. $t=2.909$, $t_{crit}=2.365$. As $t>t_{crit}$, we reject hypothesis (at $\alpha=0.05$) that there is no difference in the urea concentration as determined by the two laboratories.

4. $t=1.111$, t_{crit} 3.182 (for $\alpha=0.05$ and three degrees of freedom). As $t<t_{crit}$, we cannot reject the null hypothesis, i.e. the intercept is not significantly different from zero.

5. Two tailed F test carried out at $\alpha=0.05$. $F=3.596$, $F_{crit}=4.43$. As $F<F_{crit}$, we cannot reject the null hypothesis that both populations have the same variance.

6. One tailed F test carried out at $\alpha=0.05$. $F=2.796$, $F_{crit}=3.44$. As $F<F_{crit}$, we cannot reject a null hypothesis that both populations have the same variance.

7. I chose a bin width of 0.1 s with a bin range beginning at 0.8 s and extending to 1.9 s. Where necessary, bins were combined to ensure that the frequencies were $\geqslant5$. A chi-squared test (carried out at $\alpha=0.05$) indicates that the distribution of data in table 8.31 is consistent with the normal distribution.

8. $F=2.811$, $F_{crit}=3.55$. As $F<F_{crit}$, we cannot reject a null hypothesis that the population means are equal.

9. $F=10.61$, $F_{crit}=3.885$. As $F>F_{crit}$, we reject the null hypothesis and conclude that the population means are not equal.

References

Adler H A and Roessler E B *Introduction to Probability and Statistics* 5th Edition (1972) W H Freeman and Company, San Francisco

Akaike H A new look at the statistical model identification (1974) *IEEE Transactions on Automatic Control*, Volume 19, pages 716 to 723

Barford N C *Experimental Measurements: Precision, Error and Truth* 2nd Edition (1985) John Wiley and Sons, Chichester

Bentley J P *Principles of Measurement Systems* 2nd Edition (1988) Longmans, Harlow

Bevington P R and Robinson D K *Data Reduction and Error Analysis for the Physical Sciences* (1992) McGraw-Hill, New York

Blaisdell E A *Statistics in Practice* 2nd Edition (1998) Saunders College Publishing, Fort Worth

Blattner P *Special Edition Using Microsoft Excel® 2002* (2001) Que, Indianapolis

Bloch S C *Excel® for Engineers and Scientists* (2000) John Wiley and Sons, New York

Burr-Brown IC Data Book: Linear Products (1995) Burr Brown, Arizona

Cleveland W S *The Elements of Graphing Data* (1994) Hobart Press, New Jersey

Crow E L and Shimizu K *Lognormal Distributions: Theory and Applications* (1988) Marcel Dekker, New York

Daish C B and Fender D H *Experimental Physics* 2nd Edition (1970) English Universities Press Ltd, London

Devore J L *Probability and Statistics for Engineering and the Sciences* 3rd Edition (1991) Brookes/Cole, California

Dietrich C R *Uncertainty, Calibration and Probability: Statistics of Scientific and Industrial Measurement* 2nd Edition (1991) Adam Hilger, Bristol

Doebelin E O *Engineering Experimentation: Planning, Execution, Reporting* (1995) McGraw-Hill, New York

Feynman R P, Leighton R B and Sands M *The Feynman Lectures on Physics* (1963) Addison Wesley, Reading, Massachusetts

Flowers B H and Mendoza E *Properties of Matter* (1970) Wiley, London

Graham R C *Data Analysis for the Chemical Sciences* (1993) VCH, New York

Hamilton L C *Modern Data Analysis: A First Course in Applied Statistics* (1990) Brookes/Cole, California

Hoel P G *Introduction to Mathematical Statistics* 5th Edition (1984) Wiley, New York

Kennedy J B and Neville A M *Basic Statistical Methods for Engineers and Scientists* 3rd Edition (1986) Harper Row, New York

Khazan A D *Transducers and Their Elements* (1994) Prentice Hall, New Jersey

Lyon A J Rapid statistical methods (1980) *Physics Education*, Volume 15, pages 78 to 83

Macdonald J R and Thompson W J Least squares fitting when both variables are subject to error: Pitfalls and possibilities (1992) *American Journal of Physics*, Volume 60, pages 66 to 73

McPherson G *Statistics in Scientific Investigation: Its Basis, Application, and Interpretation* (1990) Springer-Verlag, New York

Meadows R *Electrical and Engineering Mathematics* Volume 2 (1981) Pitman, London

Meyer S L *Data Analysis for Scientists and Engineers* (1975) John Wiley and Sons, New York

Middleton M R *Data Analysis using Microsoft Excel®* (1997) Duxbury Press, Belmont

Moore D S and McCabe G P *Introduction to the Practice of Statistics* (1989) W H Freeman and Company, New York

Morris A S *Measurement and Calibration Requirements for Quality Assurance to ISO 9000* (1997) John Wiley and Sons, Chichester

Neter J, Kutner M J, Nachtsheim C J and Wasserman W *Applied Linear Regression Models* (1996) Times Mirror Higher Education Group Inc., Chicago

Nicholas J V and White D R *Traceable Temperatures: An Introductory Guide to Temperature Measurement and Calibration* (1982) Science Information Division, DSIR, Wellington, New Zealand

Orvis W J *Excel® for Scientists and Engineers* 2nd Edition (1996) Sybex, Alameda

Scheaffer R L and McClave J T *Probability and Statistics for Engineers* 4th Edition (1995) Duxbury, California

Simpson R E *Introductory Electronics for Scientists and Engineers* 2nd Edition (1987) Prentice-Hall, New Jersey

Smith F G and Thomson J H *Optics* 2nd Edition (1988) John Wiley and Sons, Chichester

Spiegel M R and Stephens L J *Statistics* 3rd Edition (1998) McGraw-Hill, New York

Tompkins W J and Webster J G (Eds) *Interfacing Sensors to the IBM PC* (1988) Prentice Hall, New Jersey

Walpole R E, Myers R H and Myers S L *Probability and Statistics for Engineers and Scientists* 6th Edition (1998) Prentice Hall, New Jersey

Weisberg S *Applied Linear Regression* 2nd Edition (1985) John Wiley and Sons, New York

Young H D and Freedman R A *University Physics* 9th Edition (1996) Addison-Wesley, Reading, Massachusetts

Index